Science Stories:
Using Case Studies to Teach Critical Thinking

Science Stories

Using Case Studies to Teach Critical Thinking

Clyde Freeman Herreid
Nancy A. Schiller
Ky F. Herreid

Arlington, Virginia

Claire Reinburg, Director
Jennifer Horak, Managing Editor
Andrew Cooke, Senior Editor
Wendy Rubin, Associate Editor
Agnes Bannigan, Associate Editor
Amy America, Book Acquisitions Coordinator

ART AND DESIGN
Will Thomas Jr., Director

PRINTING AND PRODUCTION
Catherine Lorrain, Director
Jack Parker, Electronic Prepress Technician

NATIONAL SCIENCE TEACHERS ASSOCIATION
Francis Q. Eberle, PhD, Executive Director
David Beacom, Publisher

1840 Wilson Blvd., Arlington, VA 22201
www.nsta.org/store
For customer service inquiries, please call 800-277-5300.

Copyright © 2012 by the National Science Teachers Association.
All rights reserved. Printed in the United States of America.
16 15 14 13   6 5 4 3

NSTA is committed to publishing material that promotes the best in inquiry-based science education. However, conditions of actual use may vary, and the safety procedures and practices described in this book are intended to serve only as a guide. Additional precautionary measures may be required. NSTA and the authors do not warrant or represent that the procedures and practices in this book meet any safety code or standard of federal, state, or local regulations. NSTA and the authors disclaim any liability for personal injury or damage to property arising out of or relating to the use of this book, including any of the recommendations, instructions, or materials contained therein.

PERMISSIONS
Book purchasers may photocopy, print, or e-mail up to five copies of an NSTA book chapter for personal use only; this does not include display or promotional use. Elementary, middle, and high school teachers may reproduce forms, sample documents, and single NSTA book chapters needed for classroom or noncommercial, professional-development use only. E-book buyers may download files to multiple personal devices but are prohibited from posting the files to third-party servers or websites, or from passing files to non-buyers. For additional permission to photocopy or use material electronically from this NSTA Press book, please contact the Copyright Clearance Center (CCC) (*www.copyright.com*; 978-750-8400). Please access *www.nsta.org/permissions* for further information about NSTA's rights and permissions policies.

Library of Congress Cataloging-in-Publication Data
Herreid, Clyde Freeman.
Science stories : using case studies to teach critical thinking / by Clyde Freeman Herreid, Nancy A. Schiller, and Ky F. Herreid.
    p. cm.
Includes bibliographical references and index.
ISBN 978-1-936137-25-1
1. Science--Study and teaching--Case studies. 2. Critical thinking--Study and teaching. I. Schiller, Nancy A., 1957- II. Herreid, Ky F., 1965- III. Title.
Q181.H394 2011
507.1--dc23
                              2011036547

eISBN 978-1-936959-91-4

Contents

Introduction
Clyde Freeman Herreid vii

Section I: The Nature of Science

Chapter 1   **The Scientific Method Ain't What It Used to Be**
Clyde Freeman Herreid 1

Chapter 2   **Learning About the Nature of Science With Case Studies**
Kathy Gallucci 11

Chapter 3   **Can Case Studies Be Used to Teach Critical Thinking?**
Clyde Freeman Herreid 21

Chapter 4   **The "Case" for Critical Thinking**
David R. Terry 25

The Case Studies

Section II: Historical Cases 35

Chapter 5   **Childbed Fever: A 19th-Century Mystery**
Christa Colyer 39

Chapter 6   **Mystery of the Blue Death: John Snow and Cholera**
Susan Bandoni Muench 45

Chapter 7   **Salem's Secrets: On the Track of the Salem Witch Trials**
Susan M. Nava-Whitehead and Joan-Beth Gow 57

Chapter 8   **The Bacterial Theory of Ulcers: A Nobel-Prize-Winning Discovery**
Debra Ann Meuler 69

Section III: Experimental Design 81

Chapter 9   **Lady Tasting Coffee**
Jacinth Maynard, Mary Puterbaugh Mulcahy, and Daniel Kermick 85

Chapter 10   **Memory Loss in Mice**
Michael S. Hudecki 95

Chapter 11   **Mom Always Liked You Best**
Clyde Freeman Herreid 99

Chapter 12   **PCBs in the Last Frontier**
Michael Tessmer 109

Chapter 13   **The Great Parking Debate**
Jennifer S. Feenstra 113

Chapter 14   **Poison Ivy: A Rash Decision**
Rosemary H. Ford 121

Section IV: The Scientific Method Meets Unusual Claims 129

Chapter 15   **Extrasensory Perception?**
Sarah G. Stonefoot and Clyde Freeman Herreid 133

Chapter 16   **A Need for Needles: Does Acupuncture Really Work?**
Sarah G. Stonefoot and Clyde Freeman Herreid 141

Chapter 17   **Love Potion #10: Human Pheromones at Work?**
Susan Holt 147

Chapter 18   **The "Mozart Effect"**
Lisa D. Hager 159

Chapter 19   **Prayer Study: Science or Not?**
Kathy Gallucci 167

Chapter 20   **The Case of the Ivory-Billed Woodpecker**
Kathrin Stanger-Hall, Jennifer Merriam, and Ruth Ann Greuling 173

Contents

Section V: Science and Society 183

    Chapter 21   **Moon to Mars: To Boldly Go ... or Not**
                            Erik Zavrel 187

    Chapter 22   **And Now What, Ms. Ranger? The Search for the Intelligent Designer**
                            Clyde Freeman Herreid 197

    Chapter 23   **The Case of the Tainted Taco Shells**
                            Ann Taylor 207

    Chapter 24   **Medicinal Use of Marijuana**
                            Clyde Freeman Herreid and Kristie DuRei 213

    Chapter 25   **Amanda's Absence: Should Vioxx Be Kept Off the Market?**
                            Dan Johnson 219

    Chapter 26   **Sex and Vaccination**
                            Erik Zavrel and Clyde Freeman Herreid 225

Section VI: Science and the Media 233

    Chapter 27   **Tragic Choices: Autism, Measles, and the MMR Vaccine**
                            Matthew Rowe 237

    Chapter 28   **Ah-choo! Climate Change and Allergies**
                            Juanita Constible, Luke Sandro, and Richard E. Lee Jr. 247

    Chapter 29   **Rising Temperatures: The Politics of Information**
                            Christopher V. Hollister 253

    Chapter 30   **Eating PCBs From Lake Ontario**
                            Eric Ribbens 261

Section VII: Ethics and the Scientific Process 267

    Chapter 31   **Mother's Milk Cures Cancer?**
                            Linda L. Tichenor 271

    Chapter 32   **Cancer Cure or Conservation**
                            Pauline A. Lizotte and Gretchen E. Knapp 281

    Chapter 33   **A Rush to Judgment?**
                            Sheryl R. Ginn and Elizabeth J. Meinz 289

    Chapter 34   **How a Cancer Trial Ended in Betrayal**
                            Ye Chen-Izu 295

    Chapter 35   **Bringing Back Baby Jason: To Clone or Not to Clone**
                            Jennifer Hayes-Klosteridis 303

    Chapter 36   **Selecting the Perfect Baby: The Ethics of "Embryo Design"**
                            Julia Omarzu 309

    Chapter 37   **Studying Racial Bias: Too Hot to Handle?**
                            Jane Marantz Connor 315

    Chapter 38   **Bad Blood: The Tuskegee Syphilis Project**
                            Ann W. Fourtner, Charles R. Fourtner, and Clyde Freeman Herreid 329

                            List of Contributors 341

                            Selected Bibliography on Case Study Teaching in Science 345

                            Appendix A: Case Summary and Overview 361

                            Appendix B: Alignment With National Science Education Standards
                                      for Science Content, Grades 9–12 369

                            Appendix C: Evaluating Student Casework 371

                            Index 379

Introduction

Clyde Freeman Herreid

*Critical thinking is like Mark Twain's quip about the weather—
everybody talks about it, but nobody does anything about it.*

Teachers are fascinated by facts, esoteric minutiae that beguile, tantalize, and titillate their fancies. They spend their time in the classroom trying to convince students to appreciate the same ideas. Yet when you ask teachers what they prize most, they will say critical thinking. They claim they want this most of all in their students—the ability to reason (Yuretich 2004). But teachers love their Krebs cycle, Henderson-Hasselbalch equations, tooth formulae, digestive enzymes, hormones, bones, and scientific nomenclature too much to give them up. Nor would I want them to. But let's face the facts: They are not teaching critical thinking.

Most teachers cannot define critical thinking. To take only one example, in 1995, the California Commission on Teacher Credentialing and the Center for Critical Thinking at Sonoma State University initiated a study of college and university faculty throughout California to assess current teaching practices (Paul, Elder, and Bartell 1997). Of the faculty surveyed, 89% said critical thinking was a primary objective in their courses, but only 19% were able to explain what critical thinking is and only 9% were teaching critical thinking in any apparent way. Furthermore, 81% of the faculty believed that graduates from their departments acquired critical-thinking skills during their studies, but only 9% could articulate how they would determine if a colleague's course actually encouraged critical thinking.

Experts do not really agree on a precise definition of critical thinking. But I like Moore and Parker's (2004) approach that critical thinking is the careful, deliberate determination of whether one should accept, reject, or suspend judgment about a claim and one's degree of confidence about one's position. So, critical thinking involves evaluating evidence and examining relevant criteria for making a judgment. It involves logic and clarity, credibility, accuracy, precision, relevance, depth, breadth, significance, and fairness in dealing with an argument. These are the topics of textbooks on critical thinking. Major emphasis is placed on informal logic (often said to be equivalent to critical thinking). Informal logic deals with analyzing and evaluating arguments and addresses how to avoid many of the major mistakes that humans can make. These qualities are said to constitute critical thinking.

To varying degrees, all of these qualities are desirable for anyone, not just scientists. But can they be taught, and how best to do so? We make the argument in this book that such habits of mind can be taught, and case studies are one avenue to achieve this end. In the chapter "The 'Case' for Critical Thinking," David R. Terry

Introduction

brings the critical thinking literature to bear on this issue.

Another approach to critical thinking is seen in Bloom's 1956 taxonomy of "learning domains" in approaching problems. The cognitive domain is especially relevant. Bloom and his team ranked learning in a hierarchy, starting with simple knowledge at the bottom, then comprehension, application, analysis, synthesis, and evaluation at the top of a pyramid. Bloom's original domain arrangement is shown on the pyramid in Figure 1. Anderson et al. (2001) revisited the categorization and produced the arrangement on the right in Figure 1. The differences are small. The new version has translated the original terms into action verbs and switched the order of the top two domains.

Where does critical thinking fit into these schemes? For our purposes, critical thinking corresponds to the learning categories on the upper part of both diagrams. In contrast, in traditional science courses taught by the lecture method, the focus is on the lower part of the pyramids. Students are asked to remember facts, terms, and concepts—hardly critical-thinking exercises. In contrast, the upper part of the pyramid—which deals with application, analysis, synthesis, and evaluation—fits squarely in the critical-thinking camp. So the first goal of this book is to provide a way for teachers to enhance student skills in these areas. But as cognitive scientist Daniel Willingham (2009) points out, critical thinking cannot be taught with abstract exercises. It must be taught in the context of a discipline. Critical thinking in art, music, English literature, or history is not the same as it is in natural science.

What, if any, are the unique features of critical thinking in science, and how do case studies help teach these skills? I argue that it boils down to that hoary chestnut, the scientific method. Recall the steps of the classic scientific method (sometimes called the hypothetico-deductive method): We ask a question, propose a hypothesis to answer the question, devise a test or experiment to test the hypothesis, collect the data from the test, and reach a conclusion. With repeated iterations of the process, the question is solved and science marches on. In Kathy Galluci's chapter "Learning About the Nature of Science With Case Studies," she examines the nuances of the method in detail. You will notice that much of this process is not special. People ask questions and make guesses all the time. But few of us ever do much testing and retesting to see if the data we collect are consistent with our hypotheses. This single feature is the essence of the scientific enterprise and the essence of critical thinking in science. Yes, science is a collection of facts and principles about the physical world, but what is essential to us is that it is a *way of knowing*.

Consequently, if we want to teach students about science, we need to do two things: Give them science content, and teach them the critical-thinking skills that scientists use. We need our students to have a good grounding in science content so they will be able to ask intelligent and relevant questions, suggest hypotheses, and know how to interpret data to reach reasonable conclusions that are consistent with what we already know. But if we are to teach our students about how scientists really

Introduction

go about their business, we need to help them learn about experimental design. We must look closely at the third step in the scientific method.

In short, we want them to know about controlled and uncontrolled experiments, the importance of replicability, the fact that correlation is not the same as cause and effect, falsification, prospective versus retrospective tests, the differences between historical and experimental science, blind and double-blind tests, the placebo effect, human error, fraud, and wishful thinking, as well as how the peer-review process helps identify and prevent flaws and mistakes. So throughout this book, there is an emphasis on experimental design—in fact, it is the dominant theme. If we can get students to be respectful skeptics—the major emphasis of my chapter "Can Case Studies Be Used to Teach Critical Thinking?"—we have gone a long way toward achieving our goal of developing scientifically literate citizens.

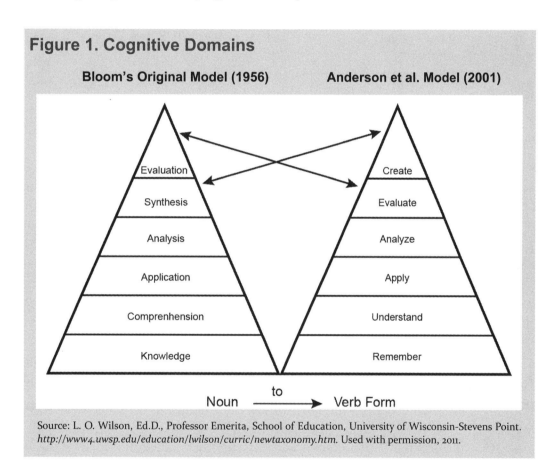

Figure 1. Cognitive Domains

Source: L. O. Wilson, Ed.D., Professor Emerita, School of Education, University of Wisconsin-Stevens Point. *http://www4.uwsp.edu/education/lwilson/curric/newtaxonomy.htm.* Used with permission, 2011.

Introduction

The Case Study Approach

There is a price to pay for our typical approach in science, technology, engineering, and math (STEM) education, and for our devotion to facts, the lecture method, and multiple-choice tests. Let's mention the obvious: We lose large numbers of excellent students who choose to major in other fields. Sheila Tobias (1990, 1992) reported that our beloved lecture is the prime culprit. Science majors have a high tolerance for boring material that seems to have nothing to do with their lives—a kaleidoscope of facts without apparent rhyme or reason. These presumptive scientists have faith that these facts will eventually become relevant in time. Nonscientists have no such faith, and, without context, they decide to leave the game. They are not intrigued by the detailed structure of the atom or the cell. It is not that they do not think it is important; they simply do not see why it should be important to them. We have not shown them how the cavalcade of facts relates to global warming, the debate over creationism versus evolution, natural disasters, cancer, AIDS, sex, or anything else they might care about.

We certainly have not helped them understand the scientific process either. They leave our STEM classrooms filled with revealed truths, believing that is what science is all about—a grab bag of oh-my-gosh facts. They certainly do not have a clue about the long, hard struggle that is involved in digging out the answers to even minor questions. Jon Miller's (1988) pronouncement that only 5% of the American public is scientifically literate did not help the situation. To illustrate the point, he used as his example of American science illiteracy that few people know what DNA is or how the seasons of the year are produced. These might be important factoids, but it is hardly sufficient to gauge scientific literacy. Far more relevant is to have a public that can make intelligent decisions about issues such as oil spills, vaccinations, alternative medicines, health insurance, cloning, and energy. Facts are important, but critical thinking is essential too, and we are doing a poor job of teaching it.

Teaching with case studies can make a difference. Why? Because whatever else they may do, they put learning into context. Case studies tell stories, and people love stories. That's why we have novels, movies, reality shows, and bedtime stories. Stories entertain us, and the best of them leave us memories and scaffolding on which to hang our facts. Jesus told parables. Aesop told fables. Homer told heroic adventures, and the brothers Grimm told fairy tales. And these stories have been remembered for centuries.

This is a storybook of case studies that show critical thinking in action in science. We have chosen these particular cases because they emphasize how science is really done. Our aim is to put flesh and bones on the scientific method, but not the classic method that schoolchildren memorize and parrot back: observation, question, hypothesis, experiment, data collection, and conclusion. A moment's reflection will reveal that not all scientists proceed along these lines. Astronomers and paleontologists have difficulty doing experiments (they are historical scientists), but they do test their ideas against the data. Theoretical physicists do not run experiments

Introduction

themselves, as they work with pen and paper or computers. Yes, the hypothetico-deductive method (the "scientific method") is the *modus operandi* for the scientific enterprise for part of the process. But it is also much more than that.

Our overarching purpose is to show scientists in action—to show how they operate and the rules that they live by. The first chapter, "The Scientific Method Ain't What It Used to Be," as the title suggests, aims to correct the common misconception that many first-year science textbooks perpetuate—namely, that science proceeds by a linear series of steps. The chapter emphasizes how scientists interact with each other and society at large. We want students to recognize how the discoveries and failures of science can affect the general public. We want them to know how results can be manipulated, distorted, and reinterpreted by folks with different experiences and agendas. We deeply care about ethical concerns. We want scientists to keep their part of the bargain to follow the canons and traditions of the discipline. We want students to know how changing social mores can force a reinterpretation of what is ethical. We want them to learn how our technological *tours de force* have created new ethical dilemmas that future generations must solve. Case studies perhaps can do all of this better than most other ways of dealing with the material because they place the students in the position of making decisions about real-world problems; they put learning into context.

The cases in this book are drawn from our website for the National Center for Case Study Teaching in Science (*http://sciencecases.lib.buffalo.edu/cs*). In this book, we present each case and an abbreviated version of its teaching notes, including a section on misconceptions that the case addresses. These misconceptions, common among our students, reveal flaws in critical thinking. More complete teaching notes for the cases can be found on our website, along with answer keys where available. The book is designed for college and high school teachers (and may also be of interest to middle school science teachers), who can select case exercises for use in their classrooms. We expect that teachers who plan to use the cases will download the individual case PDFs from the website and distribute them to students in class rather than directing students to the website itself, where the teaching notes are displayed.

To help teachers choose cases for their classes, Appendix A provides summary information about each case; Appendix B shows how each case is aligned with the National Science Education Standards for grades 9–12; and Appendix C is a discussion of ways to evaluate student work.

To give students a real sense of how scientists go about their work, we first present four historical cases in section II. These cases involve real events and people, such as Ignaz Semmelweis, whose simple observation about hand-washing and mortality rates among women in a hospital maternity ward saved countless lives, and John Snow, another physician, who set out to discover the source of cholera in Victorian London. We also meet in this section Nobel Prize winners J. Robin Warren and Barry Marshall, who discovered that ulcers were caused by a bacterium that could

Introduction

be treated with antibiotics. There is also a case that explores the Salem witch trials that took place in Massachusetts in the late 1600s. What caused the hysteria? Could there be a scientific explanation for what happened? Students will attempt to find a plausible explanation for a mystery that remains unsolved.

Douglas Allchin (from the University of Minnesota) has written extensively on the subject of historical cases. His article "How *Not* to Teach Historical Cases in Science" is a valuable resource (Allchin 2000). He is also the co-author of *Doing Biology* (Hagen, Allchin, and Singer 1996), a set of 17 historical case studies, and he maintains a website devoted to historical problem-based cases (*http://ships.umn.edu*). It is important to note that his focus is on the historical features of the stories. In this book, we focus on the discovery process and how the elements of scientific methodology are highlighted by these cases.

Next, in section III, we present six cases of scientific inquiry. When most people think of a scientist, they think of someone in a lab cooking up experiments. A lot of scientists do just that. They pose questions, make hypotheses and predictions, think up ways to test them, collect the resulting data, and draw conclusions. The cases we have selected put students through many of these same steps, all in one or two class periods. In this section, you will find cases such as the one based on the true story of statistician Ronald Fisher, who designed an experiment to test a woman's claim that she could always tell if milk was added before or after her tea had been poured into the cup. Other case studies challenge students to re-create experiments probing such matters as a possible cause for Alzheimer's disease, preferential feeding behavior in coots, how PCBs wind up in remote Alaskan lakes, whether drivers really leave their parking spaces faster if others are waiting, and whether a traditional Native American remedy relieves the skin's allergic reaction to the toxin found in poison ivy.

In section IV, we examine six cases that verge on the edges of pseudoscience, claims that are not in the mainstream of science, claims of questionable validity. Of course, we know that many of the world's most important discoveries were initially met with skepticism. With repeated testing, some of these ideas were vindicated—but not all. As Carl Sagan reminds us, "The fact that some geniuses were laughed at does not imply that all who are laughed at are geniuses. They laughed at Columbus, they laughed at Fulton, and they laughed at the Wright Brothers. But they also laughed at Bozo the Clown" (Sagan 1980, p. 34). Certainly claims of extrasensory perception and the healing power of prayer, for example—just two of the topics explored in the cases collected here—require close and careful scrutiny.

In sections V and VI, we present cases that explore the effects of science on society and society on science and show some of the ways that the news media puts spin on the scientific process. We finish in section VII with some of the ethical dilemmas that confront scientific researchers. All along we try to present cases that examine real data even as we deal with the sociology of science.

Long after students have left the classroom and forgotten glycolysis, Avogadro's

Introduction

number, Fresnel's transmission coefficient, biostratigraphy, and the ideal gas law, they will need to read newspapers and blogs and listen to CNN, or the future equivalents. After college, for the next 60 years of their lives, they will be bombarded by problems infused with science. They need to be able to consider claims that will be made and ask the first and most important questions of a critical thinker: Is it true? Why should I believe this? What is the evidence? Is there counterevidence that should be considered? And then they will need to look carefully at the logic of the argument, identify *ad hominem* attacks when they occur, and consider the consequences of their (and others') actions. We hope this book helps them on the way.

References

Allchin, D. 2000. How not to teach historical cases in science. *Journal of College Science Teaching* 30 (1): 33–37.

Anderson, L. W. et al., eds. 2001. *A taxonomy for learning, teaching, and assessing: A revision of Bloom's taxonomy of educational objectives.* Boston: Allyn & Bacon (Pearson Education Group).

Bloom B. S. 1956. *Taxonomy of educational objectives, handbook I: The cognitive domain.* New York: David McKay.

Bloom, B. S., and D. R. Krathwohl. 1956. *Taxonomy of educational objectives: The classification of educational goals, by a committee of college and university examiners. Handbook I: Cognitive domain.* New York: Longmans, Green.

Hagen, J. B., D. Allchin, and F. Singer. 1996. *Doing biology.* New York: HarperCollins College Publishers.

Miller, J. 1988. The five percent problem. *American Scientist* 72 (2): iv, 18.

Moore, B., and R. Parker. 2004. *Critical thinking.* New York: McGraw-Hill Co.

Paul, R. W., L. Elder, and T. Bartell. 1997. *California teacher preparation for instruction in critical thinking: Research findings and policy recommendations.* California Commission on Teacher Credentialing, Sacramento California, 1997. Dillon Beach, CA: Foundation for Critical Thinking.

Sagan, C. 1980. Broca's brain: Reflections on the romance of science. New York: Ballantine.

Tobias, S. 1990. *They're not dumb, they're different: Stalking the second tier.* Tucson, AZ: Research Corporation.

Tobias, S. 1992. *Revitalizing undergraduate science: Why some things work and most don't.* Tucson, AZ: Research Corporation.

Willingham, D. 2009. *Why don't students like school?* San Francisco, CA: Jossey-Bass & Sons.

Yuretich, R. F. 2004. Encouraging critical thinking: Measuring skills in large introductory science classes. *Journal of College Science Teaching* 33 (3): 40–46.

Chapter 1

The Scientific Method Ain't What It Used to Be

Clyde Freeman Herreid

Remember the time when all you had to do was memorize these five steps: ask a question, formulate a hypothesis, perform an experiment, collect data, and draw conclusions? And you received full credit for defining the "scientific method."

Well, those days are gone. Or they should be. Because the folks at the University of California, Berkeley (Roy Caldwell, David Lindberg, Judy Scotchmoor, Anna Thanukos, and their team) have put together an outstanding website called Understanding Science: How Science Really Works (*http://undsci.berkeley.edu*).

This site is a treasure trove of information for K–16 teachers, especially those of us who are interested in using case studies in the classroom. The reason? The use of cases puts science in context. And the context is the real world, and we care how science really works in that world.

Among the many gems on the site, we find a scientific checklist that helps us evaluate fields of study. Science does the following:

- Aims to explain the natural world
- Uses testable ideas
- Relies on evidence
- Involves the scientific community
- Leads to ongoing research
- Leads to benefits for society

The site goes on to compare case studies of Rutherford's claims about the nature of the atom to claims of astrology and intelligent design (ID). It quickly becomes apparent how astrology and ID fail to live up to scientific ideals. For example, although ID does focus on the natural world and aims to explain it, its ideas are generally untestable; it does not specify who or what the "designer" is or how it operates. When

Science Stories: Using Case Studies to Teach Critical Thinking

Chapter 1

ID proponents do come up with testable claims, such as the doctrine of irreducible complexity, the evidence refutes the predictions. Moreover, the ID advocates resist modifying their ideas in response to the scrutiny of the scientific community. In short, it is clear we are not dealing with science.

Another section of the site deals with science's limitations—a good thing to keep in mind when dealing with the interminable wars between science and religion (a favorite topic of evolutionist Richard Dawkins) or controversial topics of any stripe, including stem cells, cloning, genetically engineered foods, or global warming. Here is the site's list of limitations to keep in mind:

- Science does not make moral judgments.
- Science does not make aesthetic judgments.
- Science does not tell you how to use scientific knowledge.
- Science does not draw conclusions about supernatural explanations.

The site also deals with how the average citizen should approach science stories when they do not know what to believe. The site does not give the reader a blueprint for evaluating sensational claims, but it raises fundamental questions that everyone should ask themselves when they hear a statement about the latest diet fad, cancer cure, or global warming: What is the evidence? Who says so? What does the scientific community at large say?

Not surprisingly, the site also offers a fine description of peer review: what it is, how it works, and why it is essential. I have always felt that the essence of peer review is captured in the Russian proverb that Ronald Reagan used to quote during the Cold War: "Trust, but verify." This is especially true if the claims are sensational, such as the assertion that an asteroid wiped out the dinosaurs. As Marcello Truzzi, co-founder of the Committee for the Scientific Investigation of Claims of the Paranormal, has put it, "An extraordinary claim requires extraordinary proof" (Truzzi 1978, p. 11).

How Science Works

The highlight of the Berkeley website is the model showing that science is not a linear process. The hypothetico-deductive method of question, hypothesis, experiment, data collection, and conclusion is just too simplistic. It is not that these elements do not play a role (they surely are the stock and trade of the working scientist), but leaving the process at this point leads to these serious misconceptions, as enumerated by the Berkeley folks:

- Science is a collection of facts.
- Science is complete.

- There is a single scientific method that all scientists follow.
- The process of science is purely analytic and does not involve creativity.
- When scientists analyze a problem, they must use either inductive or deductive reasoning.
- Experiments are a necessary part of the scientific process. Without an experiment, a study is not rigorous or scientific.
- "Hard" sciences are more rigorous and scientific than "soft" sciences.
- Scientific ideas are absolute and unchanging.
- Because scientific ideas are tentative and subject to change, they cannot be trusted.
- Scientists' observations directly tell them how things work (i.e., knowledge is "read off" nature, not built).
- Science proves ideas.
- Science can only disprove ideas.
- If evidence supports a hypothesis, it is upgraded to a theory. If the theory then garners even more support, it may be upgraded to a law.
- Scientific ideas are judged democratically based on popularity.
- The job of a scientist is to find support for his or her hypotheses.
- Scientists are judged on the basis of how many correct hypotheses they propose (i.e., good scientists are the ones who are "right" most often).
- Investigations that do not reach a firm conclusion are useless and unpublishable.
- Scientists are completely objective in their evaluation of scientific ideas and evidence.
- Science is pure. Scientists work without considering the applications of their ideas.
- Science contradicts the existence of God.
- Science and technology can solve all our problems.
- Science is a solitary pursuit.
- Science is done by "old, white men."
- Scientists are atheists.

Chapter 1

The Berkeley Model

Take a look at the diagram in Figure 1.1, and you will see a cluster of four interlocking spheres: Exploration and Discovery, Testing Ideas, Community Analysis and Feedback, and Benefits and Outcomes. The traditional steps of the "scientific method" fall neatly into the two spheres labeled Exploration and Discovery and Testing Ideas. But there is more to the scientific process. There are all sorts of inputs and outputs from

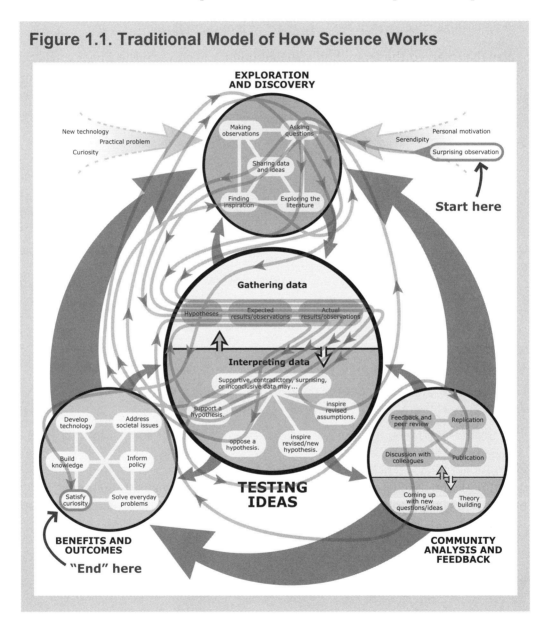

Figure 1.1. Traditional Model of How Science Works

these domains, represented by arrows going and coming. Noticeably absent in the traditional model is the influence of the scientific community and society at large—represented by the two other spheres. Let me provide a case study from the Berkeley website and see how the model plays out.

Asteroids and Dinosaurs: Unexpected Twists and an Unfinished Story

Walter Alvarez, an American geologist, was working with other scientists to look for certain rock patterns that would help to explain part of Earth's history. He examined rocks of the mountains in Italy. As he explored, he kept finding an unusual layer of clay that marked the 65-million-year-old boundary between the Cretaceous and Tertiary periods (referred to as the KT boundary). He noticed that there were many different sorts of marine fossils below the layer of clay, but few above. He asked: Why the reduction in marine fossils? What had caused this apparent extinction of many types of marine life that seemed to happen so suddenly? And could it be related to the extinction of dinosaurs that occurred at the same time on land?

Alvarez wanted to know how long it took for the mysterious clay layer to be deposited because then he would know how quickly the marine life disappeared. He discussed the question with his father, the physicist Luis Alvarez, who suggested using a chemical element called beryllium-10. Like some other substances, beryllium-10 can act as a timer because it is laid down in rocks at a constant rate. The more beryllium in the clay layer, the longer it must have taken for the layer to be deposited. Unfortunately, this investigation was a dead end. Luis suggested trying another element that acts as a timer: iridium. Iridium is often found in meteorites, and meteorite dust "rains down" on Earth's surface at a slow but constant rate.

The father-son team recruited scientists Helen Michel and Frank Asaro to help them look for iridium in the clay layer. Their results were a complete surprise! The team found more than 30 times the amount of iridium that regular meteorite dust might have caused. What could have caused this spike in iridium? Did the iridium spike occur in rock layers around the world? Now Alvarez and his team had even more questions.

Alvarez began looking through published studies to find the location of other KT boundary rock layers that might have the iridium spike. He eventually found one in Denmark and asked a colleague to check. The results were positive—the big spike in iridium was there too. So, whatever happened at the end of the Cretaceous period must have been widespread. Now, another new question: What could have happened to cause these sky-high iridium levels?

It turned out that almost 10 years earlier, two other scientists had proposed the idea that a supernova (an exploding star) at the end of the Cretaceous period had caused the extinction of dinosaurs. Since supernovas throw off heavy elements like iridium, the Alvarez discovery seemed to support this hypothesis. To further test the supernova hypothesis, the team needed other lines of evidence. Alvarez realized that

Chapter 1

if a supernova had occurred, the team should find other heavy elements, such as plutonium-244, at the KT boundary. At first, the team thought they had found the plutonium! It looked as if a supernova had occurred, but after double-checking their results they found that the sample they used had been contaminated. There was no plutonium in the sample after all—and no evidence for a supernova.

So, what could explain these different observations (plenty of iridium, but no plutonium) and tie them together so they made sense? The team came up with the idea of an asteroid impact. That would explain the iridium because asteroids contain a lot of iridium but no plutonium-244. The hypothesis made sense, but also led to a new question: How could an asteroid impact have caused the dinosaur extinction?

After talking with colleagues, Luis Alvarez suggested that a really large asteroid striking Earth would have blown millions of tons of dust into the atmosphere. According to his calculations, this amount of dust would have blotted out the Sun around the world, stopping photosynthesis and plant growth. This would have caused a worldwide collapse of food webs, and therefore many animals would have gone extinct.

In 1980, Alvarez's team published their hypothesis linking the iridium spike and the dinosaur extinction for other scientists to consider. This caused a huge debate and more exploration. Over the next 10 years, more than 2,000 scientific papers were published on the topic. Scientists in the fields of paleontology, geology, chemistry, astronomy, and physics joined the argument, bringing new evidence and new ideas to the table.

Alvarez's team was trying to learn about an event that happened 65 million years ago—when no one was around to see what happened. Many different scientists studied many lines of evidence to help test hypotheses about this ancient event. They studied the following possibilities:

- Extinctions: If an asteroid impact had actually caused a worldwide environmental disaster, many groups of plants and animals would not have survived. Therefore, if the asteroid hypothesis were correct, we would expect to find a large increase in the number of extinctions at the KT boundary. We do.

- The Impact: If a huge asteroid had struck Earth at the end of the Cretaceous, it would have flung off particles from the site where it hit. So, if the asteroid hypothesis were correct, we should find these particles at the impact site in the KT boundary layer. We do.

- Glass: If a huge asteroid had struck Earth at the end of the Cretaceous, it would have caused a lot of heat, melting rock into glass and flinging glass particles away from the impact site. So, if the asteroid hypothesis were correct, we would expect to find glass at the KT boundary. We do.

- Shockwaves: If a huge asteroid had struck Earth at the end of the Cretaceous, it would have caused powerful shockwaves. So, if the asteroid hypothesis is correct, we would expect to find evidence of these shockwaves (such as deformed quartz). We do.

- Tsunamis: If a huge asteroid had struck one of Earth's oceans at the end of the Cretaceous, it would have caused tsunamis, which would have moved ocean sediments around and deposited them somewhere else. So, if the asteroid hypothesis were correct, we would expect to see signs of these deposits at the KT boundary. We do.

- The Crater: If a huge asteroid had struck Earth at the end of the Cretaceous, it would have left behind a huge crater. So, if the asteroid hypothesis were correct, we would expect to find a gigantic crater somewhere on Earth dating to the end of the Cretaceous. We do—the Chicxulub crater on the Yucatan peninsula.

Scientists agreed that the evidence was strong—dinosaurs had gone extinct and there was a widespread iridium spike at the KT boundary. However, scientists did not all agree that the evidence supported a connection between the two.

Scientific ideas are always open to question and to new lines of evidence, so although many observations support the asteroid hypothesis, the investigation continues. The end of the Cretaceous seems to have been a chaotic time on Earth. We have found evidence of massive volcanic eruptions that covered about 200,000 square miles of India with lava. We have found evidence of changes in climate: a general cooling trend and at least one intense period of global warming. We have also found evidence that sea levels were changing and continents were moving around. With all this change going on, ecosystems were surely disrupted. These factors could certainly have played a role in triggering the mass extinction—but did they? Scientists are still studying these questions and many more.

* * * * *

Even a cursory look at the Walter Alvarez case reveals how this was not a one-man operation. Sure, we can identify the basic elements of the hypothetico-deductive method. There are questions, hypotheses, experiments, data, and conclusions, but these seem intertwined higgledy piggledy in the story and keep changing as evidence accumulates and ideas are modified. Most important, the role of the entire scientific community is revealed. Without the help of thousands of other people, the Alvarez ideas would never have been any more than a pipe dream.

Chapter 1

Figure 1.2. How Science Really Works

The scientific journey taken by Walter Alvarez and his colleagues:

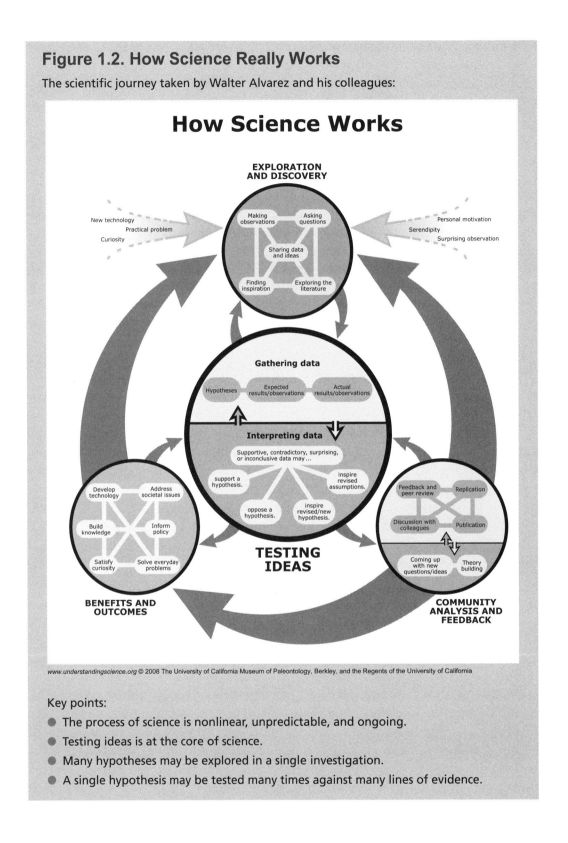

Key points:
- The process of science is nonlinear, unpredictable, and ongoing.
- Testing ideas is at the core of science.
- Many hypotheses may be explored in a single investigation.
- A single hypothesis may be tested many times against many lines of evidence.

To follow the unpredictable timeline of the process and see the elements of the model, the authors of the Berkeley website suggest that students put numbers on each statement in the text where they can spot the key elements in the model (visit Introducing the Understanding Science Flowchart on the Berkeley website for the full lesson). Then they should put the numbers in the correct sphere. Finally, they should connect the numbers in numerical order to see the flow of the journey of Walter Alvarez. Take a look at the flow diagram in Figure 1.2 and see if you think this is a better way to show students how science really works. As teachers, we have a responsibility to break away from the simplistic notion of "question, hypothesis, experiment, data collection, and conclusion" and the myth that science proceeds from a solitary genius toiling away in the laboratory; nothing could be further from the truth.

Note: This chapter originally appeared as Herreid, C. F. 2010. The scientific method ain't what it used to be. *Journal of College Science Teaching* 39 (6): 68–72.

Reference

Truzzi, M. 1978. On the extraordinary: An attempt at clarification. *Zetetic Scholar* 1 (1): 11–22.

Chapter 2
Learning About the Nature of Science With Case Studies

Kathy Gallucci

Case studies are an effective way to help students understand how science works and, perhaps even more important, how science knowledge is constructed. Yet often when we teach the content of science, we overlook the nature of science, and in particular how knowledge claims of science are justified (Abd-El-Khalick, Bell, and Lederman 1998; Duschl 1990). The nature of science (NOS) is defined as "the values and assumptions inherent in science" (Lederman 1992, p. 331) or the "sum total of the 'rules of the game'" (McComas 2004, p. 25). NOS includes both the process of science—that is, the scientific enterprise—as well as what we know about the natural world from science, or scientific knowledge. The scientific enterprise is the "context of discovery" that students often explore with inquiry activities. Scientific knowledge is the "context of justification," in which science is held responsible for its knowledge claims (Duschl 1990). Even when NOS is taught explicitly, it often does not include concepts about the "context of justification," such as hypothesis testing, tentativeness, the distinction between observation and inference, science as a way of knowing, the importance of empirical evidence, and the social and cultural embeddedness of science.

Case studies are particularly useful for teaching these NOS concepts. Case studies integrated into the curriculum early in a course can be the foundation for understanding NOS throughout the semester. Because subsequent topics will illustrate scientific knowledge acquisition, at least implicitly, these foundational cases can be revisited for their NOS concepts later in the semester.

I will describe seven examples of case studies that I have incorporated into my introductory biology course for nonscience majors. The NOS concepts that can be taught with these cases are shown in Figure 2.1 (p. 12). Sample discussion questions for each case are listed in Table 2.1 (p. 13).

Chapter 2

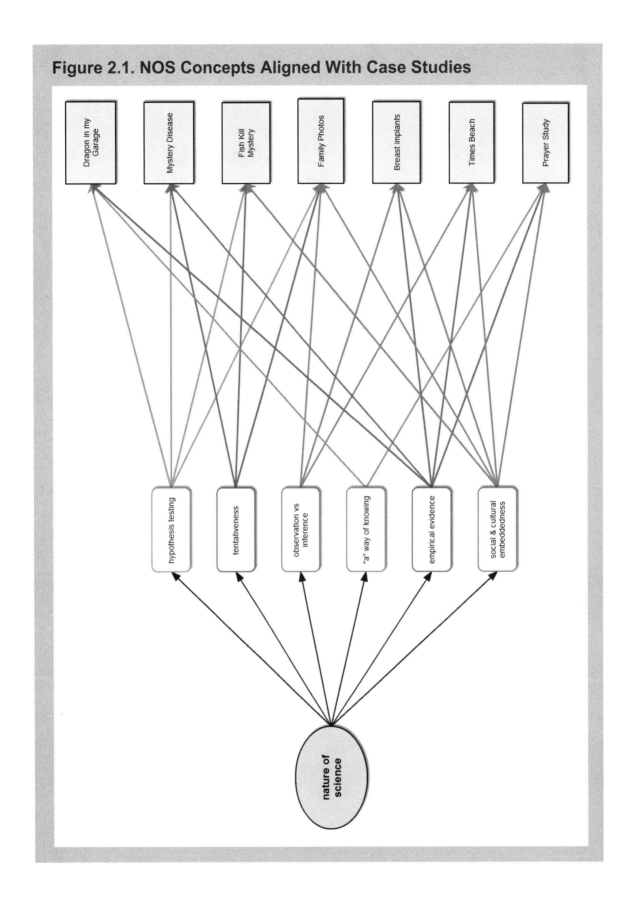

Figure 2.1. NOS Concepts Aligned With Case Studies

Table 2.1. Case Study Questions for NOS "Scientific Knowledge"

Case Study	Suggested Questions for Students

The Dragon in My Garage	1. What is this statement called: A dragon lives in my garage.
	2. How can a hypothesis be tested?
	3. What do we mean by a falsifiable hypothesis?
	4. What do we mean by a testable hypothesis?
	5. Determine if these statements represent valid scientific hypotheses:
	   a. The Earth is flat.
	   b. Life exists in other solar systems.
	   c. The world was created by an intelligent being.

Mystery Disease	1. Make a list of the possible causes of amyotrophic lateral sclerosis-Parkinson dementia complex (ALS-PDC). 2. Write a hypothesis about what you think is the most likely cause. 3. How might you test your hypothesis?
The Fish Kill Mystery	1. What are some of the possible causes of the fish kill? 2. What are some of the observations the women made as they were helping with the cleanup? 3. What inferences did they make? How were most of these possibilities eliminated as the probable cause? 4. What evidence would be required to resolve the debate between the research groups?
Family Photos	1. What is the approximate age of this woman today? 2. Determine the geographic range of this woman. 3. What can you conclude about the cultural background of this woman? 4. Can you account for anomalous observations? 5. What are some of the inferences you made based on observations?
Breast Implants	1. What inference was made by the women who had connective tissue disease? 2. What was the control in the breast implant study done by the Mayo Clinic? 3. Is there evidence to support the hypothesis that silicone breast implants cause connective tissue disease? 4. What was the social and cultural context of the decision made against Dow Corning?
Times Beach	1. Why was the dioxin-contaminated soil in Seveso, Italy, treated differently than in Times Beach, Missouri? 2. What was the prevailing public opinion about dioxin at the time of the cleanup? 3. What is the evidence to support the hypothesis that dioxin in the environment may cause disease in humans?
Prayer Study	1. Is prayer a valid area of study for science? 2. What evidence can be collected to show that intercessory prayer can result in fewer complications for a heart patient? 3. What assumptions did the researchers make?

Chapter 2

The Dragon in My Garage

I start off the first day of class by reading aloud a statement from Carl Sagan's essay "The Dragon in My Garage" (1996). In my most serious tone, I announce, "A fire-breathing dragon lives in my garage." I wait for the room to become quiet and for the students to glance at me with puzzled or cynical looks. Next statement: "Do you think there could be a dragon in my garage?" This is not your everyday open-ended question! I tell the students to ask me any questions about the dragon and I use Sagan's essay (Figure 2.2) to guide my responses. Of course, some acting ability helps prompt the discussion, but if you prefer, you can project or hand out portions of the essay for class discussion.

> **Figure 2.2. The Dragon in My Garage (Sagan 1996)**
>
> "A fire-breathing dragon lives in my garage."
>
> Suppose I seriously make such an assertion to you. Surely you'd want to check it out, see for yourself. There have been innumerable stories of dragons over the centuries, but no real evidence. What an opportunity!
>
> "Show me," you say. I lead you to my garage. You look inside and see a ladder, empty paint cans, an old tricycle—but no dragon.
>
> "Where's the dragon?" you ask.
>
> "Oh, she's right here," I reply, waving vaguely. "I neglected to mention that she's an invisible dragon."
>
> You propose spreading flour on the floor of the garage to capture the dragon's footprints.
>
> "Good idea," I say, "but this dragon floats in the air."
>
> Then you'll use an infrared sensor to detect the invisible fire.
>
> "Good idea, but the invisible fire is also heatless."
>
> You'll spray paint the dragon and make her visible.
>
> "Good idea, but she's an incorporeal dragon and the paint won't stick." And so on. I counter every physical test you propose with a special explanation of why it won't work.
>
> Now, what's the difference between an invisible, incorporeal, floating dragon who spits heatless fire and no dragon at all?

The dragon story usually elicits some interesting discussion. By the end of the class, students generally agree that a scientific hypothesis must be capable of being disproved (or falsified), and that it must be tested in some way. If it is testable, we must be able to collect evidence to support or reject it. This is what makes science a unique way of knowing. To assess student understanding, I ask students to review a list of statements and determine if each can be characterized as testable and/or falsifiable. This case has a lot of staying power. Later in the semester, sometimes a student will even refer to a different bogus hypothesis as a "dragon in the garage."

Mystery Disease

A good example of a current and exciting scientific study is the Mystery Disease case, which investigates the high incidence of amyotrophic lateral sclerosis-Parkinson dementia complex (ALS-PDC) in the Chamorro people of Guam. The case is progressively disclosed to intensify student interest (Gallucci 2006). Students propose hypotheses to explain the prevalence of the disease in Guam, and these usually reflect the list of hypotheses that have been considered by the scientists and how each was rejected for lack of evidence. I usually present this case in a PowerPoint format, with several built-in pauses for group and class discussion. The case has an exciting "O' Henry" ending when students find out which hypothesis is supported. Through this case, students come to understand the tentativeness of science as each hypothesis is rejected. Students also find out that the cause of the disease has relevance far beyond this small community on an isolated island.

The Fish Kill Mystery

This case study is about a fish kill in the Pamlico River of North Carolina (Kosal 2004). Three tourists offer to help with the cleanup, and they ask questions of the biologists about the possible causes of the fish kill. One of the three asks, "What caused this fish kill? Was it the heat? It seems to be affecting a lot of animals these days" (Kosal 2004, p. 36).

One of the biologists explains the relationship between high temperatures and low dissolved oxygen, but the biologist eliminates high temperature as a probable cause of the fish kill. This is an example of how data are collected and used to reject a hypothesis. The biologist proposes a second hypothesis, which is also rejected. Then the biologist describes the notorious dinoflagellate *Pfiesteria piscicida*: "Nutrient influx doesn't only have the potential to stimulate the overpopulation of photosynthesizers, it can also stimulate some plankton species to produce toxins and may indirectly stimulate some to change body forms" (Kosal 2004, p. 37).

This is new knowledge that scientists have acquired by collecting data in the field and doing experiments in the laboratory. But so far, scientific consensus has not been achieved. Another biologist interrupts the discussion: "Don't lead these ladies into thinking that scientists know for sure what is happening with this interesting species. ... Wayne Litaker and his colleagues have evidence supporting a much more simplistic life cycle for *Pfiesteria*, one without all the amoeboid forms" (Kosal 2004, p. 37).

This is a good example of the tentativeness of science. Each research camp has collected evidence, but has come up with a different conclusion. The first biologist continues: "It's still not resolved ... Burkholder and Glasgow ... contend that the amoeba stage wasn't found ... because they did not use a toxic strain of *Pfiesteria* ... [and] the 10-year-old sample ... was too old.... In this sense, the argument is that the absence of an amoeba form does not prove such forms never occur ..." (Kosal 2004, p. 38).

Here, students learn that absence of evidence cannot be the basis for making a conclusion. The second biologist reiterates his point, again referring to the tentativeness of science: "The point is that science doesn't always have a clear-cut answer to questions immediately. The *Pfiesteria* controversy is far from over ..." (Kosal 2004, p. 38).

The take-home lesson for students is that scientists must continue to collect evidence to acquire knowledge. The first biologist continues: "When nutrients such as nitrates come in huge quantities into the estuary ... more phytoplankton are produced .... [and] provide more food for zooplankton. More ... zooplankton provide more food for fish. When large schools of fish are present, they can excrete or secrete substances into the water that act as a cue for *Pfiesteria* to produce a deadly toxin, which results in fish kills" (Kosal 2004, p. 38).

This is the real answer to their question about the fish kill. The students have learned how science accumulates knowledge and builds on that knowledge to answer questions.

Even though this case is classified as an environmental case study on the National Center for Case Study Teaching in Science website, it is evident from these excerpts that this can also be used as an NOS case. The students learn how data can be used to support or reject many of the hypotheses, and they also learn how difficult it is to test hypotheses about complex systems.

Finally, the social and cultural embeddedness of science is also explored in this case. Factories and pig farms are the likely culprits responsible for the nutrient enrichment that promotes blooms of *Pfiesteria*. These fish kills affect not only human health but also the livelihoods of fishermen, as well as those in the tourism industry. These economic and social issues loom large as scientists are pressured to provide answers, even before they have collected enough evidence to have confidence in their recommendations.

Family Photos

This case study is designed to help students understand how scientists can acquire knowledge from the fossil record, by using photos as examples of evidence (Dickey 2000). It is also an excellent example of an NOS case study that provides practice with testing hypotheses by making observations and drawing inferences and illustrates the tentativeness of conclusions.

In Dickey's scenario, students in groups are asked to make hypotheses about the subject in the photos: What is her age, what do they think they know about her family, where does she live, and what is her sociocultural status? Then students are asked to test these hypotheses by using information from the family photos, which are presented in sequential order (with the most recent photos first), in fossil-record fashion.

The groups then report back to the class. They are instructed to list their observations and inferences in separate columns. The photos are filled with hints about the woman's habitat, lifestyle, and social customs, but often students do not have the knowledge to interpret everything they observe. They learn from this experience that inferences made from observations are only as good as their knowledge bases. "Experts" (people who can interpret what they observe in the photos) can contribute new insights about the evidence (Dickey 2000). Moreover, even small clues can be extremely valuable if they are seen in the proper context.

Students act as scientists by making their observations of this "fossil record." However, because of its context, they sometimes have a difficult time distinguishing between observations and inferences. Dickey suggests following this activity with a segment from any available human evolution video showing the observations and inferences paleontologists make when they study the human fossil record (Dickey 2000).

Breast Implants

Students in college today may not be aware that the multinational corporation Dow Corning filed for bankruptcy in 1995, precipitated by the court settlement claiming that silicone breast implants caused disease. More than 10 years later, people may think there was evidence that showed breast implants cause connective tissue disease when, to date, no such evidence has been found. Nevertheless, the connective tissue disease of many women was blamed on their breast implants, a conclusion that was based on anecdotal evidence and fueled by public hysteria.

This case is one of several presented in the ABC News Special Report *Junk Science: What You Know That May Not Be So* (ABC News 1997). The data described in the report clearly exonerate Dow Corning, but the jury chose to believe hired "experts" who blamed the company. This case study is an example that shows how the social and cultural context may influence the interpretation of data. In *Science on Trial*, Marcia Angell describes how the tentativeness of science was used against Dow Corning because the research studies cited could not prove that breast implants were safe (Angell 1997).

Not only can the case study be used to discuss the social and cultural context of science, but it also illustrates the difference between observation and inference, and the importance of collecting evidence. The case could also be presented with news stories, journal articles, or excerpts from Angell's book.

Chapter 2

Times Beach

When an explosion in a dioxin plant in Seveso, Italy, contaminated the nearby environment, government officials decided to bury the contaminated soil and build a city park on the site. In Times Beach, Missouri, a lesser dioxin contamination precipitated the U.S. Environmental Protection Agency's (EPA) decision to close the entire town, relocate the inhabitants, and destroy the "contaminated" homes (ABC News 1997). This case highlights the lack of evidence correlating health problems with dioxin exposure, in both Italy and the United States. Rather than a careful consideration of the evidence, the driving force behind the EPA's decision was the assumption that dioxin threatens human health.

In addition to illustrating the distinction between evidence and inference, the case is also a good example of the "precautionary principle" in action. John Stossel (ABC News 1997) denigrates the EPA's action as a waste of money, while an EPA official claims to be protecting not only the public's health but also the public trust. The dangers of dioxin exposure are a part of our cultural paradigm; the public believes that if science cannot prove something false, it must be true.

Prayer Study

Researchers have attempted to study the effects of intercessory prayer on patients' recoveries for more than 100 years. Recently, large grants have been awarded to prestigious institutions to collect evidence to test this controversial hypothesis. The Prayer Study case (Gallucci 2004) begins with a newspaper article. Students are asked to consider if the effect of prayer is a valid area for science to investigate. Is the hypothesis testable and falsifiable? Can science study the effects of other than natural causes? How are the current public misconceptions of science responsible for the endorsement of this type of research? The case opens up a lively discussion of the differences between science and religion, which is useful later in the semester during the evolution discussion.

This case study is interesting because both scientists and people of faith object to the scientific study of prayer. Scientists claim that the hypothesis is not testable and that there is no way that a "mechanism" (i.e., God) to improve recovery can be measured. Also, if some people routinely pray for all of the sick, there cannot be a reliable control group. People of faith regard the case as an insult to their belief in God. Furthermore, they claim that one cannot measure the sincerity of prayer and that the measurement is not reliable because one prayer may be more "effective" than a thousand.

The current religious climate of the United States and the belief that "anything can be science" (Moss, Abrams, and Robb 2001) lend support to these studies that legitimize personal beliefs. Many times my class has concluded that the prayer study is a lose-lose proposition—bad science and bad religion.

Summary

Undergraduate students typically have difficulty understanding what scientific knowledge is and why it is such a powerful force in modern life. The cases presented here attempt to contextualize NOS concepts focused on learning about how scientific knowledge is acquired. By choosing to teach how scientific knowledge is acquired ("the context of justification") with authentic case studies, as well as the how the scientific enterprise goes about its business ("the context of discovery"), we can promote a more balanced view of NOS in the classroom.

Note: This chapter originally appeared as Gallucci, K. 2009. Learning about the nature of science with case studies. *Journal of College Science Teaching* 38 (5): 50–54.

References

ABC News. 1997. Junk science: What you know that may not be so. A special 1-hour report hosted by John Stossel. *www.abcnewsstore.com*.

Abd-El-Khalick, F., F. R. Bell, and N. G. Lederman. 1998. The nature of science and instructional practice: Making the unnatural natural. *Science Education* 82: 417–436.

Angell, M. 1997. *Science on trial.* New York: W. W. Norton.

Dickey, J. 2000. Using family photographs to introduce human evolution. Paper presented at the National Association of Biology Teachers Annual Convention, Orlando, Florida.

Duschl, R. A. 1990. *Restructuring science education: The importance of theories and their development.* New York: Teachers College Press.

Gallucci, K. 2004. Prayer study—Science or not? *Journal of College Science Teaching* 33 (4): 32–35.

Gallucci, K. 2006. Learning concepts with cases. *Journal of College Science Teaching* 36 (2): 16–20.

Kosal, E. 2004. The fish kill mystery. *Journal of College Science Teaching* 33 (4): 36–40.

Lederman, N. G. 1992. Students' and teachers' conceptions of the nature of science: A review of the research. *Journal of Research in Science Teaching* 29 (4): 331–359.

McComas, W. F. 2004. The keys to teaching the nature of science. *The Science Teacher* 71 (9): 24–27.

Moss, D. M., E. D. Abrams, and J. Robb. 2001. Examining student conceptions of the nature of science. *International Journal of Science Education* 23 (8): 771–790.

Sagan, C. 1996. The dragon in my garage. In *The demon-haunted world: Science as a candle in the dark,* pp. 169–173. New York: Random House.

Chapter 3
Can Case Studies Be Used to Teach Critical Thinking?

Clyde Freeman Herreid

Everyone says they want to teach critical thinking. I have seen these words used as talismans on untold numbers of grant proposals. They fall from the lips of curriculum reformers of every stripe. It has to be the number one phrase David Letterman would put at his top 10 list of clichés of grant entrepreneurs. It is the equivalent of the Holy Grail for educators.

Yet most professors don't appear to give a hoot about the term. They are content to go into the classroom and get on with the job of ripping through their lectures so they can get back to their labs and do the university's real work—research—which ideally will bring fame and fortune to themselves and their universities.

Like many of us, I have given a lot of thought to the use of the term *critical thinking*. Just what does it mean? There are whole books written on the subject, and they haven't helped me one bit—except to make me feel guilty that I may not be putting enough emphasis on critical thinking in my classroom, even if I don't know what it is.

Let me think out loud here for a minute. Critical thinking can't be just the content of a discipline, can it? It sounds more important than that. Yet, certainly, content must be involved; otherwise, one can't really think about a subject about which he knows nothing. But then, how would that explain my daughter, who was a television reporter and often knew little about the subject she was covering? She had an uncanny ability to ask great questions and to pull information out of even the most irascible academician. But still, content knowledge must be in there somewhere. If this is true, then every teacher in some way must automatically be teaching critical thinking.

Surely, we must mean more than "pedagogical content knowledge" (a favorite phrase of educator Lee Schulman); otherwise critical thinking would be a trivial phrase. And our colleagues cannot be accused of pursuing trivial chimeras, can they? So, this leads me to think about process. Critical thinking must have something to do with the way we think—the way we go about problem solving and asking questions. But I struggle with this, too.

Chapter 3

More Than a Mind-Set

I am currently a consultant to a drug company, even though I know little about the pharmaceutical business, except what I read in *Reader's Digest* and on the back of pill bottles. The company has asked me to develop case studies to help their employees acquire a "drug hunter mind-set." After long discussions, the only thing I can get out of this phrase is the obvious point—they want a streamlined way to avoid all of the pitfalls and cost of going off in the wrong directions as they search the pharmacopoeia for miracle cures for aging, baldness, cancer, impotence, sleepwalking, and mean-spiritedness.

I have not found any magic wand to do this; if it existed, others would have been there long ago. But I do believe there are better ways to solve problems—by developing habits of mind that speed things along. They include problem solving, skepticism, flexibility, and seeing alternative strategies when others see only one way.

Let's take problem solving. Once again, this seems tied to specific content. I know that there are problem-solving exercises some authors recommend—the "thinking out of the box" thing—but I don't know of any evidence to support that they improve one's approach to problems. Maybe the data exist, but I don't know of any. It is hard to imagine that working crossword puzzles, reading an advice columnist's opinions about personal crises, or letting your inner child out to play with finger paints helps you achieve the drug hunter mind-set. But maybe it does. Frankly, I think if you want to improve someone's ability in chemistry, you ought to have them do chemistry and grapple with chemical problems.

Pardon me, but I am back to the idea that critical thinking is discipline-specific. What you really want in someone is creativity, curiosity, and skepticism in problem solving—someone who really wants to know the answers. I don't know, really, how you achieve this. We all know some people who really want to know answers, and we know hundreds who don't. Perhaps it is a genetic or a learned trait that is bestowed upon us early. But even if it is, I think we can train people to be more inquisitive.

If I had to choose one general characteristic that smart people share, it would be skepticism—the ability to ask oneself and others if the conclusions and data are correct. Smart people silently or openly say, "What is the evidence for this or that idea? Why should I believe this? Are there other explanations for the data? Is there another way to explain the data? What do you mean when you say this?" If you routinely ask such questions, even when dealing with subjects out of your own area of expertise, you will be well off. Certainly, this is true in the political arena. We have just had a terrible brouhaha—fiasco, is more like it—over the war in Iraq. Assumptions and hearsay, rather than evidence, dominated the debate.

Would that we could imbue skepticism into the American public about UFOs, psychic healing, astrology, creation "science," and a host of other paranormal claims. This goes for TV infomericals touting hair replacements, exercise equipment, vitamin therapy, and so forth. And it goes for supposed experts in our own disciplines as well.

Asking to see the evidence is a good thing. It helps if you have a little background in statistics, too!

Now, how can we develop this habit of mind in our students? The best way is to model it ourselves. Constantly, in lectures and discussions, we should openly ask: "Why should we believe this?" But this isn't enough. Most of us only got good at this in our careers as graduate students. It happened as we gained experience, read original literature, and attended journal clubs where articles were repeatedly attacked. Then we rose eagerly to the challenge. Soon we were emulating our mentors and sneering at claims of authors and doubting everything. There was probably even a stage in many of our lives when we were apt to be hypercritical and saw nothing of worth in even excellent papers because of some trivial transgression in procedure. If this scenario is correct, then skepticism can be taught!

This brings me to case studies. If reading, arguing, and challenging are the hallmarks of critical thinking, then case studies are the poster children for the process. Most of them are discipline specific, certainly. But they all grapple with the essence of critical thinking—asking for evidence—thereby developing a habit of mind that should permeate everyday life. Many case studies deal with real social problems such as global warming, pollution, environmental degradation, and medical issues. I like the ones that deal with general problems: How should one develop a dossier when seeking a job? Does prayer help heal the sick? Does acupuncture work? Such cases can be used in many different disciplines because of their general nature.

Best-Case Scenario

The best case technique that I know is one called the "Interrupted Case Method." Readers can see a version of it in Chapter 11, "Mom Always Liked You Best." The method begins when the teacher gives students (ideally working in groups) a problem faced by real researchers. The teacher asks the students to come up with a tentative approach to solving the problem. After students work for about 15 minutes, the professor asks them to report their thoughts. Then the teacher provides some additional information about the problem saying that the real scientists who struggled with the problem decided to do it in a certain way.

The professor tells of additional difficulties and asks students to brainstorm solutions. Again, they report after discussions. Then, perhaps the teacher provides additional data for their interpretations. Students consult with their teammates and report out. Again, the instructor gives them the interpretation offered by the original authors. And so on.

The interrupted case has enormous virtues. Students struggle with a real research problem and challenge each other and the data. Most important, they see different groups offering alternative approaches to the problem, and they see model behavior from the experts. I love this method because it is the way real science works—we have to work with incomplete data, make tentative hypotheses, collect more information, refine our hypotheses, make more predictions, get more data, and so on. In fact, this interrupted method is the very one that I use in workshops with the pharmaceutical industry, training folks there to attack problems. They like it, too. So, I would argue that the case method has the real potential to develop the same skepticism that we all developed in graduate school when we analyzed research papers and saw what went on in the collection of data. The trouble with the lecture method is that it seldom exposes students to what really happens in the process of collecting data. Once students see this, they are forever changed. They rapidly recognize that there are alternative ways of attacking a problem and alternative interpretations of the data. They begin to doubt.

Most textbooks and lectures give purported facts as if they were received wisdom—wisdom that is certain and irrefutable. This is a great disservice. Students are not likely to question how we know a particular fact if we speak *ex cathedra*. We cannot develop a drug hunter mind-set this way, or any other type of inquiring mind.

William Perry, the Harvard psychologist famous for outlining the Perry model of student development, pointed out that the earliest stage in the maturity of students is the "dualist." The student at that stage sees the teacher and parents as absolute authority figures and is convinced that everything in the textbook is correct. There are always right and wrong answers to questions. This student accepts that what teachers say is truth and regurgitates it back on the tests. Unfortunately, the lecture method perpetuates this stage in students. Further, it distorts the actual way that science is accomplished. Students are left with the idea that Newton, sitting under an apple tree, was bonked on the head and gravity was born—it was all "eureka!"

Case studies don't do this. They show the messy, get-the-hands-dirty approach that is real science. Cases demand skepticism, flexibility, and the ability to see alternative approaches. Problem solving is its *sine qua non*. In short, cases demand critical thinking.

Chapter 4
The "Case" for Critical Thinking

David R. Terry

In 1990, after considerable discussion and debate, a panel of 46 experts from a variety of fields devised a statement regarding critical thinking and the ideal critical thinker:

> We understand critical thinking to be purposeful, self-regulatory judgment which results in interpretation, analysis, evaluation, and inference, as well as explanation of the evidential, conceptual, methodological, criteriological, or contextual considerations upon which that judgment is based. Critical thinking is essential as a tool of inquiry. As such, critical thinking is a liberating force in education and a powerful resource in one's personal and civic life. While not synonymous with good thinking, critical thinking is a pervasive and self-rectifying human phenomenon. The ideal critical thinker is habitually inquisitive, well-informed, trustful of reason, open-minded, flexible, fair-minded in evaluation, honest in facing personal biases, prudent in making judgments, willing to reconsider, clear about issues, orderly in complex matters, diligent in seeking relevant information, reasonable in the selection of criteria, focused in inquiry, and persistent in seeking results which are as precise as the subject and the circumstances of inquiry permit. Thus, educating good critical thinkers means working toward this ideal. It combines developing critical thinking skills with nurturing those dispositions which consistently yield useful insights and which are the basis of a rational and democratic society. (Facione 1990, p. 21)

This certainly reads like it was drawn up by a committee, and it's hard to quarrel with any of it; the experts seem to be trying to please everyone, and that includes a lot of people. To get a sense of the magnitude of the literature, my search of Google using the phrase "critical thinking" on July 9, 2011, returned 36,200,000 listings. Let's be a little more selective and take a look at some of the relevant research literature.

Chapter 4

The first point to make is that most experts in the field do not regard critical thinking as a body of knowledge to be delivered as a separate subject in school; instead, it is like reading and writing, with applications in all areas of learning (Facione 1990). Various approaches for teaching critical thinking have been proposed. The most controversial issue involves whether it should be taught separately (the "general" approach) or as an aspect of domain-specific instruction (Ennis 1989). Recent research suggests that the epistemic culture of various disciplines influences the ways in which critical thinking is understood (Jones 2007) while also acknowledging that the content of thought does not strongly determine its process, which is driven more by the type of task being addressed, resulting in important commonalities across fields (Smith 2002).

One other point emerges clearly: Every definition of critical thinking seems to require that students engage in a deeper processing of information than what occurs in traditional science education (Furedy and Furedy 1985; Morgan 1995). This is ironic because most science instructors claim that one of their key goals is to teach critical thinking.

The origin of critical thinking as a goal for education dates back at least as far as the ancient Greeks, when Socrates, Plato, and Aristotle encouraged their students to realize that things often are not what they seem to be on the surface (Burbach, Matkin, and Fritz 2004). Many educators have continued to stress the importance of critical thinking, following John Dewey, who indicated in *Democracy and Education* (1916, p. 179) that "all which the school can or need do for pupils ... is to develop their ability to think."

Practical understanding tells us the following:

Learning that does not involve thinking is nothing but the memorization of facts not understood, resulting in the formation of mere opinions, not the possession of genuine knowledge and understanding. (Adler 1987, p. 11)

Critical thinking has found its way into influential science education reform documents. The National Research Council (NRC) emphasizes throughout the *National Science Education Standards* (1996) that a primary goal of science education is to strengthen problem-solving and critical-thinking skills. In addition, *Science for All Americans* contends that

[e]ducation should prepare people to read or listen ... critically, deciding what evidence to pay attention to and what to dismiss, and distinguishing careful arguments from shoddy ones. Furthermore, people should be able to apply those same critical skills to their own observations, arguments, and conclusions, thereby becoming less bound by their own prejudices and rationalizations. (AAAS 1989, p. 139)

In *Shaping the Future: New Expectations for Undergraduate Education in Science, Mathematics, Engineering, and Technology* (1996), the National Science Foundation (NSF) advises educators to "devise and use pedagogy that develops skills for communication, teamwork, critical thinking, and lifelong learning in each student" (p. iii). Meaningful education requires an understanding of the essential concepts of science, but it is just as important that scientifically literate persons use critical thinking as they apply scientific understanding to their lives. Literate individuals must be able to use scientific information appropriately to make wise choices and effectively solve problems they encounter in life. They must be able to make well-informed judgments about the reliability and accuracy of scientific information that is presented to them. People who are scientifically literate do not simply provide information about scientific concepts in a quiz-show context. Scientifically literate individuals must use science skillfully while working through the often complex thinking tasks encountered in both their personal and professional lives (Swartz 1997).

Active Learning

How do we achieve this desirable end? The answer is through active learning. Critical thinking requires students to be actively involved in their learning (Browne and Freeman 2000) as they attempt to understand and apply the information to which they are exposed (Ahern-Rindell 1998/1999). The research suggests that instructional techniques that include high-level questioning, authentic investigations, and small-group learning might be the most valuable for improving critical-thinking skills among students. Instructional techniques that encourage passivity in a learner are probably not going to support and may even impede critical thinking (Browne and Freeman 2000). In short, getting students to actively do something in the classroom rather than sit passively taking notes is likely to produce the best results.

Physics professor Richard Hake persuasively demonstrated this essential point when he published his study on 6,000 physics students. Students who merely listened to lectures on the topic of mechanics performed much more poorly on exams dealing with the topic than students who were taught the same material via active learning strategies (Hake 1998). Richard Paul, an expert in the field of critical thinking, recommended in *Critical Thinking: What Every Person Needs to Survive in a Rapidly Changing World* (1992) that activities and assignments be designed so that students must think their way through them. To develop students' thinking skills in the science classroom, instruction should require students to hypothesize, speculate, generalize, create, and evaluate while providing opportunities for identifying and solving problems, especially problems that are real and of interest and concern to students (Pizzini, Abell, and Shepardson 1988). For students to improve their critical-thinking skills, they must engage in critical thinking itself (van Gelder 2005).

Chapter 4

Here is more of the same: Lauer (2005) states that it is far more productive to have students comprehend the essence of critical thinking through active learning techniques such as case study teaching than for the teacher to describe critical thinking in a lecture and ask for a definition on a test. And Johnson and Johnson (1991) make the persuasive argument that interpersonal exchange within cooperative learning groups promotes critical thinking. According to the American Association for the Advancement of Science (AAAS) (1989), learning science is an active process dependent on social interaction. Active conversations with instructors and peers are necessary to develop a complete understanding of scientific concepts as well as to encourage critical thinking. The strength of active learning strategies is that they facilitate personal involvement with the material, provoking students into discussion and evaluation. Thinking begins when a state of doubt exists about what to do or what one believes (Browne and Freeman 2000).

Storytelling in Science Classes

Storytelling in the form of case-based teaching was introduced into Harvard University's law and business school curricula 100 years ago. These real-world problems, taught by a professor via the discussion method, introduced students to actual scenarios that they were likely to face once they graduated. The method attracted widespread admiration and is actively used today. In a Canadian medical school (McMaster University) 50 years later, faculty members introduced another formal version of storytelling called problem-based learning (PBL). In this instance, small groups of medical students met regularly with faculty facilitators to try to diagnose patients' ailments given the limited information they received piecemeal. Both case-based teaching and PBL have been successful because they are based on stories that put learning in context and actively engage students in the learning process (Herreid 2006).

If we define case studies as "stories with an educational message," it is evident that there are many ways to tell the story. Accordingly, the case study approach can be categorized into several major types depending on how the story is used in the classroom: (1) by lecture, (2) by whole-class discussion, (3) by small groups (as in PBL, for example), (4) through individual analysis, or (5) via lecture in large classes using personal response systems (clickers) (Herreid 1998). Most of these case methods depend heavily on active learning strategies, which are linked to all of the desired attributes of critical thinking. Cases encourage students to think through scientific problems with a skeptical eye, asking them to see if the conclusions are justified by the evidence. "If reading, arguing, and challenging are hallmarks of critical thinking, then case studies are the poster children for the process" (Herreid 2004, p. 13). A large percentage of case study teachers would agree. In a survey of 1,634 case study teachers, fostering critical thinking was one of the top two reasons they used the case method (Herreid et al. 2011).

Why Case-Based Teaching Works

For most students, learning by doing provides far better and more lasting results than learning through lectures (Naumes and Naumes 1999). Education policy in the United States is moving toward such teaching strategies—including group problem solving during lectures, problem-based learning, case studies, inquiry-based labs, and interactive computer learning—to more actively involve students in the learning process (Handelsman et al. 2004).

AAAS advocates the use of case studies, noting that understanding a case requires students to integrate knowledge in multiple frameworks, including historical, philosophical, social, political, economic, and technological contexts. Furthermore, there is great potential for motivating students with cases, increasing their understanding, expanding independent learning, and promoting information assessment skills (AAAS 1989). Cases in science are usually fact-driven but often are also open-ended because the data are inadequate or emotions are involved and ethical or political decisions are at stake. The newest scientific ideas are often by their very nature contentious, and different people often view evidence differently in different contexts, sometimes leading to very different conclusions (Herreid 1997).

The case study method allows students to use their prior knowledge and interests related to the case to construct new knowledge. Cases facilitate active and reflective learning by exposing learners to complex situations, allowing them to discuss and debate courses of action and providing them with the opportunity to create and discover new ideas. Good cases are realistic and generate intrinsic motivation by encouraging teamwork and accountability (Tomey 2003).

Many of the methods of case study teaching involve small groups, and we know a great deal about these well-studied collaborative or cooperative learning strategies. Peer collaboration encourages metacognition as students work together to identify what they already know about a case and determine what they need to know. By discussing their ideas with others, students come to appreciate other viewpoints and also discover their own misconceptions (Waterman and Stanley 2004). A meta-analysis of more than 300 studies that examined the relative efficacy of cooperative learning on achievement in college settings found that cooperative learning promotes higher individual achievement than either competitive or individualistic learning. The same research demonstrated that college students learning cooperatively perceive greater academic and personal social support from peers and instructors than do students working competitively, and they become more socially skilled (Johnson, Johnson, and Smith 1998).

Chapter 4

Problem-Based Learning (PBL)

Problem-based learning (PBL) is one type of case study method, but it has received more serious assessment than perhaps any other educational technique. It has been identified as one of the most promising instructional methods for promoting critical-thinking skills (Pithers and Soden 2000; Tsui 1999). According to Gijbels et al. (2005), there are six core characteristics of PBL:

1. Learning is student centered.
2. Learning occurs in small student groups.
3. A tutor is present as a facilitator or guide.
4. Authentic problems are presented at the beginning of the learning sequence, before any preparation or study has occurred.
5. The problems encountered are used as tools to achieve the required knowledge and the problem-solving skills necessary to eventually solve the problem.
6. New information is acquired through self-directed learning.

These characteristics are aligned with important aspects of critical-thinking research. The general goal of PBL is for students to develop successful problem solving in both the acquisition and application of knowledge (Gijbels et al. 2005). Students work in cooperative groups to solve deliberately ill-defined or open-ended real-world problems that engage their curiosity and drive their learning (Mierson 1998). They learn to identify the problems of a specific discipline, analyze them, and contribute to solutions (Gijbels et al. 2005). Students learn to ask critical questions and identify what they need to know to answer their questions, as well as where to find the answers. These features of problem-based learning emphasize the kinds of critical-thinking skills that are essential in science education.

Generally, one or more faculty members facilitate the problem-solving process, acting as coaches and role models for the students. According to Sheella Mierson (1998, p. S16), "The problems *lead* the students to learn basic concepts rather than being presented as applications of concepts they have already learned." Problem-based learning enables students to learn a body of essential knowledge, use knowledge effectively in the context of problem situations both in and out of school, and extend or improve that knowledge to develop strategies for dealing with future problems (Delisle 1997). As students experience problem-based learning, they progress to the level of an expert in a discipline, characterized by the coherent and flexible use of knowledge to describe and solve novel problems (Gijbels et al. 2005).

Developing critical-thinking skills in science classrooms requires the active engagement of students in problem-solving activities as well as a reorganization toward more student-centered activities in which conceptual understanding rather than memorization is taught (Narode et al. 1987). The lectures, textbooks, and perfunctory laboratory activities that are typical of science education often leave students with incomplete or incorrect knowledge of scientific principles, underdeveloped intellectual skills, and little awareness of the influence of science on their lives. Students are often successful when solving formal textbook problems, but incorrectly interpret the same scientific principles when asked to solve problems posed in real-world contexts (AAAS 1989). Problem-based learning is seen as a promising solution to some of the lack of critical-thinking and problem-solving ability in education today (Barrows 1996).

A Bit of Theory: The Constructivist Model and the Case Method

Problem-based learning derives from the theory that learning is a process in which the learner actively constructs knowledge, with instruction playing a role only to the extent that it enables and fosters constructive activities (Gijselaers 1996). Successful learning requires a continuous reworking of ideas through an individual's experiences. It is crucial that learners observe experiences from multiple perspectives, reflect on those experiences, and then use new knowledge to make decisions and solve problems (DeMarco, Hayward, and Lynch 2002). Most researchers agree that problem-based learning achieves a positive effect because the experience of working through a problem activates a mental model that facilitates performance. Capon and Kuhn (2004) believe that this activated model has the potential to allow for superior acquisition of new material (due to connections to previous knowledge structures), superior recall of new material (because of increased retrieval pathways), and superior integration of new material with existing knowledge (leading to restructuring and enhanced conceptual coherence). Their research supports the last possibility as the most likely, describing the benefit of problem-based learning as its power to promote sense making.

The general shift within the education community toward constructivist learning perspectives has led to the development of several theories regarding student thinking and learning, including that of situated cognition (Herrington and Oliver 1999), which supports many of the ideas related to the benefits of the case study method for promoting critical thinking. Situated cognition exists in various forms and is also referred to as cognitive apprenticeship, situated learning, and legitimate peripheral participation. All forms share the idea that learning and doing are inseparable and that learning is a social process (Hendricks 2001).

The theoretical framework of situated cognition relies greatly on the work of L. S. Vygotsky (1978), who proposed that learning results from complex social interactions, with individuals most often interacting and cooperating within the context of some shared task. Certainly the group problem-solving aspect of the case method of teaching fits this description. More recent research in the area of situated learning has emphasized the role of the learner as a "cognitive apprentice" who gains knowledge through imitation and practice in cooperative, authentic activities, entering at the periphery of the community and gradually becoming more active and engaged (Brown, Collins, and Duguid 1989; Lave and Wenger 1991). In this manner, learners are able to gain motivational support, participate in shared thinking and expertise, and engage in conflicts stimulating further debate. An additional benefit is that students are exposed to different models of thinking and learning strategies (Gieselman, Stark, and Farruggia 2000).

From this brief sortie into the literature, it seems abundantly clear that no matter how we look at the case method of teaching, whether it is through the practical lens of a classroom teacher or with the eye of an educational theorist, the case method is a formidable weapon against apathy and sloth in the classroom. Moreover, with regard to the major thrust of this book, problem-based learning and other case study methods encourage the development of independent critical thinkers more effectively than other methods of instruction (Arambula-Greenfield 1996; Grunwald and Hartman 2010; Quitadamo et al. 2011). This book provides an instructional resource that should encourage the development of critical-thinking skills in our science students.

References

Adler, M. J. 1987. "Critical thinking" programs: Why they won't work. *The Education Digest* 52 (7): 9–11.

Ahern-Rindell, A. J. 1998/1999. Applying inquiry-based and cooperative group learning strategies to promote critical thinking. *Journal of College Science Teaching* 28 (3): 203–207.

American Association for the Advancement of Science (AAAS). 1989. *Science for all Americans: A project 2061 report on literacy goals in science, mathematics, and technology*. Washington, DC: AAAS.

Arambula-Greenfield, T. 1996. Implementing problem-based learning in a college science class. *Journal of College Science Teaching* 26 (1): 26–30.

Barrows, H. S. 1996. Problem-based learning in medicine and beyond: A brief overview. In *Bringing problem-based learning to higher education: Theory and practice*, ed. L. Wilkerson and W. H. Gijselaers, pp. 3–12. San Francisco: Jossey-Bass.

Bloom, B. S. 1956. *Taxonomy of educational objectives, handbook 1: Cognitive domain*. New York: Addison Wesley.

Brown, J. S., A. Collins, and P. Duguid. 1989. Situated cognition and the culture of learning. *Educational Researcher* 18 (1): 32–42.

Browne, M. N., and K. Freeman. 2000. Distinguishing features of critical thinking classrooms. *Teaching in Higher Education* 5 (3): 301–309.

Burbach, M. E., G. S. Matkin, and S. M. Fritz. 2004. Teaching critical thinking in an introductory leadership course utilizing active learning strategies: A confirmatory study. *College Student Journal* 38 (3): 482–493.

Capon, N., and D. Kuhn. 2004. What's so good about problem-based learning? *Cognition and Instruction* 22 (1): 61–79.

Delisle, R. 1997. *How to use problem-based learning in the classroom*. Alexandria, VA: Association for Supervision and Curriculum Development.

DeMarco, R., L. Hayward, and M. Lynch. 2002. Nursing students' experiences with and strategic approaches to case-based instruction: A replication and comparison study between two disciplines. *Journal of Nursing Education* 41 (4): 165–174.

Dewey, J. 1916. *Democracy and education: An introduction to the philosophy of education.* New York: Macmillan.

Ennis, R. H. 1989. Critical thinking and subject specificity: Clarification and needed research. *Educational Researcher* 18 (3): 4–10.

Facione, P. A. 1990. *Critical thinking: A statement of expert consensus for purposes of educational assessment and instruction. Research findings and recommendations.* Newark, DE: American Philosophical Association.

Furedy, C., and J. J. Furedy. 1985. Critical thinking: Toward research and dialogue. In *Using research to improve teaching*, ed. J. G. Donald and A. M. Sullivan, pp. 51–69. San Francisco: Jossey-Bass.

Gieselman, J. A., N. Stark, and M. J. Farruggia. 2000. Implications of the situated learning model for teaching and learning nursing research. *Journal of Continuing Education in Nursing* 31 (6): 263–268.

Gijbels, D., F. Dochy, P. Van den Bossche, and M. Segers. 2005. Effects of problem-based learning: A meta-analysis from the angle of assessment. *Review of Educational Research* 75 (1): 27–61.

Gijselaers, W. H. 1996. Connecting problem-based practices with educational theory. In *Bringing problem-based learning to higher education: Theory and practice*, ed. L. Wilkerson and W. H. Gijselaers, pp. 13–21. San Francisco: Jossey-Bass.

Grunwald, S., and A. Hartman. 2010. A case-based approach improves science students' experimental variable identification skills. *Journal of College Science Teaching* 39 (3): 28–33.

Hake, R. R. 1998. Interactive-engagement vs. traditional methods: A six thousand-student survey of mechanics test data for introductory physics courses. *American Journal of Physics* 66 (1): 64–74

Handelsman, J., D. Ebert-May, R. Beichner, P. Bruns, A. Chang, R. DeHaan, J. Gentile, S. Lauffer, J. Stewart, S. M. Tilghman, and W. B. Wood. 2004. Scientific teaching. *Science* 304: 521–522.

Hendricks, C. C. 2001. Teaching causal reasoning through cognitive apprenticeship: What are results from situated learning? *Journal of Educational Research* 94 (5): 302–311.

Herreid, C. F. 1997. What is a case? Bringing to science education the established teaching tool of law and medicine. *Journal of College Science Teaching* 27 (2): 92–94.

Herreid, C. F. 1998. Sorting potatoes for Miss Bonner: Bringing order to case-study methodology through a classification scheme. *Journal of College Science Teaching* 27 (4): 236–239.

Herreid, C. F. 2004. Can case studies be used to teach critical thinking? *Journal of College Science Teaching* 33 (6): 12–14.

Herreid, C. F. 2006. *Start with a story: The case study method of teaching science.* Arlington, VA: NSTA Press.

Herreid, C. F., N. A. Schiller, K. F. Herreid, and C. Wright. 2011. In case you are interested: A survey of case study teachers. *Journal of College Science Teaching* 40 (4): 76–80.

Herrington, J., and R. Oliver. 1999. Using situated learning and multimedia to investigate higher-order thinking. *Journal of Educational Multimedia and Hypermedia* 8 (4): 401–421.

Johnson, D. W., and R. T. Johnson. 1991. Collaboration and cognition. In *Developing minds, vol. 1: A resource book for teaching thinking*, ed. A. L. Costa, pp. 298–301. Alexandria, VA: Association for Supervision and Curriculum Development.

Johnson, D. W., R. T. Johnson, and K. A. Smith. 1998. Cooperative learning returns to college:

Chapter 4

What evidence is there that it works? *Change* 30 (4): 26–35.

Jones, A. 2007. Multiplicities or manna from heaven? Critical thinking and the disciplinary context. *Australian Journal of Education* 51 (1): 84–103.

Lauer, T. 2005. Teaching critical thinking skills using course content material: A reversal of roles. *Journal of College Science Teaching* 34 (6): 34–37.

Lave, J., and E. Wenger. 1991. *Situated learning: Legitimate peripheral participation*. New York: Cambridge University Press.

Mierson, S. 1998. A problem-based learning course in physiology for undergraduate and graduate basic science students. *Advances in Physiology Education* 20 (1): S16–S27.

Morgan, W. R. 1995. "Critical thinking"—What does that mean? Searching for a definition of a crucial intellectual process. *Journal of College Science Teaching* 24 (5): 336–340.

Narode, R., M. Heiman, J. Lochhead, and J. Slomianko. 1987. *Teaching thinking skills: Science*. Washington, DC: National Education Association.

National Research Council (NRC). 1996. *National science education standards*. Washington, DC: National Academies Press.

National Science Foundation (NSF). 1996. *Shaping the future: New expectations for undergraduate education in science, mathematics, engineering, and technology*. Document NSF 96-139. Washington, DC: NSF.

Naumes, W., and M. J. Naumes. 1999. *The art and craft of case writing*. Thousand Oaks, CA: SAGE Publications.

Paul, R. W. 1992. *Critical thinking: What every person needs to survive in a rapidly changing world*. Santa Rosa, CA: Foundation for Critical Thinking.

Pithers, R. T., and R. Soden. 2000. Critical thinking in education: A review. *Educational Research* 42 (3): 237–249.

Pizzini, E. L., S. K. Abell, and D. S. Shepardson. 1988. Rethinking thinking in the science classroom. *Science Teacher* 55 (9): 22–25.

Quitadamo, I. J., M. Kurtz, C. N. Cornell, L. Griffith, J. Hancock, and B. Egbert. 2011. Critical-thinking grudge match: Biology vs. chemistry—Examining factors that affect thinking skill in nonmajors science. *Journal of College Science Teaching* 40 (3): 19–25.

Smith, G. 2002. Are there domain-specific thinking skills? *Journal of Philosophy of Education* 36 (2): 207–227.

Swartz, R. J. 1997. Teaching science literacy and critical thinking skills through problem-based literacy. In *Supporting the spirit of learning: When process is content*, ed. A. L. Costa and R. M. Liebmann, pp. 117–141. Thousand Oaks, CA: Corwin Press.

Tomey, A. M. 2003. Learning with cases. *Journal of Continuing Education in Nursing* 34 (1): 34–38.

Tsui, L. 1999. Courses and instruction affecting critical thinking. *Research in Higher Education* 40 (2): 185–200.

van Gelder, T. 2005. Teaching critical thinking: Some lessons from cognitive science. *College Teaching* 53 (1): 41–46.

Vygotsky, L. S. 1978. *Mind in society*. Cambridge, MA: Harvard University Press.

Waterman, M. A., and E. D. Stanley. 2004. Investigative case-based learning: Teaching scientifically while connecting science to society. In *Invention and impact: Building excellence in undergraduate science, technology, engineering and mathematics (STEM) education*, ed. S. Cunningham and Y. S. George, pp. 55–60. Washington, DC: American Association for the Advancement of Science.

Section II:
Historical Cases

Historical case studies involve real people caught in famous moments in time. The reason we have chosen the cases in this section is simple: The discoveries recounted in them have affected the lives of millions of people, and these cases illustrate for students how science was practiced in earlier times. How people went about solving important problems in the past was limited by the knowledge of the time, the equipment then available, prevailing scientific and cultural biases, and political pressures. These same forces exist today. We don't know what causes arthritis, autism, schizophrenia, or aging, or if there is life on other planets. As we seek answers to these and other questions, our views are conditioned and constrained by the same forces that affected our predecessors. The path to discovery is usually convoluted and not obvious.

In the first case, "Childbed Fever: A 19th-Century Mystery," we encounter Ignaz Semmelweis, a Hungarian physician in the mid-1800s who was dismayed by the high death rate among women shortly after they gave birth. We see Semmelweis proceed through the steps of "the scientific method"—asking a question, proposing an answer, testing that proposition, and reaching a conclusion. But the case is much more than that. In the broader sense, we see the special conditions of the time: the skepticism of his colleagues and their ignorance of the germ theory of disease, which would not be formulated for another 50 years with the pronouncements of Koch's postulates regarding the causal relationship between a specific microbe and an infectious disease. We note the lack of basic hospital hygiene as well as the resistance of Semmelweis's colleagues to accepting his findings—resistance that subsided only after his death. Even today in hospitals, the lack of hand washing is blamed for many deaths and infections. The lessons of Semmelweis are still to be learned.

The case of John Snow and the cholera epidemics in England in the 1830s follows a similar path: Here we see yet another physician-scientist on the trail of a killer disease. In "Mystery of the Blue Death: John Snow and Cholera," we learn how Snow, one

Section I

of the founders of the field of anesthesiology, applied his experience in anesthesiology to the cholera epidemic. We might say that Snow was the world's first epidemiologist. He found himself on the front lines of a public health crisis, the cholera outbreak in London. He collected data on the history of the disease. He convinced William Farr to share information on where cholera outbreaks had occurred in the city. He looked for patterns in the data. He published his suspicions and received feedback from a peer revealing that he had overlooked an important possible cause of the epidemic. He took to heart a suggestion on how he could test his hypothesis, leading to a crucial natural experiment. We learn that Reverend Henry Whitehead, who originally was skeptical of Snow's hypothesis that water from a street pump was contaminated, carried out his own investigation and found that people drinking from this pump were nine times more likely to become sick than those drinking water from other sources. This finding ultimately led to the closing of the pump, the rebuilding of London's water system, and Snow's vindication. This is a telling story of how epidemiologists approach public health problems even today when new and exotic diseases threaten the public. And, of course, cholera is still a scourge upon human populations whenever disasters such as hurricanes, earthquakes, floods, and tsunamis disrupt sanitary and water-supply systems. Witness the Haiti cholera epidemic of 2011.

The case "Salem's Secrets: On the Track of the Salem Witch Trials" recalls an infamous time in early American history in 17th-century Massachusetts, when witchcraft was accepted as an explanation for undesirable actions or events. Young girls inexplicably began accusing citizens of the town of witchcraft, a crime subject to trial and often punished by death. As students work through the case study, they become sleuths trying to unravel the reason for the girls' apparent irrational behavior. Was it witchcraft, mass hysteria, or something else? Students are confronted with the historical background and asked to consider the evidence against the accused citizens. They then contemplate the possible explanations for the behavior, including a recent interpretation provided by modern scientists. The case ends with students having to grapple with whether the new hypothesis is adequately supported by the evidence. The game is afoot!

The final case in this section, "The Bacterial Theory of Ulcers: A Nobel-Prize-Winning Discovery," documents how our understanding of gastric ulcers recently has changed dramatically. Prior to the 1990s, people with ulcers were believed to be the victims of poor diet, stress, and unfortunately sensitive stomachs that secreted too much acid. Treatment consisted of a bland diet, no alcohol, and plenty of pleasant thoughts. Today, we know the culprit is *Helicobacter pylori*, a bacterial infection readily cured by a course of antibiotics. The primary message of the case once again is how a physician-scientist made an observation that aroused his curiosity but was inconsistent with the prevailing ideas of the time. Pathologist J. Robin Warren of Australia began finding bacteria in the stomach lining of people with ulcers, reaching the tentative conclusion that the bacteria might be the cause of the ailment. A young resident physician, Dr. Barry Marshall, was assigned to Warren, and together they began a quest to solve the problem. They started a pilot study of 100 patients, collecting data on their diets and looking for the presence of bacteria and a possible correlation with ulcers. Convinced that they had found the culprit, they tried to culture the bacterium but failed repeatedly. But chance intervened. Because of a holiday, technicians who routinely threw out cultures after two days left the cultures for five days; when they returned, they discovered that the bacteria were growing! It turned out that the bacteria required unusually long incubation periods before growth occurred. Serendipity is a common springboard to scientific discovery. However, as Louis Pasteur famously said, "In the field of observation, chance favors only the prepared mind" (Pasteur and Pasteur Vallery-Radot 1939, p. 131).

Once Warren and Marshall had their hands on the suspect bacteria, they began serious studies to discover if the organism would infect pigs. No such luck. Frustrated, Marshall resorted to self experimentation. He drank a healthy slug of a bacterial cocktail and got a dreadful infection, which he promptly cured with antibiotics. Exhilarated by their success, Warren and Marshall tried to publish their results, but their work was rejected because their discoveries were in conflict with the prevailing view of the times. This was to be expected. Most unusual claims ultimately are found to be wrong, and it requires compelling evidence to overthrow a major doctrine. But vindication was around the corner. A respected microbiologist, Dr. Martin Skirrow, was able to repeat Warren's and Marshall's work. After that, their manuscript was accepted for publication—albeit grudgingly—and the importance of their work was gradually recognized and appreciated. Warren and Marshall were awarded the Nobel Prize for Medicine in 2005. Would that all stories of frustrated scientists turn out so well!

Reference

Pasteur, L., and L. Pasteur Vallery-Radot. 1939. *Oeuvres de Pasteur*. Vol. 7. Paris: Masson and Co.

Chapter 5
Childbed Fever: A 19th-Century Mystery

Christa Colyer

The Case

Part I

Ignaz Semmelweis, a young Hungarian doctor working in the obstetrical ward of Vienna General Hospital in the late 1840s, was dismayed at the high death rate among his patients. He had noticed that nearly 20% of the women under his and his colleagues' care in Division I of the ward (the division attended to by physicians and male medical students) died shortly after childbirth. This phenomenon had come to be known as "childbed fever." Alarmingly, Semmelweis noted that this death rate was four to five times greater than that in Division II of the ward (the division attended by female midwifery students).

Questions

1. What were Semmelweis's initial observations?
2. What was the problem at hand?
3. What possible explanatory story might Semmelweis come up with?
4. How might Semmelweis test his suspicions?

Part II

One day, Semmelweis and some of his colleagues were in the autopsy room performing autopsies, as they often did between deliveries. They were discussing their concerns about death rates from childbed fever. One of Semmelweis's friends was distracted by the conversation and punctured his finger with the scalpel. Days later,

Science Stories: Using Case Studies to Teach Critical Thinking

Semmelweis's friend became quite sick, showing symptoms not unlike those of childbed fever. His friend's eventual death from the illness strengthened Semmelweis's resolve to understand and prevent childbed fever.

Questions

1. What might Semmelweis now propose as an explanatory story?
2. How could Semmelweis test his new hypothesis?

Part III

In an effort to curtail the deaths in his ward due to childbed fever, Semmelweis instituted a strict hand-washing policy among his male medical students and physician colleagues in Division I of the ward. Everyone was required to wash their hands with chlorinated lime water prior to attending to patients. Mortality rates immediately dropped from 18.3% to 1.3%, and in fact not a single woman died from childbirth between March and August 1848 in Semmelweis's division.

Questions

1. What conclusions can be drawn from Semmelweis's experiment?
2. How might Semmelweis revise his original hypothesis or experiments to gain additional information?

Part IV

Despite the dramatic reduction in the mortality rate in Semmelweis's ward, his colleagues and the greater medical community greeted his findings with hostility or dismissal. Even after presenting his work on childbed fever (technically named puerperal sepsis) to the Viennese Medical Society, Semmelweis was not able to secure the teaching post he desired, so he returned to Hungary. There, he repeated his successful hand-washing attack on childbed fever at St. Rochus Hospital in Budapest. In 1860, Semmelweis finally published his principal work on the subject of puerperal sepsis, but this too was dismissed. It is believed that the years of controversy and repeated rejection of his work by the medical community caused him to suffer a mental breakdown. Semmelweis died in 1865 in an Austrian mental institution. Some believe that his death was, ironically, caused by puerperal sepsis.

Questions

1. When presented with what appears to be unequivocal evidence in support of hand washing, why might Semmelweis's colleagues have dismissed his ideas?
2. How else might Semmelweis have approached the problem of disseminating his research findings to ensure their acceptance?
3. What, if any, role did serendipity play in Semmelweis's story of childbed fever?

Teaching Notes

Introduction and Background

This case provides a brief, factual account of the pioneering work of Ignaz Semmelweis and his efforts to remedy the problem of childbed fever in mid–19th-century Europe. The case was designed to be used in a freshman seminar class called Scientific Serendipity as a concrete example of the scientific method in practice.

It is important for students to understand that although there is no single sequence of events that must always transpire for scientific discoveries to be made, a set of commonly used strategies can facilitate such discoveries. If we simply present a didactic list of steps to teach the scientific method, then we fail to allow students the freedom to explore alternative methods and pathways. Instead, this case elicits from students the key aspects of the scientific method in a discovery-based format, without resorting to memorization or rote learning. It is likely that the students may not even be aware that the questions associated with this case are designed to guide them to conduct the scientific method through Semmelweis's eyes. As such, this case is best used as an introduction to the scientific method, even before an explicit discussion of the method has taken place. Such a discussion could be held subsequently and would be facilitated by referring back to this case. This case could be used successfully in any introductory science course in which the scientific method is discussed or practiced.

Objectives

The overall purpose of this case is to allow students to learn about the scientific method by "dissecting" the various steps involved in an important historic medical breakthrough. More specifically, the objectives of this case are

- to learn about the importance of observation when conducting scientific experiments and encourage observations beyond those expected or anticipated;
- to define a problem or a question given a set of observations;
- to formulate an "explanatory story," or hypothesis, to solve the problem at hand;
- to design a suitable experiment to evaluate the validity of the proposed hypothesis;
- to be able to draw logical conclusions based on experimental results; and
- to understand the importance of the dissemination of scientific information and establishing credibility within the scientific community.

Common Student Misconceptions

- All hypotheses are derived from carefully planned studies or purposeful observations.

Chapter 5

- The outcomes of carefully conducted scientific investigations will be accepted and acted on by the scientific community.
- Doctors have always known about bacterial infections but haven't always had the means to treat them.

Classroom Management

This is a short and effective case that requires only about 30–40 minutes to conduct in a freshman seminar class of 15 students. However, the use of this case should not be limited to small, discussion-based classes. It would be equally successful in a large (60 or so students) introductory science class, although an entire 50-minute class would likely be required to facilitate discussion among that large of a class.

Progressive Disclosure Method

For students to figure out for themselves the important steps that constitute the scientific method, this case should be presented by the method of progressive disclosure. That is, only one piece of information at a time should be distributed to the students, followed by discussion, before moving on to the next piece of information. If students are already familiar with the history of childbed fever, they will be tempted to jump ahead and suggest experiments that are not warranted by the observations presented in the earlier parts of the case.

The instructor distributes Part I of the case, accompanied by its questions. He or she instructs students to read the brief text provided and consider the questions before reconvening for a whole-class discussion. It has proven effective at this point to ask students to work in pairs with their nearest neighbor when considering the questions. Not only will the pairs come up with more possible solutions to the questions than any given individual might, but the pairs also will be more willing to share their solutions with the class as a whole. Be sure to emphasize that there is a range of possible answers to these questions; often science students are hesitant to write down an answer that might not be correct, and this hesitancy will severely stifle discussion.

After a brief time, ask the class to come together again to share their answers to the questions (see "Directing the Discussion: Expected Outcomes" below). Maintain a class list of the suggested explanatory stories and possible tests on one of the blackboards in the room (or on an overhead transparency) because the class will refer back to this list after completing Part II of the case.

Next, distribute Part II of the case, with the same directions to students as for Part I: Read the text and, in pairs, consider solutions for the accompanying questions. After a brief time, the class will come together again to create a new list of possible explanatory stories and tests. Students should compare the new list with the original list and attempt to explain any differences between the two.

The case proceeds in this fashion of progressive disclosure, moving next to Part III and its questions and, finally, Part IV and its questions. The final question in Part

IV, which refers to the role of serendipity in this case study, need not necessarily be addressed by an introductory science class that is focusing only on the scientific method. However, even a class that will not intentionally discuss serendipity in science might well enjoy a brief exposure to it in this context. Then the instructor can highlight serendipitous aspects of regular course material throughout the remainder of the semester.

Finally, the class should retrace the steps through this case to gain an appreciation for the "big picture." That is, an open-class discussion should follow after completion of Parts I–IV of this case to encourage the students to recognize that they have just witnessed an example of the scientific method in action.

Directing the Discussion: Expected Outcomes

This case was specifically designed to elucidate the steps of the scientific method, defined as follows: (1) observation, (2) statement of the problem, (3) hypothesis, (4) experiment [reiteration of steps (1)–(4), as necessary], and (5) conclusion. By establishing this set of events as a frame of reference, students are later able to compare future cases to these benchmarks and will come to appreciate the variability within this thing known as the scientific method. It seems more instructive for students to deduce the "standard scientific method" from this case rather than simply to recite a list of five steps as defined by a textbook or other source, since such a list tacitly implies invariability. Other instructors may, of course, choose to define the standard scientific method differently from the way in which it is defined here, such as, for example, citing experimentation as the first step or not including the statement of the problem as a distinct step. This case should be adaptable enough to teach other definitions of the scientific method too.

To ensure that the class is able to ultimately conclude that their analysis of this case highlighted the pertinent steps of the scientific method, it is important to direct the discussion each time the student pairs return to the class as a whole.

Web Version

This case and its complete teaching notes, references, and answer key can be found on the website of the National Center for Case Study Teaching in Science at *http://sciencecases.lib.buffalo.edu/cs/collection/detail.asp?case_id=429&id=429*.

Chapter 6
Mystery of the Blue Death: John Snow and Cholera

Susan Bandoni Muench

Image credit: Dr. John Snow (1813-1858), British physician. *http://commons.wikimedia.org/wiki/File:John_Snow.jpg*

The Case

Part I: Beginnings

John Snow was born in 1813 in a poor neighborhood in York, England, one of 10 children in a working-class family. He obtained a scholarship to a local school to learn to read and write and, with some extra money that his parents provided, learned arithmetic as well. Snow's wealthy and well-connected uncle, Charles Empson, arranged an apprenticeship for him with a surgeon-apothecary, one of the two types of health-care providers in 19th-century London. Physicians were graduates of the medical programs at Oxford or Cambridge, while surgeon-apothecaries went through a longer apprenticeship, attending classes part-time at smaller medical schools. Snow moved to Newcastle at age 14 to apprentice with William Hardcastle, dividing his time between classes and assisting Hardcastle with routine tasks. Snow stayed in Newcastle for several years. It was in Newcastle, near the end of his apprenticeship, that he first encountered Asiatic cholera when it arrived in England in 1831.

In his medical studies, Snow learned the prevailing humoral model of disease, which held that health depended on the balance of four humors: blood, phlegm, black bile, and yellow bile. The tools of the surgeon-apothecary were limited, consisting of bleeding patients with leeches and using purgatives and emetics to balance a patient's humors. Cholera was treated in various ways depending on how physicians interpreted the humoral imbalance. Sometimes they recommended soup or thin gruel. Often, they prescribed the use of purgatives to cause diarrhea and emetics to induce vomiting, believing that these "purges" would help the body achieve a balance among the humors. During the 1831 cholera outbreak, Hardcastle sent Snow, then only 19, to the Killingworth Colliery, a coal mine, to treat the cholera outbreak there.

Chapter 6

In the coal mines of 19th-century England, women, children, and men worked 12-hour shifts under unsanitary conditions. In the village where the mine was located, 330 of the 550 residents were stricken with cholera, and 65 died. Snow worked tirelessly caring for the ill, but could do little to help.

On his own, away from his mentor, Snow had the opportunity to think about the cholera outbreak, and his later work suggests that it was during this time that he first began to doubt the conventional medical wisdom about the cause of cholera. Some biographers have suggested that these doubts may have played a role in his decision to leave Newcastle for London and pursue further medical studies there.

Sanitation and Victorian London

In the middle of the 19th century, London contained 2.5 million people housed in 30 square miles, a population density greater than present-day Manhattan. During the Industrial Revolution, thousands of people moved from the countryside to London to take factory jobs, working long hours for little pay. Many people also lacked formal employment and worked as day laborers for very low wages or were self-employed, including many in the recycling trades. Nearly 100,000 people earned a living by collecting and reselling bone, rags, bits of copper, lumps of coal, human and animal wastes, and other salvageable materials. Mid-19th-century London was crowded and dirty. Sanitation was generally poor, as the city lacked a sewer system except for draining rainwater. Cesspools were used to collect used water from washing, and human excreta were collected in pit latrines. The cesspools and latrines often overflowed, and some buildings had several feet of accumulated waste in basements or courtyards.

Because of the problems of waste disposal, few Londoners had a source of drinking water uncontaminated by human sewage. At that time, a total of nine different water companies supplied Londoners with water, obtained from either shallow wells or the Thames River. In general, neighborhoods south of the Thames obtained water from the river, whereas neighborhoods north of the Thames had a wider range of sources. Some of the companies servicing these neighborhoods obtained their water farther upstream than others. Though only rainwater was supposed to be in the sewer system, often human waste also ended up in the river. Water companies that had their intake pipes farther downstream were more likely to obtain water contaminated with human waste.

According to the medical orthodoxy of the time, cholera was an example of an epidemic disease, a category of diseases thought to be explained by exposure to a toxic gas or miasma. The name *malaria* comes from the Italian for "bad air," reflecting this belief. The gases were believed to result from the fermentation of organic material. The sanitarian movement, early promoters of public health, focused on reducing miasmas by calling for the removal of the cesspools and piles of composting feces. They planned to build the network of sewer pipes that would carry wastes out to the Thames. One early proponent of public health, Edwin Chadwick, declared, "All smell is disease" (Finer 1952, p. 298). Chadwick sought to have sewers constructed for waste disposal in the river to rid communities of overflowing pit latrines. His plan also called for moving the water-intake pipes upstream once the system of sewer pipes was constructed. Thus, although Chadwick was a leader in declaring that health issues could be addressed through infrastructure at the community level, his focus on the miasma model as the cause of disease may have led to the deaths of many Londoners from waterborne diseases as a result of drinking contaminated water.

John Snow and the Origins of Anesthesiology

After John Snow's initial experience with cholera in 1831, there were no further cholera outbreaks in England until 1847. In the meantime, Snow went to London for further medical studies, probably again with the financial assistance of his uncle. After he finished his studies, he practiced surgery in a London hospital. At that time, surgery was done without anesthesia. Snow read the account of the first use of anesthesia in Boston and witnessed the first demonstration of a surgical procedure employing anesthesia in England. He immediately realized the value of general anesthesia, although he recognized a critical problem with the procedure, specifically the lack of precise control over the dosage of the anesthetic. He immediately set out to develop devices for administering gases as well as for measuring and controlling dosages, which he succeeded in doing quickly. Snow had not been a particularly successful surgeon, but he went on to develop one of the very first anesthesiology practices. This led to greater financial success as he cultivated a wealthier clientele, eventually including Queen Victoria, who sought relief from the discomfort of childbirth.

As his anesthesiology practice grew, Snow continued to do research in a lab he had built in his home. In addition to designing gauges for controlling dosages of gases and masks for precise delivery of the drugs, he experimented with different drugs using a variety of animal models, carefully recording dosages and effects. As a result of his extensive studies of anesthetics, he began to question conventional wisdom about miasmas, as epidemics did not seem consistent with what he had learned about toxic gases.

Questions

1. Models are analogies that allow us to clarify hypotheses—proposed explanations of relationships between causes and effects. What roles do models play in testing hypotheses?

2. What did the humoral model of disease propose as the cause for cholera?

3. What did the miasma model of disease propose as the cause for cholera?

4. Unlike Snow's later work on cholera, his research on anesthesia was experimental in nature. What general skills of experimental design were necessary to plan effective experiments to (a) test dosage measuring and delivery systems for anesthesia and (b) investigate the properties and effects of different drugs?

5. Why are experiments considered strong tests of hypotheses?

Chapter 6

Part II: The Mystery of the Blue Death

Cholera returned to London in 1847 and rapidly spread through some neighborhoods, causing numerous deaths and widespread fear. John Snow began to turn his attention to investigating the outbreak, seeing fewer patients in order to make time for his research. Having spent a great deal of time studying gases for anesthesiology, Snow recognized that the patterns of cholera cases were inconsistent with the patterns one would expect if people were being poisoned by a toxic gas; he reasoned that it was unlikely that a toxic gas would kill some members of a household and not others. Snow had no way of inducing cholera in an animal model, and he realized that studies of cholera would have to take a very different form.

The symptoms that cholera patients presented also seemed to Snow to be inconsistent with exposure to a toxic gas. Often called "blue cholera" or "the blue death," cholera usually caused death by respiratory failure, giving the skin a bluish tinge. To most of Snow's contemporaries, this observation seemed consistent with the miasma model, but Snow thought that the initial gastrointestinal symptoms might be more significant and suspected that the causative agent was ingested rather than inhaled. He wondered about the "animalcules" identified by early microscopists, although the prevailing scientific wisdom was that these were harmless. He became particularly interested in the possibility that cholera was transmitted in water. He confided his suspicions to some of his friends from the medical society. They were skeptical, and he resolved to collect additional data.

Snow recognized that an experimental test of his hypothesis was not possible and began to look for other ways to test his theory. He combed through all of the available records of cholera outbreaks in England to search for patterns. In studying an outbreak in an affluent neighborhood, he found that cesspools that should have contained only water from washing contained partially digested food. He hypothesized that the cesspools were contaminated with human feces and that somehow this contamination had spread to drinking water.

In 1849, Snow published a monograph titled *On the Mode of Communication of Cholera*. It was poorly received, with negative reviews in *The Lancet* and the *London Medical Gazette*. However, a reviewer in the latter journal made a helpful suggestion in terms of what evidence would be compelling: The crucial natural experiment would be to find people living side by side with lifestyles similar in all respects except for the water source.

Questions

1. What causes cholera, and how is it transmitted?
2. Why were Snow's ideas about cholera not accepted at this early date?
3. Explain why cholera outbreaks are more consistent with contamination of water than of air.

4. Given that cholera outbreaks are more consistent with contamination of water than of air, why did the miasma model persist?

5. How did Snow's experimental research on anesthesia help him design a new model for the cause of cholera?

6. Why would evidence of cholera in people living side by side, differing only in their water supplies, provide critical evidence?

7. When was the germ theory of disease proposed, and on what basis?

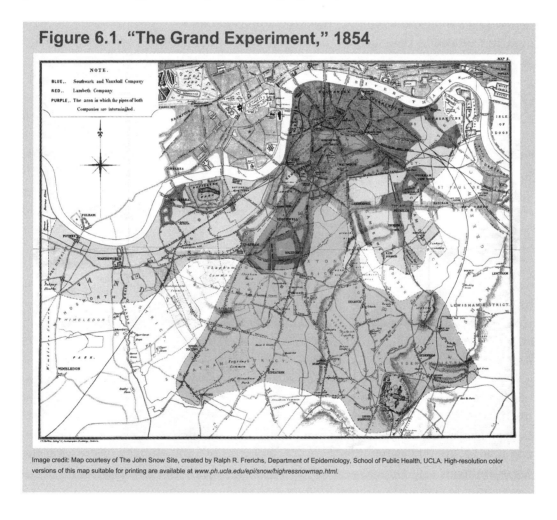

Figure 6.1. "The Grand Experiment," 1854

Image credit: Map courtesy of The John Snow Site, created by Ralph R. Frerichs, Department of Epidemiology, School of Public Health, UCLA. High-resolution color versions of this map suitable for printing are available at *www.ph.ucla.edu/epi/snow/highressnowmap.html*.

Chapter 6

Part III: Solving the Mystery of the Blue Death

After Snow's monograph was rejected, he sought ways to strengthen his argument by carrying out the crucial experiment sought by the *Medical Gazette*'s reviewer. He went door to door interviewing families of cholera victims. He interviewed physicians who had treated cholera patients. In addition, he looked at the geographic distribution of other cholera outbreaks in England as compared to the drinking water sources.

At the same time that Snow was beginning to consider questions of public health, so was William Farr, a contemporary of Snow's. Farr kept records of births and deaths for the city of London. He encouraged physicians to write down the causes of deaths and began looking for patterns. Although a miasmatist, Farr, in response to Snow's inquiries, agreed to keep records of the water companies used by each neighborhood along with records of cholera deaths.

Snow began marking cholera deaths on city maps, and patterns began to emerge. One of the first patterns he noticed was in the mortality statistics from prisons and insane asylums. Residents of those institutions were not able to leave and were forced to use the same water source. He found that they often had mortality rates that were either higher or lower than those of the surrounding communities. From this, he concluded that he might be able to make a comparison of cholera rates in people with different water supplies. He soon realized that during the period when some water companies were moving their intake pipes upstream and others were not, there would be an opportunity to distinguish between the miasma model and his own hypothesis of water-borne transmission.

Snow studied maps of the city, initially trying to find areas served by a single water company. He was unsuccessful in finding such a neighborhood from the maps, but soon he recognized that neighborhoods in which two water companies were competing might be more valuable. There, neighbors would breathe the same air and would resemble each other more in variables other than water source (see Figure 6.1 [p. 49]).

Snow went door to door asking people about their water source. He found that often people did not know, so he sought a way to verify the source. He recognized that the tidal flow in the Thames held the key, with water obtained in London having a far higher salt content than water from sources upstream. He began taking water samples and testing them to determine the salt concentration, and thereby identifying the water company and the location of the source along the Thames. By then, it was winter and cholera had subsided for another year.

Questions

1. Why was it useful to be able to verify the source of the water?
2. Why would a neighborhood served by two different water companies be more useful for testing Snow's hypothesis than two neighborhoods each with its own source?

3. Epidemiologists often draw causal webs to illustrate the interrelationships among biological, social, and environmental variables that contribute to disease outbreaks. Based on what you have learned so far, what variables should be included in a causal web for cholera?

4. Snow considered his conclusions about cholera to be inferences from observations, whereas the reviewer from the medical journal considered these to be conjectures. What is the difference between inference and conjecture?

Part IV: The Broad Street Pump

Late in the summer of 1854, cholera broke out once again, with a cluster of cases in the vicinity of Broad Street. This was a neighborhood that Snow knew well, having lived there for several years when he first arrived in London. By knocking on doors, Snow determined that all of the cases of cholera involved people who obtained their water from the Broad Street pump. Most of the victims lived closest to this pump.

At that time, the local parish, St. James Church, was responsible for taking care of poor families in the neighborhood. A parish board was responsible for making decisions on how to provide charitable assistance. Snow went to the board and made his case, arguing that the handle on the pump should be removed so that people couldn't get water from the pump, thus preventing further cases of cholera. Although those on the board were skeptical, they decided that Snow's proposal would do no harm, and the handle was removed. Cases of cholera dropped.

Reverend Henry Whitehead, the reverend of St. James Church, was skeptical of Snow's explanation and saw a critical flaw in Snow's work. Snow had focused on interviewing those households affected by the epidemic. As the parish cleric, he had ministered to the sick and dying during the epidemic. Whitehead decided to interview people in households in which no one died from cholera. Much to his chagrin, he found that households using the Broad Street pump were nine times more likely to have cholera victims than those not using it. Through his careful interviewing, Whitehead also succeeded in identifying an earlier case, an infant living in a house a few feet from the Broad Street pump who died from diarrhea two days before the cholera outbreak was officially recognized.

Although Whitehead had initially sought to criticize Snow's work, the two became allies. Together, Whitehead and Snow asked the Board of Public Health to excavate the area around the pump. There, they found that water from the cesspool under the building where the first infant victim died was seeping out and into the water from the pump. On further inquiry, they learned that the baby's mother had washed the soiled diapers of her sick infant in the basement of her building, dumping the water in the cesspool.

Although many in the community and in the scientific and medical professions continued to reject Snow's explanation, some began to give it grudging acceptance, often without acknowledging his contribution. Snow's vindication came at a meeting of the London Medical Society, where a member stood up after such a presentation insisting that Snow be given credit. Sadly, not long after this, Snow suffered a series of strokes, dying at the age of 45.

The following summer, a drought compounded the problems of contamination of the Thames, resulting in pollution so severe that it was called the Great Stink. Members of Parliament were forced to meet in a remote location because of the stench, where they voted to initiate a vast construction project to move the sewer pipes far downstream. After the construction of the new sewer system, there were no more cholera outbreaks in England.

Questions

1. The basic questions of epidemiology focus on time and place: Why here, and why now? What are the answers to these questions for the Broad Street outbreak?

2. Epidemiology relies on nonexperimental tests of hypotheses. What was Snow's hypothesis, and how did he test it?

3. How did Whitehead improve on Snow's test of this hypothesis?

4. What is the difference between correlation and causation?

Teaching Notes
Introduction and Background

This case study introduces students to John Snow, considered to be one of the founders of both epidemiology and anesthesiology, and a remarkable figure in the history of science. Although historical case studies are often less popular with students than case studies exploring contemporary issues (Herreid 1997), several aspects of this case make it attractive to students. First, students have told me that they find "detective stories" about important medical discoveries to be inherently appealing. Second, the questions and methods that Snow used to demonstrate the causes of cholera outbreaks are the same as those used in contemporary epidemiological investigations. Third, although we now know the causes of cholera outbreaks, there continue to be cholera outbreaks with causes similar to those studied by Snow.

This case was written for an undergraduate course in global health taught in a biology department. The course is taught at the 200 level, with a full year of introductory biology for biology majors as the prerequisite. The class size ranges from 30 to 50 students and contains both biology majors and nonmajors, most of whom are biology minors or elementary education majors with a science concentration.

The global health course fulfills a general education requirement for a course in non-Western traditions and issues (multicultural core) because of its focus on health in Asia, Africa, and Latin America. Three areas of study make the course multicultural:

(1) non-Western scientific traditions, (2) contemporary issues where science affects policy, and (3) contemporary science in non-Western cultures. In teaching this course, I have observed that students do not generally have a clear understanding of the ways in which their own culture shapes their perceptions of health and disease. Although this case study is concerned only with Western science, it introduces students to a period of time in which Western ideas about disease were still forming. Considering alternative models of disease within Western culture sets the stage for understanding the models of health and disease in other medical traditions.

This case study can also be used in a variety of other contexts, including courses in microbiology and introductory biology for either majors or nonmajors. Because it addresses the nature of science, it is also appropriate for courses in the history, philosophy, or sociology of science.

Objectives

- To apply terminology and concepts from epidemiology and public health from a text to a discussion of the case
- To discuss several aspects of the nature of science using cholera as a case study. These features may include the role of models in hypothesis testing, nonexperimental tests of hypotheses, and populational thinking
- To discuss the relationship between science and the surrounding culture, as well as cultural and class influences on the practice of science
- To write a summary essay on one of the topics discussed, such as
    - history of models of disease in Western culture,
    - nonexperimental tests of hypotheses,
    - populational thinking in epidemiology and in biology in general,
    - cultural and social context of science, or
    - health inequities and contemporary cholera outbreaks

Common Student Misconceptions

- Experiments are the only rigorous method for testing hypotheses.
- Good ideas in science will immediately and automatically receive recognition and replace poorer ideas.
- The germ theory quickly replaced the inferior humoral model of disease.

Classroom Management

This case is designed to be taught in the first or second week of the semester to introduce basic concepts in epidemiology and public health as well as models of disease.

Chapter 6

I use a team-based learning format (Michaelson, Knight, and Fink 2004), beginning each topic with a quiz on the reading assignment in the text. A chapter in a textbook (Jacobsen 2008) is assigned ahead of time that introduces John Snow and his experimental removal of the handle of the Broad Street pump. This chapter also introduces populational thinking and the basic questions of epidemiology, including the evaluation of possible causal factors and development of causal webs. A second text reading (Sherman 2006) describes cholera in the history of Western culture and was assigned for the discussion of cholera in the previous year the course was taught.

I recommend that instructors who are teaching the case assign the first section before class and assign the questions for homework. Begin class with individual and team quizzes on the assigned readings from the textbook, and then progress to discussion of the questions from the first part of the case study. Over the remainder of that 75-minute class period and all of the next, introduce the three additional parts of the case using small-group discussion of the student questions interspersed with segments of full-class discussion of some of the additional questions. Afterwards, students write summaries of the discussion of one or more issues.

In a microbiology course, the case could be used at the beginning of the semester to introduce the history of the discipline, but placing a different emphasis on the topics. In introductory biology courses, it might fit either with the introduction of micro-organisms or with the discussion of the nature of science, with very different selections of questions for class discussion. For some classes, especially for beginning students, it might be more appropriate to stretch the discussion over four class periods with shorter discussions of each part during each class period. For those who wish to extend the case study into the laboratory, the data and maps are available online (Frerichs 2008).

An underlying goal of the case study is to move students toward critical thinking. Nelson (1999) has reviewed Perry's work on cognitive development in the college years and identified the challenges for teaching critical thinking in science to students at varying stages. Students in the earliest stage of Perry's scheme (dualism) tend to view science as a march toward truth and learning science as a matter of learning right answers. For students in the next (multiplicity) stage, there is no reason to prefer one model over another, and the instructional task at this stage is to spell out criteria for fair tests of different hypotheses and then examine the evidence. For students at later stages of development, placing science in a context of policy or ethics is vital to helping students see that scientific evidence matters and is not simply a question of following a set of disciplinary conventions.

The story of John Snow contains many elements that support the instructional strategies identified by Nelson (1999). There are critical and unavoidable ambiguities in the story that create uncertainty and set the stage for the shift from dualism to multiplicity. There are multiple models to explain disease, but some are better supported by evidence than others. This variation in models allows students to make judgments about the quality and quantity of supporting evidence and helps them move from multiplicity to understanding how decisions are made in science. Finally, Snow's epidemiological study of cholera offers an opportunity to explore the role of science in informing decisions about policy in multiple social and political contexts.

Web Version

This case and its complete teaching notes, references, and answer key can be found on the website of the National Center for Case Study Teaching in Science at *http://sciencecases.lib.buffalo.edu/cs/collection/detail.asp?case_id=258&id=258*.

References

Finer, S. E. 1952. *The life and times of Edwin Chadwick*. London: Methuen.
Frerichs, R. R. 2008. John Snow and cholera. *www.ph.ucla.edu/epi/snow.html*.
Herreid, C. F. 1997. What makes a good case? *Journal of College Science Teaching* 27 (3): 163–165.
Jacobsen, K. H. 2008. *Introduction to global health*. Sudbury, MA: Jones and Bartlett Publishers.
Michaelsen, L. K., A. B. Knight, and L. D. Fink. 2004. *Team-based learning: A transformative use of small groups*. Westport, CN: Praeger.
Nelson, C. E. 1999. On the persistence of unicorns: The trade-off between content and critical thinking revisited. In *The social worlds of higher education*, ed. B. A. Pescosolido and R. Aminzade, pp. 168–184. Thousand Oaks, CA: Pine Forge Press.
Sherman, I. W. 2006. *The power of plagues*. Washington, DC: ASM Press.

Chapter 7
Salem's Secrets: On the Track of the Salem Witch Trials

Susan M. Nava-Whitehead and Joan-Beth Gow

The Case

Part I: Salem's Secrets

There was a chill in the courtroom that day—a chill colder than could be explained by the unbearable winter. It was a cold that started at the back of the neck and lodged deep in the spine. Something evil was afoot. The question was, to whom did that evil belong?

"She killed Goodwife Betty's baby. She killed it with those evil eyes. I saw her staring, as in a trance, at Betty's house at sunset one evening last week. Then her cow and her baby died. She also makes poisons in her house. When people won't take her poison, she sends her spirit to force them by choking them until they swallow it. I see her spirit here now. It is over near Abby. Oh Abby, Abby! Be careful, Abby, she has pins and they are red hot! Stop her, she is pricking me! Help me, I am burning. Help me."

The courtroom hummed with whispers as the spectators watched two young girls—Elisabeth, the speaker, and Abby, her best friend—tear and swat at their arms and legs as if swarmed by invisible bees. Their contortions escalated into convulsive fits, which were so grotesque and violent that witnesses agreed they could not be manufactured. Soon, as if on cue, other girls from Elisabeth and Abby's circle of friends joined in. The girls collapsed in exhaustion. Dr. William Griggs, the village physician, examined the girls and, finding only bruised skin, made a diagnosis: "The evil hand is upon them. They are bewitched."

Chapter 7

Hathorne, the magistrate, directed his attention to Sarah Good, the latest woman to be accused of witchcraft in Salem in 1692, and in a powerful voice demanded, "Goodwife, why do you torture these girls so?"

"Sir, I do not hurt them."

"Who do you employ then to do it?"

"I employ nobody."

"And what say you of the poisons you keep at your home?"

"They are nothing more than good broths. When a child is to be born to a woman of this village ... my broths bring them ease."

"What evil spirit directs you in the making of these broths?"

"No spirit, good sir. I am falsely accused."

This dialogue is fabricated from translations of actual depositions given in the Salem Witchcraft Papers (Boyer and Nissenbaum 1977). Scenes such as this were not uncommon in New England in the early colonial days. Most students of American history know that a group of young girls in Salem were the initial catalyst that led to accusations of witchcraft against more than 200 people. These accusations resulted in the execution of 20 persons.

Question 1: What do you think caused the girls to behave this way?

To understand the phenomena at Salem, it is necessary to understand the culture and community of the time. In general, colonial life was hard. Rich, farmable land was scarce, and any food it yielded was a result of strenuous physical labor. Diets were poor and deficient in essential nutrients and vitamins. Colonists, including young children, often worked from first light until after dark. Disease and death rates were high. It was not uncommon for families to suffer the loss of children.

The year spanning 1691–1692 was not a particularly good one for the Puritans, with an unusually severe winter and a rainy spring. As a result, the harvest that year was extremely poor. In addition, many families escaping the Indian Wars of Maine had moved into the northwest side of the area known as Salem, into an area called Salem Village. This situation forced farmers to "utilize their swampy, sandy, marginalized land" (Matossian 1982, p. 357) for rye production and families in the community to share crops. Compounding these ills was a perceived imbalance in social status and power, with the "haves" on the east side of town in direct conflict with the "have nots" of the village. Salem villagers were disgruntled by having to pay taxes to and serve in the militia for Salem Town without receiving any direct benefit, namely, a church of their own. In addition, Salem was factionalized by the leadership of two strong men (Parris of the village and Proctor of the town). The original, core group of girl accusers were kin of the Parris family.

Puritans were fervently religious and believed strongly in the balance of good and evil. To them, the devil was "a physical being who was incarnate, there to seduce them from the path of righteousness" (Woolf 2000, p. 458). When the march to the gallows struck Salem in 1692, it struck hard. The litmus test for bewitching was not substantial: The mere accusation of spectral evidence (victims would "see" a witch touching, pinching, or otherwise harming them) was sufficient to place a citizen in jeopardy. Likewise, one could earn the title of witch when the "passing by" of a person's house or "fixing a gaze" upon someone correlated with the stillbirth of a child or the death of a domestic animal.

Question 2: In the opening passage, what "evidence" did the girls provide for the presence of witches/witchcraft? (List this information in column 1 of your data management sheet.)

Question 3: Assume you are living in Salem in 1692. Develop a hypothesis based on your observations. (Remember that a hypothesis must be supported by scientific evidence.)

Question 4: Reflect for a moment on this concept of evidence. How do we define *evidence* in science? Does the girls' evidence pass scientific muster?

Part II: Mass Hysteria

For more than 100 years, the prevailing belief was that the Salem tragedy was a direct result of mass hysteria, a condition in which a large group of people exhibit similar physical or emotional symptoms not attributable to any physiological cause. The Salem girls as a group experienced an array of unusual symptoms. In the absence of a clear medical diagnosis, and based on the limited technology of the time, the doctor who examined the girls pronounced them bewitched.

Collective human behaviors, however, are more common than many people realize. Some are simply the crazes and fads that often affect teenagers and other social groups. Others are bizarre, such as the example given in Table 7.1 (p. 60) that occurred in the early 1900s when students in Szechuan, China, were convinced their penises were shrinking. Many, though, are less innocuous and involve severe symptoms of illness. On the next page are listed some selected examples of mass hysteria events spanning several centuries.

Chapter 7

Table 7.1. Mass Hysteria Events Throughout History

Mass Hysteria Event Location	Year	Summary
Southern Europe, especially Italy	1200s–1800s	Symptoms such as headache, giddiness, twitching, and delusions, culminating in frenzied dancing, in response to perceived bites from a tarantula spider.
Milan, Italy	1630	Several people executed after being pronounced guilty of spreading poison throughout the city in cooperation with the devil.
Szechuan, China	1907	Twenty students convinced their penises were shrinking.
Newark, NJ	1938	Following radio broadcast of H.G. Wells's *War of the Worlds,* dramatizing a "gas raid from Mars," mass panic occurred, involving thousands. Several were treated at hospitals for shock.
West Bank, Jordan	1983	Nearly 1,000 people, mostly young females, afflicted with headaches, fainting, dizziness, and abdominal pain. Initially attributed to poison gas.
Kosovo, Yugoslavia	1990	Outbreak of flulike symptoms such as headaches, dizziness, and respiratory distress that persisted for weeks among thousands of mostly adolescent Albanians. Initially attributed to poisoning by Serbs.
Central Falls, RI	1991	Seventeen middle school students and four teachers with rapid onset of an array of symptoms such as dizziness, pain, vomiting, and chills. Initially attributed to chemical spill or toxic gas exposure.
A large Midwestern university	1996	Sixty-nine college students and workers treated for shortness of breath, eye and skin irritation, and general feelings of illness. Initially attributed to a dusty substance in the snack bar.
McMinnville, TN	1998	Following a "gasoline-like" smell detected in a classroom, close to 200 students and staff members experienced headaches, nausea, shortness of breath, and dizziness.
Amman, Jordan	1998	More than 800 students in grades 1–10 displayed a variety of symptoms such as fever, chest tightness, chills, and feeling faint following tetanus-diphtheria vaccination.

Have your thoughts regarding the events at Salem changed after examining this table? Reflect on the observations you listed in column 1 of your data management sheet.

Part III: Ergot: A Toxic Fungus

Question 1: Incidences of witchcraft are found universally among cultures during this time, but none had the devastating impact that Salem's had. What other factors may have contributed to the phenomena at Salem?

Claviceps purpurea is the genus and species name of a toxic fungus that grows as a parasite on many grains, particularly rye. In rainy, wet weather, all plants become more vulnerable to fungi. Because rye also grows best in damp weather, it is particularly susceptible to fungal growth. When *Claviceps* spores germinate, they form distinct dark, hard structures called sclerotia. These sclerotia are commonly known as ergots, thus the term *ergot poisoning*, or *ergotism*. Within the ergots is produced a poisonous brew of fungal toxins, ingestion of which can lead to severe illness or death. Ergot fungal toxins are particularly stable and are not destroyed by boiling or baking (Bennett and Bentley 1999).

Two forms of ergotism exist: convulsive and gangrenous. The convulsive form has the greatest effect on the central nervous system, leading to seizures, insomnia, and insatiable appetite. In the gangrenous form, blood flow to the extremities is restricted; in severe cases, this can lead to blackened tissue and subsequent loss of bloodless limbs. Either form of ergotism may cause tingling, itching, alternating perception of hot and cold temperatures, hallucinations, perceptual disturbances, and gastrointestinal upset. In addition, people under the influence of ergot derivatives are known to be highly suggestible (Matossian 1982).

Several factors play a role in the severity of ergot poisoning. Nutritional status—in particular, vitamin A deficiency—is one such factor (Bennett and Klich 2003). Age and sex play a role as well, and ergotism seems to affect teenage females much more than it does males (Bennett and Bentley 1999; Caporael 1976). Humans are not the only species to suffer from ergot poisoning; farm animals can be affected as well.

Question 2: In column 2 of your data management sheet, list evidence that the events at Salem could have been caused by ergot poisoning.

Question 3: After reading Parts II and III of the case study, develop a second hypothesis, different from your first, explaining the events at Salem. Record this hypothesis on your data management sheet.

Part IV: Data Interpretation

Table 7.2 (p. 62) is extracted from an article written by Nicholas Spanos and Jack Gottlieb on ergotism and the Salem witch trials published in 1976 in *Science*. Spanos and Gottlieb collected these data by reading through the *Records of Salem Witchcraft* (Woodward [1864] 1969) and making note of the frequency of symptoms suffered by witnesses outside the original group of girls. At the time of the trials, the adult

Table 7.2. Frequency of Symptoms

Reported Sufferers	RSW* (vol. & page)	A	B	C	D	E	F	G	H	I	J	K	L	M	N	O	P	Total
W. Allan	I:38	0	0	0	0	0	0	0	0	0	0	1	0	0	0	0	0	1
J. Bayley	I:113	0	0	0	0	0	0	0	0	0	0	1	0	1	1	0	1	4
S. Bittford	I:108	0	0	0	0	0	0	1	0	0	0	1	0	0	0	0	0	2
A. Booth	II:180	0	0	0	0	0	0	0	0	0	0	1	0	0	0	0	0	1
J. Childen	I:92	0	0	0	0	0	0	0	0	0	0	1	0	0	0	0	0	1
G. Cory	I:55	0	0	0	0	0	0	0	0	0	0	0	0	0	0	0	1	1
J. Doritch	I:262 and II:179	0	0	0	0	0	0	0	?	0	0	1	0	1	1	0	0	3
B. Gould	II:178	0	0	0	0	1	0	0	0	0	0	1	0	1	0	0	0	3
J. Holton	I:71	0	0	0	0	0	0	0	?	0	0	0	0	0	0	0	0	1
J. Hughes	I:38	0	0	0	0	0	0	0	0	0	0	1	0	0	0	0	0	1
J. Indian	I:64	0	0	0	0	0	0	0	1	0	0	1	0	0	0	0	0	2
T. Indian	I:44	0	0	0	0	0	0	0	?	0	0	1	0	0	0	0	0	2
E. Keysar**	**	0	0	0	0	0	0	0	0	0	?	1	0	0	0	0	0	2
M. Pope	I:59	0	0	0	0	0	0	0	1	0	0	1	0	0	1	0	0	3
H. Putnam	I:275	0	0	0	0	0	0	0	0	0	0	1	0	0	0	0	0	1
J. Putnam	I:95	0	0	0	0	0	0	0	?	0	0	0	0	0	0	0	0	1
W. Putnam	I:96	0	0	0	0	0	1	1	0	0	0	0	0	0	0	0	0	1
D. Wilkins	II:7	0	0	0	0	0	0	0	?	0	0	0	0	0	0	1	0	2
R. Wilkins	II:5	0	0	0	0	0	0	0	0	0	0	1	0	0	0	1	0	3
S. Wilkins	II:3	0	0	0	0	1	0	0	0	0	0	1	0	0	1	0	0	3
E. Woodwell	II:178	0	0	0	0	0	0	0	0	0	0	1	0	0	0	0	0	1
Total		0	0	0	0	2	1	2	7	0	1	16	0	3	3	2	2	39

Key: Symptoms of witnesses (other than the afflicted girls) who testified against the accused witches. A: vomiting; B: diarrhea; C: livid skin; D: permanent contractures; E: pain in extremities; F: death; G: temporary muscle stiffness; H: convulsions; I: ravenous appetite; J: perceptual disturbances (not including apparitions); K: apparitions; L: sensations of hot and cold; M: skin sensations (biting and pinching); N: stomach pain; O: choking sensations; P: temporary inability to speak; 1: symptom reported; 0: symptom not reported; ?: symptom questionable.
*RSW stands for *Records of Salem Witchcraft*, compiled by W. E. Woodward in 1864–85.
** This testimony comes from Boyer and Nissenbaum, 1972, p. 75.

population of Salem was estimated to be 215 people; no estimate is provided for the child population.

After reviewing the data from this table, use the information it provides and your observations from columns 1 and 2 of your data management sheet to prepare an argument in support of either one of your hypotheses. Consider the following questions:

Question 1: What do the data suggest? What symptoms are reported in high frequency? In low frequency? What patterns exist?

Question 2: Consider who was afflicted. What other health information was reported in 1692?

Question 3: Why did the idea of witchcraft occur at this time and in this place in history? Is this situation unique?

Part V: The Societal Frame: What Is the Secret of Salem?

Review the data from your data management sheet and then answer the questions below:

Question 1: When you first read "Salem's Secrets" on pages 57–59, did you think the girls at Salem were bewitched? Faking?

Question 2: Did your group acknowledge or dismiss the idea of mass hysteria with respect to Salem? Did you consider the social dynamics of the time in your thinking?

Question 3: As you read more about the events of Salem (pp. 57–59), did your thinking about what happened change as you progressed through the material? How many times?

Question 4: If a similar set of symptoms were presented today, do you think the result would be similar? Why or why not? In other words, how does who we are and what we know affect our interpretations?

Question 5: What questions still remain? What other information would help you decide what happened at Salem?

Part VI: Classroom Extensions to Salem's Secrets

Research further one of these topic extensions from *Salem's Secrets*. Choose one of the following questions to develop into a one-page essay.

1. How does public health reporting differ today from reporting in the 1600s?

2. What can you infer from this case about the general safety of the world's food supply in 1692? What about today's food supply?

3. If you, as a member of the Department of Public Health, were informed of a group of people seemingly afflicted with a similar set of symptoms, what might be your initial thoughts and course of action?

4. Ergot derivatives today are used to treat migraine headaches and alleviate bleeding after childbirth. How is it possible that such toxic substances can have therapeutic uses?

5. If Salem had had a well-defined government in 1692, would the march to the witches gallows have occurred?

6. The testimony of the children at Salem was accepted without question. Children often make accusations of many kinds of abuse by adults. Should such accusations be accepted at face value?

7. Will it ever be possible to prove that the events at Salem were attributable to a specific cause? Why or why not? What is the difference between correlation and causation?

Teaching Notes

Introduction and Background

This case study examines the Salem witch trials that took place in Salem, Massachusetts, in the late 1600s. Although based on an actual historical event, the case study is fiction. As mentioned earlier, the dialogue is fabricated from translations of actual depositions given in the Salem Witchcraft Papers (Boyer and Nissenbaum 1977).

The case is designed to prompt students to analyze and critique data and to help them understand the scientific method. We have used the case in a nonmajors general biology course. It has led to increased comprehension and retention, as evidenced by students' self-selection of this topic when given essay options on examinations.

The case could be used in a variety of courses. The topic of mass hysteria could be integrated into a psychology or sociology course. Mycotoxins and ergot toxicity could be emphasized in a microbiology course. Mass psychogenic illness could be highlighted in epidemiology courses. Indeed, this is a perfect case study to use for interdisciplinary teaching. The case also would work well in a course in American history.

Objectives

At the conclusion of this case, the student will be able to

- apply the principles of the scientific method to analyze and evaluate evidence critically;
- define the term *evidence* using scientific terminology such as *observable*, *measurable*, and *repeatable*;
- outline the defining social dynamics of mass hysteria;
- identify the signs and symptoms of ergot poisoning;
- synthesize data to formulate and defend a conclusion; and
- appreciate and communicate that scientific understanding is contextual, as it is interpreted at the technological level and within cultural norms of the time.

In addition, enrichment opportunities exist for

- cross-curriculum exploration (history of New England in 1692);
- an examination of ecology (environmental influence);
- defining organizational schemata (fungus);
- identifying biological toxins (in particular mycotoxins);
- investigating biochemical effects of ergot on the central nervous system;
- discussing the role of nutrition, age, and gender in susceptibility to toxins; and
- examining food sanitation and the history of food poisoning.

Common Student Misconceptions

- Scientific data is not open to interpretation.
- Scientists always agree on the implications of results and are not subject to societal or cultural influence.
- The scientific method is a set of steps to be memorized because scientific research always proceeds sequentially.
- Scientific research is always experimental.
- Victims of mass hysteria do not experience physical symptoms; mass hysteria is fakery.
- Modern humans are not susceptible to mass hysteria.
- The world food supply is safe; fungal toxicity is controlled in modern agriculture.
- The Salem witchcraft outbreak is an isolated incident in historical and modern times.

Chapter 7

Classroom Management
We have taught the case using an interrupted case study method in which students receive the case in parts, or sections, that they must complete in turn. The case is devised to take place in a single 75-minute classroom period. We hand out a data management sheet to help students keep track of the information they glean from the case. Students like to complete case studies with a tangible product in hand, particularly if they are to be tested on the information. Specifically, the handout has space for them to write down their observations, first for Part I of the case, then later for Part III, as well as their hypotheses, data interpretation, and final conclusion.

Part I: Salem's Secrets
Students are asked to read Part I of the case study in advance of the class in which it will be discussed. In class, students are put into small groups, handed a copy of the data management table, and given five minutes to read Part I. The instructor then facilitates a class discussion regarding the concept of scientifically acceptable evidence. After extracting from the students their ideas of what constitutes evidence and writing these ideas on the board, the instructor leads the students to conclude that scientific evidence must be subject to some form of validation. The gold standard for evidence obtained experimentally is, of course, data that are measurable, observable, and repeatable. For observational studies, particularly those done retrospectively, these criteria cannot always be met.

Students in groups are asked to formulate a hypothesis based on what they have read so far that explains the events at Salem. Then, using their newly learned definition of evidence, they are asked to list in the data management sheet all the pieces of data that support the hypothesis they developed.

Part II: Mass Hysteria
As students list pieces of evidence from Part I on their data management sheet, the theme of mass hysteria will emerge. The instructor facilitates a brief discussion on the concept of mass hysteria and then distributes Part II, which includes a table describing selected mass hysteria events throughout history. Many students equate mass hysteria with fakery. The instructor should emphasize that victims of mass hysteria really do experience a set of symptoms, although there is no identifiable organic cause.

Part III: Ergot: A Toxic Fungus
The instructor next hands out Part III of the case study, essentially an information sheet about ergot. After the students read this new information, the instructor asks them to list in column 2 of their data management sheet the evidence that ergot could have been responsible for the events at Salem.

At the bottom of column 2 in the data management sheet, students are asked to write a hypothesis different from their first one to explain the Salem events. The students are polled to see how many believe mass hysteria caused the events in Salem, how many believe ergot caused the events in Salem, and how many believe there is an altogether different explanation. Students like to go off on tangential discussions here, and it is important to assist them in staying on task.

At this point, students are wrangling with the two conflicting hypotheses they likely have developed. Both seem viable, based on observations they have made. It is appropriate now to expose students to some retrospective, quantitative data gleaned from information published during the Salem witch trials.

Part IV: Data Interpretation

The instructor now shares with students how several scientists have examined some of the same data they have and also disagree about the interpretations. Part IV of the case study (original data from the Spanos and Gottlieb 1976 *Science* paper) is then handed out. The instructor explains how the data are represented in a binomial table, examining whether or not certain symptoms were experienced by affected persons other than the initial group of girls. These data are important, as they show that the signs and symptoms of being bewitched extended further than the initial 11 girls. Students are asked to use the data interpretation section of their data management sheet to incorporate information to make a defensible argument for either mass hysteria or ergot toxicity.

The instructor should circulate in the classroom playing devil's advocate, challenging the students' ideas at this point. This section of the case in particular provides an excellent opportunity for team teaching—additional data from the disciplines of history and environmental science provide a more complete consideration of the cultural and environmental conditions of the time.

Once groups have had adequate time to work, they should share their position with the class, with a survey taken of the various groups' positions. The instructor should facilitate a final discussion by intimating that majority rules; if a majority of the class selected and defended ergot toxicity, surely that must be the answer? Expect a lively discussion to follow. Instructors may choose to end the case study here or continue on to Parts V and VI.

Parts V and VI: Wrap-up: The Societal Frame and Classroom Extensions

Here the instructor should emphasize the iteration of the scientific method. It is also appropriate here to be sure students understand the distinction in science between hypotheses, theories, and laws. It should be emphasized to students that science builds on knowledge. As more data are accumulated and examined by more scientists, old hypotheses are discarded and new ones formulated. New technologies make it possible to examine old data in different ways. For example, if someone were to show

Chapter 7

symptoms of ergot poisoning today, a definitive urine test is available as a diagnostic tool. This is an opportune time to ask students what methods scientists employ today and what the limitations of analysis are. Finally, although the case emphasizes an either-or approach to the question of mass hysteria versus ergot poisoning, it is plausible that a number of causes were jointly responsible. Ergot poisoning might have affected some of the girls, providing them with symptoms; friends may have been inclined to more easily believe in the symptoms due to the described schism in local politics; and mass hysteria could have chipped in further once the ball got rolling.

At the end of the class, follow-up questions are distributed that expand on various aspects of the case that are relevant to the 21st century. The case study is left unresolved and, to our great pleasure, the study discussion does not end in the classroom.

Web Version

This case and its complete teaching notes, data management sheet, references, and answer key can be found on the website of the National Center for Case Study Teaching in Science at *http://sciencecases.lib.buffalo.edu/cs/collection/detail. asp?case_id=307&id=307*.

References

Bennett, J. W., and R. Bentley. 1999. Pride and prejudice: The story of ergot. *Perspectives in Biology and Medicine* 42 (3): 333–353.

Bennett, J. W., and M. Klich. 2003. Mycotoxins. *Clinical Microbiology Reviews* 16 (3): 497–516.

Boyer, P., and S. Nissenbaum. 1977. *The Salem witchcraft papers.* Verbatim transcriptions of the court records. New York: Da Capo Press. Also available online at *http://etext.virginia.edu/ salem/witchcraft/texts/transcripts.html*.

Caporael, L. R. 1976. Ergotism: The satan loosed in Salem. *Science* 192 (4234): 21–26.

Matossian, M. K. 1982. Ergot and the Salem witch affair. *American Scientist* 70 (4): 355–357.

Spanos, N. P., and J. Gottlieb. 1976. Ergotism and the Salem witch trials. *Science* 194 (4272): 1390–1394.

Woodward, W. E. (1864) 1969. *Records of Salem witchcraft V2: Copied from the original documents.* Vol. 2. New York: Da Capo Press.

Woolf, A. 2000. Witchcraft or mycotoxin? The Salem witch trials. *Journal of Toxicology-Clinical Toxicology* 38 (4): 457–460.

Additional Resource

Salem Witch Trials Documentary Archive and Transcription Project. *www.salemwitchtrials.org/ home.html*.

Chapter 8
The Bacterial Theory of Ulcers: A Nobel-Prize-Winning Discovery

Debra Ann Meuler

The Case

Dr. Robin Warren

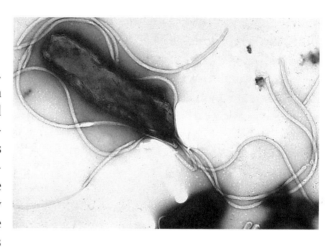

The story begins with Dr. J. Robin Warren. Dr. Warren was a pathologist at the Royal Perth Hospital (RPH) in western Australia. A pathologist is a medical doctor who examines tissues and is responsible for the accuracy of laboratory tests. Pathologists interpret the results of these examinations and tests and provide important information for a patient's diagnosis and recovery. Part of Warren's duties as a pathologist was to examine histological sections from gastric biopsies. With the arrival in the early 1970s of the fiber-optic endoscope, it was increasingly common to find good, well-fixed sections from gastrointestinal tissue. Prior to this, it was unusual for a pathologist to see good histological sections from any part of the stomach that was not postmortem. In 1979, after examining a hematoxylin- and eosin-stained section of a stomach biopsy from a man with severe gastritis, Warren noticed a thin blue line on the surface of the tissue. When he increased magnification, he thought he saw bacteria.

To get a better look at these bacteria, Warren ordered a Warthin-Starry silver stain of the histological section. Silver staining deposits silver granules on some types of bacteria, making them larger and more pronounced. The stain revealed numerous spiral-shaped bacteria. This was a bit surprising because, according to medical textbooks, bacteria were not supposed to colonize the stomach. Intrigued, Warren began requesting Warthin-Starry silver stain for all the gastric biopsies he examined.

Chapter 8

He began to see a pattern. There was a definite correlation between active, chronic gastritis and the presence of these bacteria. He also observed that the number of bacteria seemed to correlate with the degree of inflammation of the stomach lining—the more severe the inflammation, the more abundant the bacteria. But this didn't make sense. Medicine taught that the stomach was sterile. It was too acidic for any bacteria to survive for very long. Bacteria in the stomach had been reported before, but scientists had dismissed the bacteria as a contaminant or secondary to the problem. Warren, however, did not believe his spiral bacteria were simply a contaminant. Electron microscopy revealed bacteria attached to the mucosa and infiltrating between the tops of cells—an area not reached by acid. Also, because of the large numbers and homogeneity of colonization, he believed the bacteria were actively multiplying and living in the stomach lining.

What was even more peculiar to Warren was that the bacteria were associated with gastritis. Histological gastritis is diagnosed when sections of stomach biopsies show infiltration of tissue with lymphocytic-type cells, small collections of neutrophils, micro-erosions, epithelial cell damage, and a decrease in the thickness of the mucus layer. If Warren saw this type of histological presentation in biopsies, the bacteria were usually present. When Warren discussed his findings with colleagues, they were unconvinced and asked for more data. The stomach was a sterile environment, so Warren's bacteria were more than likely an anomaly. Despite the unenthusiastic response, Warren held to his belief that these bacteria were important and tried to recruit others to his study.

Dr. Barry Marshall

In 1979, Dr. Barry Marshall moved to RPH to begin three years of internal medicine training. At RPH, doctors in the College of Physicians training program were expected to do a small research project and write a paper each year. Marshall's early attempts at research were aimed at studying heat stroke. Australian summers can get very hot, and with the increased popularity of marathons, heat stroke was becoming a problem. Marshall wanted to find a way to monitor core body temperatures of runners while they were running. Toward this end, Marshall developed a mini-rectal thermometer attached to a piece of fishing line for easy retrieval. He and a resident spent many a day running stairs with the thermometer safely in place in hopes of raising their core body temperature to test their new monitoring system. He got a chance to use his invention during a marathon in 1982. It was an extremely hot day and many runners collapsed as a result of heat stroke. Marshall worked on one runner who collapsed and developed prolonged epileptic-like seizures that could be life threatening. Marshall took out his mini-thermometer and recorded a core body temperature high enough to cause brain damage. With the help of bystanders, Marshall carried the man to a nearby pool to cool him off until he regained consciousness and could be transported to the hospital, where he made a full recovery.

A Partnership Begins

In 1981, Marshall was in his second year of internal medicine training and was assigned to a rotation in gastroenterology. The chief of gastroenterology suggested that Marshall work with Dr. Warren, investigating his gastric spiral bacteria. After speaking with Dr. Warren, Marshall's interest was piqued. Marshall put aside his study of heat stroke, and together he and Warren worked to isolate and study Warren's spiral-shaped bacteria.

Marshall began his research by searching the current literature. Marshall was surprised to find that Warren's spiral bacteria had already been "discovered." In 1982, the Italian pathologist Giulio Bizzozero described the presence of spiral bacteria in the stomachs of dogs. Even earlier, a Polish clinical researcher, Dr. Walery Jaworski, described spiral organisms in sediments of gastric washings obtained from humans. In fact, over the years, many investigators had described spiral organisms in human stomachs postmortem and in fresh surgical specimens. In the early 1970s, because of the arrival of fiber optic endoscopy, researchers began to report the presence of gram-negative bacillus in 80% of patients with gastric ulcers. However, these bacteria were thought to be pseudomonas, a common contaminant in biopsy channels of endoscopes, and therefore not significant. In addition, isolating and culturing these bacteria proved difficult, making them impossible to study. Subsequently, few took these observations seriously. Most dismissed their presence as an artifact of collecting the specimen or some type of postmortem contamination.

The Pilot Study

Warren and Marshall, however, were convinced the bacteria were important. In 1982, they initiated a pilot study of 100 consecutive patients undergoing routine endoscopy. Prior to their scheduled endoscopy, each patient was given a survey to complete. In the survey, patients were asked about their symptoms, as well as their exposure to animals, travel, dental hygiene, and diet. It was important for Warren and Marshall to know about their symptoms to determine whether there was any correlation between stated symptoms and the presence of the bacteria. But why were patients asked the other questions? Warren and Marshall needed to rule out other explanations for the presence of the bacteria. For example, at the time of the study, Kentucky Fried Chicken was the most popular fast food in Perth. Marshall felt it was important to know about patients' diets, because chickens were known to possess spiral bacteria in their intestinal tracts. Dental hygiene history was important because spiral bacteria commonly inhabit dental plaque. Finally, information about the travel history of patients was important because Warren and Marshall did not know if patients contracted the bacteria while traveling to other countries.

At the time of their endoscopy, each patient had two stomach biopsies. One was sent for sectioning and the other to the microbiology lab for culture. The study was designed to answer the following questions: (1) Are the bacteria in normal stomachs?

(2) Does their presence correlate with type and severity of pathology? (3) Can the bacteria be cultured?

Isolating the Bacteria

To study a pathogen, you must first isolate it and grow it in culture. The spiral bacteria from the stomach biopsies proved difficult to isolate. Warren and Marshall believed the bacteria to be a type of Campylobacter. Warren's bacteria were similar in appearance to *Campylobacter jejuni*, a type of bacteria known to cause intestinal problems. At RPH, *Campylobacter jejuni* was routinely being isolated from stool samples sent to the microbiology lab. *Campylobacters*, however, are notoriously difficult to culture. They require a micro-aerobic technique, which requires the growth chamber to be evacuated of air and an atmosphere of less than 5% oxygen pumped in. By April 1982, Warren and Marshall had attempted to culture samples from about 30 patients without success.

Then Easter arrived. The hospital at the time was fighting an outbreak of *Staphylococcus aureus*, and the staff by mistake left the latest culture attempt in the incubator over the holidays. The normal procedure was to leave cultures to incubate for two days. If after two days no growth was evident, the cultures were discarded. But with the holiday and the *Staphylococcus aureus* outbreak, cultures were left in the incubator for five days. When the staff came back to discard the cultures, they were surprised to find growth. Apparently, Warren's spiral bacteria were slow growers, and it took five days for the colonies to emerge. Lab technicians, following standard protocol, had been throwing out viable cultures for the past six months!

Warren and Marshall were ecstatic. After months of effort, they had finally isolated and cultured their bacteria. Marshall had specimens sent to an electron microscopist. The results indicated they were similar to *Campylobacter jejuni*, but there were some key differences. The main one was the number and type of flagella. Warren and Marshall's bacteria had four sheathed flagella at one end. *Campylobacter jejuni* has only one unsheathed flagellum at one or both ends. The new isolate was named *Campylobacter pyloridis* because of its similarity to *Campylobacter jejuni*. The name was later changed to *Campylobacter pylori*. However, after additional studies, *Campylobacter pylori* was found not to be a *Campylobacter* at all and renamed *Helicobacter pylori*.

The Data

The analysis of the pilot study continued. The data were sent to a statistician to look for any relationship between the clinical and endoscopic results and the presence of gastritis and *Helicobacter pylori*. The data, as presented in the original paper published in the journal *The Lancet*, can be seen in Figure 8.1.

Figure 8.1: Data Tables Published in *The Lancet* by Warren and Marshall in 1984

Table II: Association of Bacteria With Endoscopic Diagnosis

Endoscopic Appearance	Total	With Bacteria	*p*
Gastric ulcer	22	18 (77%)	0.0086
Duodenal ulcer	13	13 (100%)	0.00044
All ulcers	31	27 (87%)	0.00005
Oesophagus abnormal	34	14 (41%)	0.996
Gastritis^T	42	23 (55%)	0.78
Duodenitis^T	17	9 (53%)	0.77
Bile in stomach	12	7 (58%)	0.62
Normal	16	8 (50%)	0.84
Total	100	58 (58%)	

Note: More than one description applies to several patients
^{T}Refers to endoscopic appearance, not histological inflammation.

Table III: Histological Grading of Gastritis and Bacteria

	Bacterial Grade				
Gastritis	Nil	1+	2+	3+	Total
Normal*	29	2	0	0	31
Chronic^T	12	9	7	1	29
Active	2	5	15	18	40
Total	43	16	22	19	100

*Gastritis grades 0 and 1 are normal.
^{T}1 case showed bacteria on gram-stained smear.

Based on their data and the statistical analysis, Warren and Marshall came to the following conclusions:

a. There were no well-defined clinical symptoms associated with the presence of the bacteria except for burping and abdominal pain.

b. There was no evidence that other sources of infections—such as animal contact, diet, and poor dental hygiene—were the source of this particular infection.

c. Gastritis, which was histologically diagnosed, was closely associated with the presence of the bacteria.

d. Bacteria were present almost exclusively in patients with chronic gastritis and were also common in those with peptic ulceration of the stomach or duodenum.

In fact, 26 out of 30 patients with ulcers had the bacteria. Those four patients with ulcers who tested negative for the bacteria had taken nonsteroidal, anti-inflammatory drugs known to induce ulcers. Warren and Marshall now felt they had sufficient evidence linking *Helicobacter pylori* with gastritis and peptic ulcer disease. The next step was to present their findings.

Presenting Their Findings

Warren and Marshall presented their findings to the College of Physicians at RPH. Their work was not well received. Most of their colleagues were skeptical of their claims. Of the many objections, the main one was the link between gastritis and duodenal ulcers. Prevailing wisdom said that gastritis was associated with gastric ulcers, not duodenal ulcers. Marshall did further research and found a study done at the Mayo Clinic in the 1950s. In this study, stomachs from either motor vehicle fatalities or partial gastrectomy patients were examined for the presence of gastritis and ulcers. Researchers found a surprisingly high number of specimens that showed evidence of gastritis. Of those with duodenal ulcers, 100% (N = 250) also had gastritis. Here was evidence that supported Warren and Marshall's hypothesis. Why had the ulcer research community skipped over the relationship between gastritis and duodenal ulcers? Marshall speculated it was because the finding did not fit with current beliefs and was therefore ignored.

Marshall continued reviewing the literature for any information pertaining to their research. He read that while acid-reducing drugs relieved the symptoms of the ulcers, they did not cure them. The relapse rate was high. Cimetidine, a common acid-reducing agent, helped heal the ulceration but did not cure the disease. Marshall was further intrigued when he read that colloidal bismuth subcitrate (CBS, which is similar to the over-the-counter drug Pepto-Bismol) greatly reduced the relapse rate in those with gastric ulcers. Marshall hypothesized that if a drug lowered the relapse rate for gastric ulcers and bacteria can cause gastric ulcers, then CBS should have antimicrobial properties. To test this, Marshall soaked a filter paper disc in CBS and then placed it in the middle of a petri dish inoculated with *Helicobacter pylori*. A few days later, much to his delight, there was a clear zone of inhibition around the paper disc, while cimetidine had no effect. This provided further evidence supporting the bacterial theory of ulcers.

Warren and Marshall needed more evidence. They needed to show a causal relationship between contracting *Helicobacter pylori* and developing gastritis. Marshall decided to use pigs as his model system. Pigs suffer from peptic ulcer disease just like humans, and they are large enough to be endoscoped. Marshall tried to infect two pigs with *Helicobacter pylori,* but neither pig developed gastritis or gastric ulcers. Also, as the pigs grew larger, it became dangerous to work with them.

Marshall needed to show that *Helicobacter pylori* caused gastritis. So, in desperation, Marshall submitted to a self-experiment. He knew that *Helicobacter pylori* were sensitive to several antibiotics, and he had successfully treated patients with gastritis using antibiotics. The culture he eventually would drink was isolated from a man who had been diagnosed with active gastritis. The organism was found to be sensitive to the antibiotic metronidazole. The man was given a two-week treatment of bismuth and metronidazole and cured of the disease. Therefore, Marshall felt confident that if he became infected with the same culture, he would be able to cure himself.

First, Marshall submitted himself to an endoscopic examination. The results, as expected, indicated no evidence of inflammation. A month later, after premedicating himself with the anti-acid drug cimetidine, Marshall drank 30 ml of peptone broth containing *Helicobacter pylori* from a three-day culture isolated from the patient mentioned above. Seven days after drinking the bacterial cocktail, Marshall became sick and began to vomit. After ten days, still quite ill, he underwent a second endoscopic biopsy of his stomach. The results, much to his satisfaction, revealed gastritis and the presence of *Helicobacter pylori*. A few days later, at the insistence of his wife, Marshall began antibiotic treatments. Soon after, his symptoms disappeared. Marshall underwent an additional endoscopy, but none of the histological sections showed the presence of *Helicobacter pylori*; it had totally disappeared. But Warren and Marshall now had evidence linking *Helicobacter pylori* infection of a normal stomach to gastric inflammation.

Publishing Their Results

Warren and Marshall had a difficult time getting their work published. They submitted an abstract of their research for presentation at a meeting of the Gastroenterological Society of Australia, only to have it rejected. The reason cited was the large number of abstracts submitted for the meeting. In the rejection letter, the selection committee stated that of the 67 abstracts submitted, they were able to accept 56. Apparently, the committee felt Warren and Marshall's paper to be in the bottom 10%, not important enough to be presented at the meeting. They rewrote their abstract and submitted it to the International Campylobacter Workshop in Brussels with more success. After the pilot study was completed, Warren and Marshall submitted their work to the medical journal *The Lancet*. Editors were initially reluctant to publish the paper because they could find no reviewers who believed their results. Dr. Martin Skirrow, chair of the Brussels Campylobacter Workshop, was contacted about their problem. He had their work repeated in his laboratory and found similar results. Dr. Skirrow contacted *The Lancet* to confirm Warren and Marshal's findings and, in June 1984, Warren and

Chapter 8

Marshall succeeded in publishing their paper titled "Unidentified Curved Bacilli in the Stomach of Patients with Gastritis and Peptic Ulceration."

In Conclusion

By 1994, after many more years of research by Warren, Marshall, and others, the medical community began to accept their bacterial theory of ulcers. The following quotation by noted gastroenterologist Dr. Ransohoff suggests how difficult it was for the medical community to give up their beliefs:

> The long-held hypothesis that duodenal ulcer disease is caused primarily by acid has, after a decade of siege by the *Helicobacter pylori* hypothesis, finally collapsed. That the acid hypothesis could even be challenged, much less toppled, appeared as unthinkable 10 years ago as the fall of Communism in the former USSR. Within the last few years, strong evidence has accumulated, however, about *Helicobacter pylori*'s importance, persuading even this previously skeptical writer. (Ransohoff 1994, pp. 62–63)

The impact of Warren and Marshall's bacterial theory of ulcer is profound. The medical community thought peptic ulcer disease was caused by excess stomach acid. Drug companies spent millions researching and marketing drugs such as Tagamet to treat the disease. Patients with extreme cases even underwent surgery to cut nerves involved in stimulating acid production. While these treatments did alleviate the symptoms, they did not cure the disease. Relapse rates were always high, prolonging patient suffering. By altering the way the medical community looked at peptic ulcer disease, patients are being spared unnecessary pain and anguish. Because of Warren and Marshall's ideas, the medical community has changed the way it looks at ulcers. Now the standard treatment regime for peptic ulcer disease is long-term antibiotics. This has resulted in a significantly reduced relapse rate for the disease, reduced medical costs, and, more important, reduced patient suffering. Drs. Marshall and Warren were awarded the 2005 Nobel Prize in Physiology or Medicine for their discovery of the bacterium *Heliobacter pylori* and its role in gastritis and peptic ulcer disease.

Questions

1. Do you think it is ethical and appropriate for Marshall to have used himself as a test subject and swallowed a sample of *Helicobacter pylori*? What precautions did he take? Would you do this? Why or why not?

2. How did the colloidal bismuth subcitrate (CBS) experiment provide evidence supporting Warren and Marshall's hypothesis?

3. Answer the following questions based on the data presented in Figure 8.1 (p. 73):

a. In Table II, of those patients with ulcers, how many were positive for *H. pylori*? Of those patients with normal endoscopic results, how many were positive for the bacteria?
b. Based on this data, Warren and Marshall hypothesized that there was a causal relationship between ulcers and bacterial infection. But there were four patients with ulcers who were negative for the bacteria. Why is this not significant?
c. If there is a causal relationship between the presence of *H. pylori* and ulcers, how might you explain that 50% of the patients with a normal endoscopic examination were infected with the bacteria?
d. In your own words, explain the results presented in Table III. What do you conclude from this data?

4. Robert Koch was a German physician who identified the bacteria causing anthrax and tuberculosis. His methods established four criteria that must be met for a specific pathogen to be considered the cause of a disease. These four criteria are listed below. For each one, discuss whether Warren and Marshall fulfilled them and, if so, how.

I. The pathogen should be found in the bodies of animals having the disease.
II. The suspected pathogen should be obtained from the diseased animal and grown outside the body.
III. The inoculation of that pathogen, grown in pure cultures, should produce the disease in an experimental animal.
IV. The same pathogen should be isolated from the experimental animal after the disease develops.

5. What role did chance, assumptions, and curiosity play in Warren and Marshall's research on *Helicobacter pylori*?
6. Describe how the story of Warren and Marshall's discovery illustrates the process of science.
7. How does this case illustrate the tentative nature of science?
8. How does this case illustrate the role of technology in scientific progress?
9. Why is this discovery significant? Do you think it is worthy of a Nobel Prize?
10. What does the *Helicobacter* story tell us? What lessons can be learned from this story?
11. Albert Szent-Györgyi, the 1937 Nobel Laureate in Physiology and Medicine, said in his Nobel Prize speech: "Discovery consists of seeing what everybody has seen and thinking what nobody has thought" (Good 1963, p. 15). Describe how this statement applies to Warren and Marshall's pioneering work on peptic ulcer disease.

Chapter 8

Teaching Notes

Introduction and Background

This case is an account of the events that led Dr. Robin Warren and Dr. Barry Marshall to the bacterial theory of ulcers. It is a tale of two physicians who refused to accept the standard explanations for what they had observed. Their alternative hypothesis transformed the way the medical community viewed peptic ulcer disease and has saved countless patients from unnecessary pain and suffering. But Warren and Marshall's bacterial theory of ulcers is also a story that provides a fascinating glimpse into the process of scientific discovery. It shows the importance of curiosity, questioning, chance, and tenacity in scientific inquiry. It shows how science is built on the work of others, and how assumptions can cloud people's views. Their story shows the tenuous nature of truth in science, and that scientific "truth" can and does change when faced with new data and new interpretations.

The case has been successfully used in both major and nonmajor classes during the unit on the nature of science. The case is appealing to today's student in that it is current, showcases the story of two scientists who are contemporaries, and is relatable, considering that many students have experienced the effects of *H. pylori* either firsthand or through relatives.

Objectives

The teaching objectives for this case are as follows:

- To trace the development of a major idea in biology/medicine
- To learn about the cultures in which the people profiled in this case story lived and worked
- To illustrate the scientific method using a real-life example
- To provide insight into how a discipline's base assumptions can affect scientific progress
- To provide an opportunity for students to analyze authentic data
- To show how scientists identify a causal relationship between a pathogen and a particular disease

Common Student Misconceptions

- Long-standing scientific knowledge is "true"—it does not or will not change.
- If the data do not fit our belief system, they are probably erroneous or an anomaly.
- Our assumptions are true.
- The human stomach is too acidic to allow for bacterial growth.
- Bacteria associated with humans all have the same incubation period.

Classroom Management

This case can be completed in a single 50-minute class period. The case is distributed during the preceding class, and students are asked to read the case and formulate answers to the questions provided at the end of the story (pp. 76–77). When the class meets to discuss the case, students are put into groups of 3 or 4 and given 20 minutes to discuss their answers to the questions. Students then come back as a class and discuss their responses. Students are then told that questions 5, 6, 7, and 8 are possible essay questions for the next exam.

Web Version

This case and its complete teaching notes, references, and answer key can be found on the website of the National Center for Case Study Teaching in Science at *http://sciencecases.lib.buffalo.edu/cs/collection/detail.asp?case_id=571&id=571*.

References

Good, I. J. 1963. *The scientist speculates.* New York: Basic Books.

Ransohoff, D. F. 1994. Commentary. *Annals of Internal Medicine (ACP Journal Club suppl.,* May/June): 62–63.

Warren, J. R., and B. J. Marshall. 1984. Unidentified curved bacilli in the stomach of patients with gastritis and peptic ulceration. *Lancet* 1 (8336): 1273–1275.

Section III:
Experimental Design

Every general biology book has an introductory chapter that outlines the steps of the scientific method. But this idealized approach to solving scientific questions in fact is not the way the scientific enterprise proceeds. Scientists seldom are so organized. They stumble onto discoveries, some may not do experiments at all (recall Snow's epidemiological approach to cholera in chapter 6), they often make innumerable false starts, and theoretical physicists leave the testing to others. But the most egregious error in thinking that the idealized steps of the scientific method are the inevitable path to discovery is that the idealized model ignores the roles of the scientific community and society. Scientists are not hermits whiling away the hours in isolated laboratories for their personal amusement. Yes, scientists are motivated by curiosity, but they have a multitude of other items on their agendas. Real scientists are, to a greater or lesser degree, motivated by the desire for peer recognition, funding, career advancement, job satisfaction, job security, and more—the whole gamut of emotions. In short, scientists are just as human as everyone else.

Still, with that said, it remains the case that the core of science is that questions are asked, guesses are made, and answers are sought. In this, scientists are not unique—everyone in the world asks questions and seeks answers dozens of times every day. What makes scientists different is that they vigorously and systematically test their ideas, repeating that process over and over again to seek generalizations about how the physical world is constructed. Asking good questions, formulating testable hypotheses, and then following through with the actual tests and generating useful data are earmarks of not only a good scientist but also a good critical thinker.

The cases in this section highlight the core steps associated with research in most areas of science. Like other cases in this book, the emphasis is not on scientific content but on the *process* of science. There is a well-worn axiom that captures the idea: Science is a way of knowing. It is our purpose to drive that point home, especially with

Section III

the cases in this section, in the hope that they will sharpen students' skills in thinking scientifically.

The first case, "Lady Tasting Coffee," retells a famous story of the statistician Ronald Fisher, who, upon hearing a woman claim that she could tell by taste if milk had been added to her cup of tea before or after the tea was poured, set out to design an experiment to test the claim. Students wrestle with a modern version of the story, this time about coffee, to see if they can come up with an appropriate experimental design to test the claim. As part of the case, students revisit Fisher's original paper to see if their approach compares favorably to his. Students learn some basic elements of statistical analysis, discovering that a statistical experiment requires careful planning and that making statistical conclusions or generalizations requires that the experiment be repeated more than once. Fisher's work has had a huge impact on how we evaluate research today and serves as an illustrative model for today's students.

The case study "Memory Loss in Mice" is unique in this collection. It consists of a single newspaper paragraph summarizing a scientific paper on Alzheimer's disease. Under the guidance of the teacher, students are asked to reverse-engineer how that research might have been conducted, reconstructing in class the experimental design that led to the claims made in the short abstract. The goals here are not only to solve the problem but also stimulate the students' creativity and compare their predictions with those of the original authors. In this way, the case demonstrates science in action while also exploring the workings of the nervous system and the importance of carefully crafted and implemented animal model systems in human disease research.

The third case, "Mom Always Liked You Best," involves an actual field study conducted in the marshes of British Columbia to determine whether bird parents favor some chicks over others as they raise them (origins of sibling rivalry?). The case proceeds through a series of steps as more information is revealed to the students. We call this method the interrupted case method. We have found it to be a favored method among science teachers for teaching a case study. Students first are given a snippet of information from a real published paper and asked to imagine the question the authors seem to have been asking. Students then are asked to guess what the hypothesis might have been. After a short time working in small groups, students are given more information and asked to devise a method for testing that hypothesis. Again, after further small-group discussion, more information is provided. Students then are given some data and asked to predict other results. The case concludes with the students receiving the real data and comparing the actual findings to their own. The beauty of this case teaching method is that it mimics in many ways how real scientists work—proposing ideas, trying them out, refining their thinking, and analyzing data.

The fourth case, "PCBs in the Last Frontier," deals with polychlorinated biphenyls, now recognized as an environmental pollutant. Students grapple with the mystery of how PCBs ever reached pristine, isolated lakes in the wilds of Alaska. Students are asked to generate a hypothesis and propose a way to test it. New observations are introduced that show that PCB levels in nearby lakes differ one from another, a fact that leads to new hypotheses and further tests. Finally, data from migrating salmon are introduced, solving the mystery. The case is another example of the interrupted case method in action. It challenges students to think like scientists—making observations, offering hypotheses, and suggesting tests—all while constantly revising their views as new data are introduced.

Unlike the previous natural history cases, the fifth case in this section deals with social psychology. "The Great Parking Debate" will resonate with anyone who has struggled to find a parking space in a crowded mall parking lot or a university parking lot as students rush to class. Students must negotiate their way through real data as they make observations, form hypotheses, devise tests, and make predictions, following the familiar strategies of both natural and social scientists.

The last case in this section, "Poison Ivy: A Rash Decision," has historical flavor. We are introduced to a student afflicted by poison ivy and learn that the American Indians and early colonists also had learned about the plant the hard way. The protagonist in the story researches poison ivy on the internet and learns of a questionable method of treatment. Data are introduced, and students must evaluate if the treatment is effective. As with the other cases in this section, students practice forming and evaluating hypotheses, making predictions, designing experiments, interpreting data, and drawing conclusions.

In summary, the goal of the cases in this section is to hone the skills of students in experimental design. More than that, the cases here (and case studies in general) require students to work in groups, enhancing their skills in teamwork and collaboration, the modern approach of today's scientists, who know that no one person has all the necessary skills or knowledge to get the job done. Finally, all of the cases refer to published scientific studies, and students are able to compare their efforts to what really happened, giving the students a chance to see how their ideas compare to the experts' lines of attack.

Chapter 9
Lady Tasting Coffee

Jacinth Maynard, Mary Puterbaugh Mulcahy, and Daniel Kermick

The Case

Part I: Coffee Shop Wager

Characters

Model: An attractive young woman with impeccable taste, working as a successful model for an advertising firm.

Escort: A tall, dark, and handsome young man, working as a marketing and survey researcher for the same advertising firm.

Older Gentleman: An adjunct professor of biostatistics at a local college and a coffeehouse regular; dressed in a slightly aged tweed jacket.

Setting

The red sports car makes a quick stop in front of the Philadelphia coffeehouse after its occupants have spent a night on the town. A sleek model and her tall, dark, handsome escort gracefully exit the car and approach the counter, where the escort purchases two cups of house coffee. At the condiments table, the escort proceeds to pour milk into the two coffees, and the following dialogue results.

Dialogue

Model: Oh, no. I'll need a fresh cup.

Escort: You like milk in your coffee, right?

Model: Yes, I like milk in my coffee, but only, and I say only, if it is added to the cup first.

Escort (laughing): Oh, come on, that's ridiculous; you can't possibly tell the difference. Coffee is coffee whether you add the milk first or second.

Chapter 9

Model: Of course I can tell the difference; you've no right to laugh at me.

Older Gentleman: Pardon me for overhearing your conversation. Actually, the lady may be able to tell the difference.

Escort: You're kidding me, right?

Older Gentleman: Such claims have been made before. A woman made that claim at an afternoon tea party in Cambridge in the 1920s. She stated that she could always tell whether milk was added before or after the tea was poured in the cup. The famous statistician Sir Ronald Fisher was at high tea that afternoon and immediately designed an experiment to test the woman's palate. Rumor has it that she shocked the guests because she correctly told Sir Fisher whether milk had been added first or second after tasting multiple cups of tea. The event became famous because Sir Fisher used it to explain the basics of experimental design in one of the first textbooks ever published on the topic of experimental design.

Escort: Well that's a good story, but it also sounds suspiciously like a story with no basis in reality, invented by an imaginative professor writing a textbook. You certainly haven't convinced me to buy the lady another cup of coffee.

Older Gentleman: From what I've read, it was a real event, and furthermore, humans have very sensitive taste buds. I think she deserves a new cup.

Model: Now that the coffee is cold, I certainly deserve a fresh cup.

Older Gentleman: Ma'am, I fully agree. In fact, I think you deserve several cups of fresh coffee! Let's set up our own experiment to determine whether this lady can or cannot discern whether milk was added to the cup before or after the coffee.

Model: How fun! Yes, let's see who is right with our own tea test, I mean coffee test.

Older Gentleman: Clever pun, ma'am. You know the real student's t-test was developed in the early 1900s by a taste tester of sorts at the Guinness Brewery in Dublin, Ireland. Only he wasn't tasting tea either. Sir, are you a gambling man? How about a wager? If our experiment shows that she can tell the difference, you will pay for the coffee. If the experiment does not demonstrate her discerning palate, I will pay for the fresh cups.

Escort: You're on! But I want the rules hashed out before she starts sipping. I mean, how many cups are we talking about? And what if she's wrong for just a few cups? I don't want to pay for the coffee just because she is a good guesser.

Older Gentleman: Fisher would agree completely. Even the smallest experiment requires forethought and planning. You must tell me, sir, just how sure do you want to be that she isn't guessing?

Questions

Imagine that you are going to design and perform the experiment described in the dialogue.

1. What is the hypothesis that will be tested in this experiment?

2. Why is it important to offer the model more than just two cups (one with the milk added first and one with the milk added second)? Explain your answer.

3. How many cups of coffee should the model taste? Explain your answer.

4. Describe exactly how the cups should be prepared. Does every cup need to be exactly the same in every way except the order in which the milk and coffee are added? Can you actually make every cup identical? Explain your answer fully.

5. In what order should the cups be presented? What method or decision rules might you use to decide which cup will be offered first, second, and so on?

6. How do you recommend that the characters decide if the model is able to tell whether the milk was added to the cup before or after the coffee? (In other words, how many cups does she have to correctly evaluate for you to conclude that she really can tell the difference?) Explain your choice.

7. Without looking it up in a textbook or online, provide your own definition of "experimental design."

Part II: Tasting Tea

Read Fisher's essay "Mathematics of a Lady Tasting Tea," available online at *http://legacy.library.ucsf.edu/tid/fqi22e00/pdf*. Then answer the questions below.

Questions

The questions below ask you to compare your answers to the questions in Part I to the explanations found in Fisher's essay.

1. Referring to your answers to the questions in Part I, was the hypothesis you chose different than the null hypothesis given by Fisher in his essay? Explain your answer.

2. What reason did you give for why it is important to offer the model more than just two cups (one with the milk added first and one with the milk added second)? Was your answer the same as Fisher's answer? Based on the essay, please describe Fisher's answer to this question.

3. How many cups did you say the model should taste? How many cups did Fisher say that the "tea lady" in the story should taste? Please describe fully Fisher's

Chapter 9

answer to this question, including any mathematical considerations. Was your answer the same as Fisher's answer? If not, how was it different?

4. Before reading Fisher's essay, did you think that every cup needed to be exactly the same in every way except the order of the addition of milk and coffee? Does Fisher believe that every cup should be prepared identically? Describe Fisher's explanation for how to deal with uncontrollable variations among cups.

5. Before reading Fisher's essay, what method did you recommend for choosing the order in which the cups should be presented? Was your answer different from what Fisher recommends in his essay? Describe Fisher's explanation for how to choose the cup order.

6. In Part I, you answered the question of how one should decide if the model is able to tell whether the milk was added to the cup before or after the coffee. (In other words, how many cups does she have to correctly evaluate for you to conclude that she really can tell the difference?) Compare your answer to Fisher's answer and then describe Fisher's answer fully.

7. Your instructor will provide you with a textbook definition of *experimental design*. Was your definition complete? Your answers to all of the questions in Part I probably differed in small or large ways from Fisher's proposed design of a tea-tasting experiment. Are there any differences that changed your perception of statistics and experimental design? If yes, describe how Fisher's essay enlightened you. Even if you did not find that Fisher's essay changed your views, make a short summary list of the important design concepts that you think are emphasized in Fisher's essay.

8. Following the guidance of your instructor, use Part III of the story or, alternatively, if your instructor has provided materials, set up a mock event similar to the tea-tasting experiment. Instead of tea, you might consider seeing if your classmates can tell the difference between 1% and 2% milk, between two brands of bottled water, between two brands of flavored diet or regular soda, or some other simple taste comparison. Adhere to the principles of Fisher's paper, and draw conclusions based on your results, using Fisher's rules.

9. Rumor has it that the real lady who had tea with Fisher on an afternoon in the 1920s was able to accurately tell whether the milk was added first or second every single time she was offered a new cup. Can you conclude from her success that most people can tell the difference between milk-first and tea-first cups? Briefly describe the design of an experiment that would test this broader question.

Part III: Tasting Coffee

The escort and the older gentleman prepared four cups of coffee to which they added the milk first and four cups of coffee to which they added the milk second. The two men attempted to make sure that the cups were identically prepared as much as possible in terms of the amount of coffee, milk, and other factors. By writing the numbers 1–8 on slips of paper and pulling the numbers out of the hat, the escort and the older gentleman randomly decided that cups #2, 3, 6, and 7 would be prepared with the milk added first and cups #1, 4, 5, and 8 would be prepared with the coffee added first. The model has just finished sipping all cups (presented in order 1–8), and she has written down her decision.

Older Gentleman: Are we all in agreement that we will use Fisher's rules for deciding the conclusion for the experiment?

Escort: Yes, we are. Please tell us which cups you think had the milk added first.

Model (smiling broadly): It is so obvious that cups #2, 3, 4, and 7 had the milk first.

Older Gentleman: Are you sure? Is this your final answer?

Model: Yes, those are the cups that had the milk added first.

Escort: Ha! Based on Fisher's rules, you can't really tell the difference! The milk was added first to cups #2, 3, 6, and 7. Hot dog, mister! *You* will have to pay for the cups!

Model: But I got 6 of the 8 correct! That's pretty good! Who is this Fisher guy to say I can't do it?

Older Gentleman: I have to agree, ma'am, that you may be able to tell the difference, but we did decide beforehand that we would use Fisher's rules, and his rules are quite strict. We can only conclude that you can tell the difference if there were less than a 5% chance that your success could be explained by good guessing.

Model: But I didn't guess! I really can tell; it seems unfair that you accuse me of guessing just because I made one mistake!

Escort: We aren't really accusing you of guessing; the 5% rule is the common cutoff that is used in many statistical analyses.

Model: Fisher's rules are too strict. Give me another two cups and, if I get it right this time, my success rate will be better than the 5% guess rate. I'll show you that I really am able to tell the difference!

Escort: Hold on! It would be cheating to change the plan now. No more cups. Please, let's quit while I'm ahead!

Chapter 9

Questions

1. Did the characters correctly follow Fisher's rules when they concluded that the results do not allow us to conclude that the woman can tell the difference between the two types of cups?

2. Do you think the lady really can tell the difference between the milk-first versus the milk-second cups?

3. Which of the following sentences do you think most accurately and clearly states the conclusion of the experiment? Propose your own statement if you find flaws in all of the statements below.
    a. The model cannot tell the difference between the milk-first and the milk-second cups.
    b. There is a 5% or greater chance that the woman guessed her answers.
    c. At the 5% level, the model cannot significantly tell the difference between the milk-first and the milk-second cups.

4. Would it be acceptable to add two more cups now? Why or why not? What is the value of deciding the experimental design before you begin an experiment and not changing it in the middle of the experiment?

5. Do you think Fisher's rules are too strict? Why or why not?
    a. Would you feel the same way about his rules if we were testing whether a monitor at a nuclear power plant can really recognize elevated levels of radiation?
    b. Would you feel the same way if we were testing whether a new children's vitamin caused increased risk of kidney dysfunction?
    c. To answer questions (a) and (b) better, read the more detailed descriptions of the two hypothetical experiments given below. For each design, state the null hypothesis. There are two types of mistakes that statisticians can make at the conclusion of a significance test: They can incorrectly reject a true null hypothesis (Type I error), or they can incorrectly fail to reject a false null hypothesis (Type II error). The 5% rule ensures that a Type I error is never made at a greater rate than 5%, and the likelihood of making a Type I error is often inversely related to the likelihood of making a Type II error. Type II error rates are usually not controlled in scientific experiments and can be considerably higher than 5%. For the following two experiments, what would be the human ethical consequence of making Type I and Type II mistakes in your conclusion at the end of the experiment? Would you recommend the 5% rule for these experiments? Why or why not?
        i. Nuclear Monitor Experimental Design: The monitor is exposed to eight environments (four high and four low radiation levels), and the experimenter records whether the monitor warning light comes on or not.

ii. Vitamin Experimental Design: Eight children are given a placebo and eight children are given the new vitamin. Urine from the children is collected and the pH in the urine is monitored and compared between the two groups. Both high and low pH are indications of kidney dysfunction.

6. Return again to the coffee-tasting experiment. Does the outcome described in the dialogue on pages 85–86 prove that the model cannot tell the difference between the two types of cups of coffee?

Teaching Notes
Introduction and Background
Students majoring in biology need to be able to recognize and design good experiments. Unfortunately, most biology curricula do not include a required statistics course that would formally train students in the design of experiments, and many statistics courses are focused more on analysis and probability than on the initial setup of an experiment. This case study was written to introduce students in a sophomore-level course in ecology and evolution to some important basic principles in experimental design.

The case is also an opportunity to teach students about the history of science. Ronald Fisher's essay is easy and fun to read despite the fact that it was written more than 70 years ago. Prior to the experimental design books and research papers published by Fisher in the 1920s and 1930s, experiments were largely designed according to the style of the particular scientist. Fisher's paper articulated some guidelines and standards that have become commonplace in experimentation in many disciplines.

Objectives
Students completing this case should be able to

- state the null hypothesis for a simple experiment and
- design a simple experiment, including
  - selecting samples and applying treatments,
  - establishing the level of significance, and
  - making appropriate conclusions based on significance levels.

Common Student Misconceptions
- A statistical experiment does not require much thought or planning.
- There should be no variation in experimental subjects.
- Experiments require that you eliminate all variations.

- Statistical conclusions and generalizations can be made after performing an experiment once on a small number of subjects.
- Statistical conclusions are always correct.

Classroom Management

Day 1: In-Class Activities

(15 minutes) Students act out the dialogue in Part I of the case. This can be done in small groups of three students with each student taking the role of one of the characters (the model, her escort, and the older gentleman), or it can be done by just three students role-playing the characters for the entire class.

(20 minutes) Students work in small groups of three (recommended) or up to five to answer the questions associated with Part I and devise an experimental design to test the question: *Can the lady tell if the milk was added before or after the coffee?*

(15 minutes) One student from each group briefly summarizes the group's design for the entire class. Because of limited time, the instructor can call on just one group to present the answer to Question 1, move to another group to answer Question 2, and so on. The instructor can ask for a show of hands as to whether other groups chose the same or different designs. The instructor should take notes on the designs so that important issues can be discussed at the next class meeting.

Day 1: Homework Assignment

As homework, students are given Part II of the case, which directs them to read Sir Ronald A. Fisher's paper "Mathematics of a Lady Tasting Tea," then answer the questions associated with it (answers to these questions may be turned in or can kept by the students to use for discussion at the next class meeting). Fisher's essay is available online at *http://legacy.library.ucsf.edu/tid/fqi22e00/pdf*.

Day 2: In-Class Activities

(20 minutes) Students discuss how (and why) their experimental designs differ from that of Fisher's. To stimulate discussion, we give students a mock set of wrong answers. The instructor can lead the discussion by saying, for example, "Here is an answer given by a student to question 1: *The null hypothesis is: The lady can tell the difference between milk-first cups and milk-second cups.* What is your reaction to this student's response?"

The instructor should make sure that by the end of the discussion the students can do the following:

1. Determine the null hypothesis

2. Select samples and assign treatments, and understand the concept that it is impossible to eliminate all variations among experimental units

3. Establish the level of significance

4. Collect and analyze the data

5. Correctly state the conclusion of a statistical test, usually based on the level of significance

(15 minutes) The instructor can use all or portions of the PowerPoint presentation (see "Web Version," below) developed for this case to give some background information on Fisher and his essay.

(30 minutes) The instructor can have students act out the experimental test in small groups of three to five students using almost any kind of beverage.
Examples:
- Skim milk versus 1%, 2%, or whole milk
- Tap versus bottled water
- One brand of soda versus another brand of soda

Or you can give the students the simulated results provided in Part III of the case.

Day 2: Homework Assignment
Students are told to turn in a report that describes (1) the null hypothesis, (2) the methods they used in the simulated experiment, including the significance level, (3) their results shown in a simple table, and (4) the correct conclusion based on the rules in Fisher's essay.

Web Version

This case and its complete teaching notes, answer key, and supplemental PowerPoint presentation can be found on the website of the National Center for Case Study Teaching in Science at *http://sciencecases.lib.buffalo.edu/cs/collection/detail.asp?case_id=414&id=414.*

Chapter 10
Memory Loss in Mice

Michael S. Hudecki

The Case

Please read the following abstract of an article in the *New York Times* and prepare a short written response addressing each of the following questions.

Memory Loss in Mice

A biochemist, Eugene Roberts, and researchers at the City of Hope Medical Ctr. [in Duarte, California] discovered that injecting fragments of a brain protein called beta-amyloid into the brains of mice caused the mammals to forget chores they had just been taught [J.F. Flood et al. Amnestic effects in mice of four synthetic peptides homologous to amyloid B protein from patients with Alzheimer disease. *Proc. Nat. Acad. Sci.*, USA, 88 (8):3363–6, Apr 91]. "This is really the first correlation between the presence of [beta-amyloid] in the brain and the loss of memory," said Rachael Neve, a molecular biologist at U. California at Irvine. (Anonymous/Associated Press 1991)

Chapter 10

Questions

1. State in concise terms the problem being investigated.
2. Describe the details, if any, of the experimental method apparently used in this study.
3. Describe any pertinent results that originate from the study.
4. What specific conclusions can you draw from this study?

Teaching Notes

Introduction and Background

Named after the German neuropathologist Alois Alzheimer (1864–1915), Alzheimer's disease is currently an incurable medical condition affecting a significant percentage of senior citizens worldwide. Experts estimate that approximately 10% of those over 65 years of age and half of those over 85 are afflicted. Moreover, research has demonstrated that about 50% of Alzheimer's disease is caused by a faulty gene(s), presumably affecting the expression of specific brain proteins such as beta-amyloid.

This discussion case explores the scientific process involved in implementing an animal model in the study of Alzheimer's disease. Students read a short paragraph describing a study in which the brains of "trained" mice were injected with beta-amyloid fragments, which subsequently caused them to forget their tasks. The paragraph is a very short *New York Times* story reporting on an experimental study originally published in the *Proceedings of the National Academy of Sciences, USA* (Flood, Morley, and Roberts 1991). Based on the short description provided, I ask students to identify relevant components of the scientific method, such as the problem, method, results, and conclusions. All students are generally able to accomplish the assignment in the allotted 20-minute time period, after which I initiate a class discussion.

I originally used this case in a weekly freshman seminar course, History of Contemporary Biology, with no more than 20 first-year undergraduate students. While most were science majors, a significant proportion of the students were arts and humanities majors. More recently, I have incorporated this case into my two-semester nonscience majors' course, Perspectives in Human Biology. This particular case comes up at the beginning of the second semester when we focus on the human nervous system. During the first semester of the course, the students develop a working knowledge of biological principles and the scientific method directed at a range of contemporary human biology issues.

Objectives

- To demonstrate the scientific method in action. Through in-class participation and discussion, pertinent components of the scientific method can be sequentially identified and analyzed.

- To explore the workings of the nervous system in health and disease, with specific attention given to the degenerative disorder Alzheimer's disease.
- To understand that important advances in human disease research often rely on carefully crafted and implemented animal model systems. According to U.S. Food and Drug Administration (FDA) guidelines, preclinical animal studies serve as the foundation for future therapeutic trials involving human subjects (including Alzheimer's patients).

Common Student Misconceptions

- Science is a straightforward process of discovery.
- Experiments are easy to devise and perform.
- Working with animal models will lead to simple conclusions.
- Scientists have not made any progress in understanding Alzheimer's disease.

Classroom Management

I initiate the discussion of the case by asking students to define the problem being investigated. Using a blank overhead projector transparency, I write "Problem" on top and wait for student responses. Students offer information on memory loss and Alzheimer's (as well as disease expressions involving amyloid brain plaques and the debilitated status of the mouse model). Very quickly, I am able to fill up the transparency with student feedback. Through dialogue and by drawing on their prior academic and personal experiences, students eventually group and refine responses into a manageable number of research problems. In broad terms, the consensus of the class is that the central problem is Alzheimer's disease (and its attendant memory loss). On the other hand, many students narrow the problem and focus more on determining the unknown etiology of the disease (i.e., amyloid protein). This part of the exercise yields both general and specific characterizations of the research problem at hand.

After spending sufficient time determining the problem, we move next to the experimental method. Using a new transparency with "Method" written on top, I ask students, "What did the scientists do to address the research problem?" and record the wide range of responses they propose. Some students immediately begin to describe the amyloid injection process, whereas others concentrate on the "training" phase of the study.

Although various methods are suggested, usually the class agrees that a "maze" is needed in which there is an incentive or reward for navigating the configuration. Consequently, the "trained" mice are thought to be suitable test subjects for the amyloid protein injections. These mice would be tested again using the maze/reward configuration to determine the amyloid's effect on memory. After further discussion, the class generally concurs that the critical result of the study is whether there is a change in the ability of the injected mice to navigate the maze successfully.

Chapter 10

Before we begin discussing the critical results of the study, the class must decide what makes a proper "control" for the study. At this juncture, I typically ask, "How do we know that amyloid injection causes memory loss in the mice?" I then begin to develop a list of control (as well as further experimental) groups of mice for the study. I find this particular phase of the discussion both enjoyable (because the students by this time are really into the analysis of the case) and challenging (because I need to pay particular attention to each suggested control and the inherent rationale of its inclusion).

As a class we expand the study to include a growing list of groups: control groups of injected but not trained mice and mice that were injected only with the watery vehicle of the amyloid (sham group). We also include an expanded range of experimental animals, including males and females, young and old, and treatment groups composed of synthetic and natural amyloid protein fragments. Another offshoot is a discussion of how many animals should be included in each group. One or two? A dozen? Hundreds? At this point we are able to discuss both statistical significance (what it is and how it is used) as well as the cost of conducting any large-scale study involving animals.

To close the case, I have often turned to the original article and shown the students how the investigators really went about their research protocol.

Most recently, with a current class enrollment of more than 300, I have made the case the subject of a written report that is due within a three-week period. In this report, I require the students to discuss (and research) in depth the four components of the scientific method evident in the mouse study. After the assignments have been turned in, I open the floor to a whole-class discussion on the case.

As a result of the success I have had using this case, I now use newspaper articles of varying lengths as both case study material in the classroom and subject matter for my formal exams. Daily newspapers are a virtual treasure trove of new science case study material. It's simply a matter of cutting, copying, and distributing, and then asking the important questions: What is the problem? What did the scientists do? What did they find? What does it mean?

Web Version

The case and its complete teaching notes can be found on the website of the National Center for Case Study Teaching in Science at *http://sciencecases.lib.buffalo.edu/cs/collection/detail.asp?case_id=194&id=194*.

References

Anonymous/Associated Press. *New York Times*. 1991. Clues to loss of memory in Alzheimer's reported. April 16.

Flood, J. F., J. E. Morley, and E. Roberts. 1991. Amnestic effects in mice of four synthetic peptides homologous to amyloid beta protein from patients with Alzheimer's disease. *Proceedings of the National Academy of Sciences* 88 (8): 3363–3366.

Chapter 11
Mom Always Liked You Best

Clyde Freeman Herreid

Figure 11.1. Mother Coot and Two Chicks

Photo by Mike Baird, www.flickr.com/photos/mikebaird/530816116

The Case

Part I: What Is the Problem?

Many birds have bright, ornamental plumage. Most often the plumage is displayed by the male of the species, who is believed to use the plumage to attract females. The female may select male breeding partners on the basis of this feather advertisement, perhaps assessing their health. According to evolutionary theory, it behooves the female to choose a strong, healthy male to be the father of her chicks, not only because the male helps feed them but also because the chicks will carry the father's healthy genes. Darwin labeled such mate choice "sexual selection."

Another possible explanation of plumage selection has been studied by biologists in Canada. They think that the parents of certain bird species may select the "prettiest" chicks out of a nest as favorites and feed them better (why they might do this is an interesting question).

American coots are birds that live in the marshes of western North America. As adults, they are grayish-black with white beaks. The chicks are unusual, for unlike most birds whose nestlings are usually drab, coot chicks are surprisingly conspicuous. They have long, orange-tipped, slender feathers; brilliant red papillae around their eyes; bright red bills; and bald red heads (see Figure 11.1). The chicks lose this colorful appearance at three weeks. The Canadian biologists speculated that the plumage may make some chicks more attractive to their parents; possibly the most "attractive" chicks might be able to successfully beg for more food from their parents and have a better chance of survival. That seems possible because about half of all chicks die from starvation. But how could the authors test such an unusual notion?

Here is your challenge:

- First, identify the specific question(s) the researchers are asking.
- Second, what is the hypothesis that they suggest?
- Third, what predictions (deductions) can you make if the hypothesis is correct?
- Fourth, how can we test the predictions; for instance, what exactly might we do if we were the researchers who had studied coots for several years?

Part II: The Biologists Find a Method to Attack the Problem

During the breeding season of 1992, Bruce Lyon, John Eadie, and Linda Hamilton studied 90 pairs of coots nesting in the marshes of British Columbia. They decided to try to alter the plumage of the chicks by dyeing them. Unfortunately, the dye made the chicks sick and removed the oils from their feathers. The scientists next tried to alter the appearance of the chicks by cutting the orange tips off their body feathers. This produced black chicks that seemed to act the same as normal orange chicks. The scientists now began a test of their hypothesis using this technique. What do you expect they might do? Design an experiment that would test the hypothesis that you think the scientists might be using as a working idea.

Part III: What Should Be Measured?

The biologists decided to set up three types of nest conditions. In the first group of nests (let's call this the experimental group), they trimmed half of the chicks and made them black and left half of the brood with orange feathers. In a second group (call this a control group), all of the chicks were trimmed so they appeared black. In a third group of nests (call this another control group), all the chicks were left their natural color, orange. In all three groups, the chicks were captured within a day of hatching and were generally handled the same way even though some were trimmed. The chicks were kept in captivity for 30 minutes before being replaced in their nest. To control for hatching order in the experimental groups, the first chick hatched was randomly assigned to be trimmed or left orange. Thereafter, treatments were alternated with hatching order. The biologists worried about what kinds of data to collect, how to collect the data, and what kinds of results to expect. What would you suggest they do?

Part IV: At Last, Some Data!

The biologists decided to compare the feeding, relative growth, and survival rates of the chicks in the different nests. Since the chicks had been individually color-marked, they could be easily observed and identified from floating blinds. To estimate growth rates, swimming chicks were photographed at known distances and their body lengths at waterline were estimated from projected slides. In pilot tests, the scientists found this measure of size is strongly correlated with body mass ($r = 0.97$, $n = 43$). This meant that they would not have to handle the chicks to collect the body mass data. Part of the data has been reproduced in Figure 11.2. Predict the results of the other values by plotting the values on the graph. Why have you made these predictions?

On the left side of the figure (panels a, c, and e) are the values for the two control groups (nests with either all orange chicks or nests with all black chicks). But the only data shown are for the nests composed entirely of orange chicks. Remember this is the normal situation that we find in the wild. The values shown are the medians, interquartile ranges, and 10–90 percentiles. With this knowledge, plot on the graph the values you would expect for the control black chicks raised with only black nest mates.

On the right side of the figure (panels b, d, and f) is a space for the expected values for the chicks in the experimental broods. Here, half of the chicks are black and half are kept orange. Now plot what you predict the values will be if

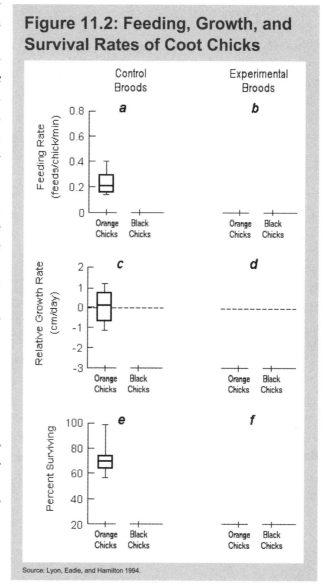

Figure 11.2: Feeding, Growth, and Survival Rates of Coot Chicks

Source: Lyon, Eadie, and Hamilton 1994.

the researchers' hypothesis was correct.

Part V: The Rest of the Story

In Figure 11.3, you will see the actual results that the researchers collected. How do the data compare with your predictions?

The biologists ran a series of statistical tests and noted that there were no significant differences between the two control groups in any of the measures—that is, when the orange and black chicks were in separate nests, they had similar feeding, growth, and survival rates.

But in the experimental group where both black and orange chicks occurred together and the parents had a true choice of which to feed, statistical analysis showed that the orange chicks fared better. Orange chicks were fed at a higher rate, had a higher growth rate, and enjoyed a higher survival rate than the black chicks in the same brood. What conclusions might the biologists make about their original hypothesis? Here are examples of questions:

Figure 11.3: Feeding, Growth, and Survival Rates of Coot Chicks

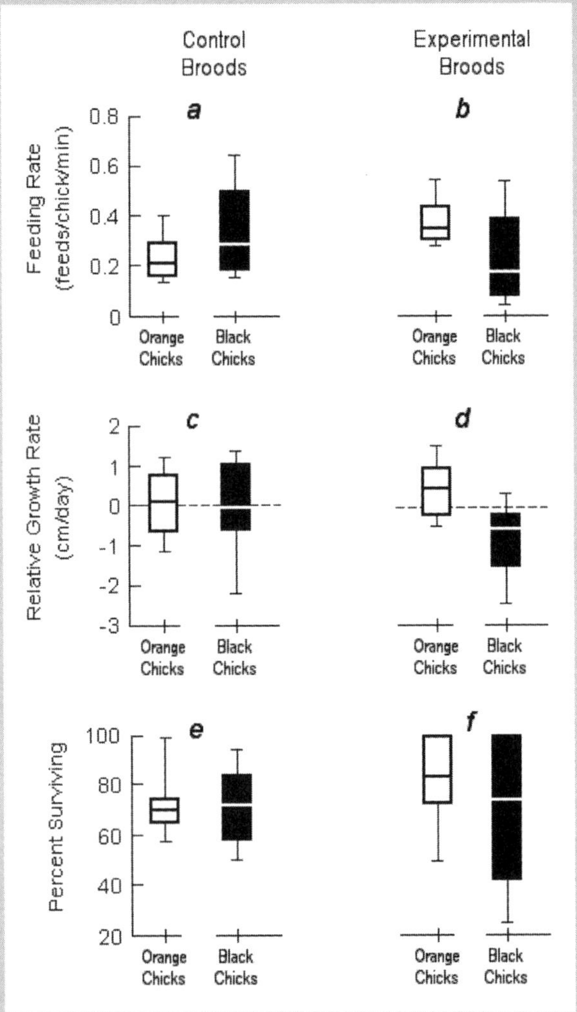

Source: Lyon, Eadie, and Hamilton 1994.

- Do black chicks have a lower survival rate simply because they are black or because they are "inferior" relative to the orange chicks?
- Do the data support the hypothesis?
- Do the data prove the hypothesis?

Teaching Notes

Introduction and Background

This case is based on a fascinating article that appeared in *Nature* titled "Parental Choice Selects for Ornamental Plumage in American Coot Chicks" (Lyon, Eadie, and Hamilton 1994) and the commentary published in the same issue, titled "Parents Prefer Pretty Plumage" (Pagel 1994).

As the case reveals, Lyon, Eadie, and Hamilton concluded that coot parents "feed ornamented chicks over non-ornamented chicks, resulting in higher growth rates and greater survival for ornamented chicks" (1994, p. 240). They sum up "that parental preference is relative, rather than absolute, an important element in the evolution of exaggerated traits. These observations provide the first empirical evidence that parental choice can select for ornamental traits in offspring" (p. 241).

I use this case in courses whenever I wish to emphasize how scientists solve problems. Obviously, the subject matter of the case is appropriate for courses in biology, especially those focusing on evolution and ecology. Yet the case is accessible to students without any background in science at all, and I have also used it in a nonscience majors' course called Scientific Inquiry. The way the case unfolds mimics the way that scientists go about their work: Scientists do not have all of the facts all at once; they get them piecemeal, as in this case.

Objectives

There are several objectives for this case beyond teaching coot biology:

- To give students practice in making predictions and interpreting data
- To give students practice in designing experiments
- To give students an explicit experience with the hypothetico-deductive method of reasoning (the "scientific method"), where a question is asked, a hypothesis is suggested, predictions or deductions are made in light of the hypothesis ("if the hypothesis is true, then such and such should occur"), tests are accomplished, and the data are evaluated as either supporting or rejecting the hypothesis

Common Student Misconceptions

- Experiments are always straightforward; scientists always know what they are doing.
- Scientists set up an experiment and then never change their minds.
- Scientists can predict what will happen in an experiment.
- Controls are not important in an experiment.
- All science is done in the laboratory.

Chapter 11

- Natural selection acts only on adults.
- Natural selection affects only structures, not behavior.

Classroom Management

This case is an example of the interrupted case method, where information is fed piecemeal to students working in small groups. The basic technique is to give the students a problem to work on in small groups (in this case, I give them a real research problem reported in *Nature*). Then, after the groups discuss the problem for a short time, I supply them with additional information and let them go back to work. This sequence is repeated several times as the problem gets closer to resolution. Just before I give them new information, I ask each group to briefly report their conclusions so that other groups can hear their progress. This has the effect of keeping the groups alert to other possible interpretations of the data and information.

It is important to note that the only parts of the case I hand out to the students are Part I and Figures 11.1, 11.2, and 11.3. I do not hand out Parts II–V, but instead read this information out loud to the entire class.

This case takes about 75 minutes for analysis, but it easily can be shortened or lengthened.

In Part I, students work in small groups; I give them 15 minutes for this. I first hand out Part I to each person in the group. I instruct the students to read the case and, as a group, respond to the bulleted items that follow the case.

After the small-group discussion of Part I is in full swing, about five minutes along, I walk around the room with a color copy of the cover of *Nature* magazine (September 15, 1994), which has a photograph of the orange baby coots. The students do not know it at the time (some hardly glance at it; others are simply enamored with how cute the babies are), but what the young coots look like is essential to what the authors did in the study. At that time, I tell them that coots can see color and they should consider there might be an average of eight chicks in a nest. I then put a color overhead of the cover of the magazine on the projection screen for the rest of the class period so that the coots are omnipresent.

When the 15 minutes have passed, I interrupt the discussion and ask for a representative from each group to tell me what they thought was the question that the researchers were addressing. I quickly have the groups report their conclusions. I do not comment at this point, but only ask questions for clarification. I summarize the discussion by saying it is not unusual to find that different researchers can look at the same situation but come up with different questions: Some groups will have been specific, focusing on coots, while others will have focused on the big evolutionary picture.

After this round robin is over, I ask the groups to quickly report on the hypothesis they think the researchers are pursuing. Again, when the reporting is done, it will be apparent that there are differences among groups. I choose not to comment as this process is occurring, except to say that clearly people differ even though they have the same information. At this point, I say that I know they have begun to consider the ways the authors might test their hypothesis. Perhaps they might like to hear a little bit more from the biologists who conducted the actual study.

I now read Part II out loud. I finish by saying that now that they know what the researchers are up to, they are to use the method of clipping the feathers and design an experimental program that will test the hypothesis. It is essential at this point to emphasize that yes, there are other ways to attack the problem, but they must use the researchers' approach. If you do not do this, some groups will go off on their own and produce radically different designs. Although this might be productive in some ways, it will surely be diversionary to the present case strategy.

I give the groups seven minutes to come up with an experimental design using the feather-clipping method from the *Nature* article. As the discussion occurs, I move from group to group, checking on their progress and making sure they are using the authors' method. I try not to answer many coot questions. I simply say that yes, I know that they do not know much about coots, but they should just make some reasonable assumptions as they design their experiments. I do reassure them that they do not need to worry about budget at this point. Their goal is simple: Design an experiment to test the hypothesis.

When the time is up, I ask each group in turn to report its findings. Most groups will now have similar experimental designs. Typically, they will have decided to have a nest with orange chicks and to have another with all black chicks. Some will stop there, not realizing that they need to have a third group, a nest condition with 50% black chicks and 50% orange chicks. It is only in this situation that the parents will have a choice between the two conditions. Also, there will be variations in which data to collect and how to collect it. Once again, I do not comment except to ask questions occasionally for clarification.

I then read Part III out loud, telling the students what the researchers really did. This passage is instructive because it reveals details about how the authors of the study handled the coots that some of the student groups will have considered. At the end of the passage, I specifically ask them to decide which data the authors should collect and how they might do that. The groups now have a chance to revise their ideas in light of other groups' comments and the further information. Also, it is essential to emphasize that they should begin to seriously think about what the data might show.

After about five minutes, I have the groups briefly report their proposals. There will be a few surprises here, as some groups will have some novel and often very expensive ways to monitor the birds.

Next, I read Part IV out loud and then hand out Figure 11.2 (p. 101) to everyone. After they have had a moment to look at the figure, I go over it with them, pointing out that the figure shows only the results for the nests where the birds are all the normal orange color. I ask them to plot what the missing data might look like if the

Chapter 11

hypothesis were correct. I suggest that they first plot the data for the experimental nests where half of the chicks are black and half are orange—that is, where the parents have a choice between black and orange chicks. When that is done, they should turn their attention to what would happen in the nests where there are only black chicks. I emphasize that anything might really occur, but they should ask themselves, What would be the ideal data that one might like to have?

The plotting of the data will take about 10 minutes, and, after some hemming and hawing, most groups end up with the appropriate predictions. I walk around the room as this is going on, and when I find a group that thinks it is finished, I ask them to consider how they would interpret the data if the black control group were different. When most groups are finished, I hand them a completed version of what happened (Figure 11.3, p. 102).

After they have had a few moments to compare their predictions with the real data, I ask the entire class to respond to the questions listed at the end of Part V. By this point, students will recognize the distinction between "support" and "prove." Also at this point, I ask students to consider how they might interpret the data if the chicks in the all-black control nest had a lower rate of survival than those in the all-orange nest. I challenge them to tell me why black chicks might not do well. It does not take long for them to suggest myriad possibilities, including temperature control, predation, parasitism, lack of parental recognition that they are chicks rather than adults, and other factors. I finish with the statement "Isn't it great that the authors used this control group? Without it, they would never know if these factors were involved in the experimental nests. It appears clear now that they were not involved. So what do you think the authors' final conclusion is for 'Why do parents prefer orange-feathered chicks?'"

Here is what the authors say:

- Ornamental plumage may be a signal of a chick's high genetic or phenotypic quality, leading the parents to invest more caregiving in them.
- Orange plumage may be a signal of age, allowing parents to selectively feed the chicks in an optimum way.
- The color preferences may not be directly related to feather color but simply due to a parent's color preference for other reasons. Perhaps it is the chick's head color that signals the offspring's need.

What experiments might resolve the issue?

Web Version

The case and its complete teaching notes can be found on the website of the National Center for Case Study Teaching in Science at *http://sciencecases.lib.buffalo.edu/cs/collection/detail.asp?case_id=505&id=505.*

References

Lyon, B. E., J. M. Eadie, and L. D. Hamilton. 1994. Parental choice selects for ornamental plumage in American coot chicks. *Nature* 371 (6494): 240–243.

Pagel, M. 1994. Parents prefer pretty plumage. *Nature* 371 (6494): 200.

Chapter 12
PCBs in the Last Frontier

Michael Tessmer

The Case

Part I: PCBs

Polychlorinated biphenyls (PCBs) are compounds that were once used as insulators in electrical transmission lines and in the production of polymers. Each PCB differs by the quantity and location of the chlorine atoms. An example of one of the many different PCBs is shown in Figure 12.1. PCB production was halted in 1977 due to the potential toxicity, but the chemicals are still found in the environment due to their stability. Studies in remote areas of Alaska have shown that PCBs can even be found in lakes untouched by humans. There is no known natural process that produces PCBs, so all of the PCBs in existence are presumed to have been produced by humans.

Mount McKinley and Wonder Lake: Denali National Park, Alaska

Figure 12.1. 2,2',4,4',5-Pentachlorobiphenyl

Questions

1. What scientific observation about PCB distribution is described above?

Science Stories: Using Case Studies to Teach Critical Thinking

Chapter 12

2. Propose a hypothesis or "explanatory story" to explain the global movement of pollutants such as PCBs. Specifically, how could they end up in the most remote Alaskan lakes?

3. Propose a method, either through observations or direct experimentation that would test your hypothesis from question 2. (Note: Your approach may be on a local scale despite examining a global phenomenon.)

Part II: Global Transport

Later studies showed that the global circulation of PCBs was due at least in part to atmospheric transport. PCBs enter the atmosphere by several mechanisms, including the burning of organic material and evaporation in warmer climates, followed by condensation at higher latitudes. This explained how chemicals made by humans could be found in areas untouched by humans.

Questions

1. Come up with a hypothesis or "explanatory story" to answer the following question: Should PCB levels differ significantly in Alaskan lakes that are near each other and at the same altitude? (Keep in mind that a hypothesis is an educated guess, so it requires a reason why you think your answer is correct.)

2. Propose a method, either through observations or direct experimentation, that would test your hypothesis from question 1.

Part III: Significant Difference?

Recent observations of PCB levels in arctic lakes have shown that the levels of PCBs are not the same in all lakes that are near each other and at the same altitude. In fact, lakes at the terminus (i.e., the start) of river systems had higher PCB levels than completely isolated lakes that were close by.

Questions

1. What possible "explanatory story" might explain the observation described above? (Hint: Think of species that leave a lake but return later in life.)

2. How would you test your hypothesis?

Part IV: Riddle Solved

Recent scientific studies have shown that sockeye salmon returning from the ocean to spawn in Alaskan lakes contain elevated levels of PCBs. After spawning, the salmon die and their contents become part of the lake sediment and/or enter the food chain. The salmon are responsible for adding approximately six times as many PCBs to remote lakes as atmospheric circulation. The types of PCBs in the salmon also match those found in the ocean.

Question
Imagine yourself as a scientist working on this issue. What would you want to look at next?

Teaching Notes
Introduction and Background
This case study was developed after reading an article by Krummel et al. (2003) in the journal *Nature* concerning the bio-accumulation and transport of PCBs by sockeye salmon from the Pacific Ocean to Alaskan lakes. It involves students reading basic background information before proposing hypotheses to explain the information. The emphasis is on making predictions and explaining the reasoning behind the predictions.

PCBs are a good example of a persistent pollutant that has a global distribution. The example shown in the case study is one of 209 possible congeners. The precise mixture of congeners depends on the original source and is quite variable. The compounds are excellent insulators and were used mainly in heavy electrical equipment. Examples of other uses include polymer manufacturing and carbonless copy paper production. The persistence of PCBs in the environment is related to their thermal stability and general resistance to biodegradation. The acute toxicity of PCBs was first recognized on a large scale in the 1960s from an accidental contamination of cooking oil in Japan. Several thousand people suffered a variety of illnesses, ranging from skin discoloration to higher mortality for infants born to exposed mothers.

This case was designed to be used early in a course such as general or introductory chemistry, general biology, or environmental science. Since little background knowledge is needed, it can be used with majors or nonmajors. The case could also be extended for use in a course such as analytical chemistry, where it could involve reading and discussing the original paper from *Nature* and subsequent work.

Objectives
- To help students review the scientific method
- To teach students how to better state hypotheses
- To encourage students to design experiments that test a hypothesis
- To give students an introduction to the scientific literature with a relatively easy-to-read article

Chapter 12

Common Student Misconceptions
- Areas with little direct human interaction are pollutant-free.
- All global transport of chemicals occurs via the atmosphere.
- Chemicals can "drift" up rivers from the ocean.
- Salmon cannot transport measureable amounts of chemicals long distances.

Classroom Management

I have used this case in a General Chemistry I class of 40 students working in groups of three. As written, the case can be completed in a 50-minute lecture session or in a lab setting. The case is broken up into four parts, and the instructor distributes one part at a time, with discussion after each part. (In large classes, this may be less practical, and the instructor may want to hand out the entire case and instruct students not to look ahead.) Students read each part and then spend about 5 minutes discussing the questions as a group. After each part, there is a short class discussion with a summary of the best answers.

Before beginning the case, I run a short class discussion on the steps of the scientific method. With input from the class, I draw a diagram showing the steps of the scientific method as a refresher. Most students have seen the steps to the scientific method enough times that the basics can be discussed as a class. Added discussion can also occur during the case study on the difference between conducting an experiment where a variable is manipulated and making observations to answer a question. This more subtle distinction can be emphasized while students are discussing ways to test hypotheses. The case presents several opportunities for students to propose experiments, and the discussion from the early part of the case will likely lead to improved answers later.

After completing the case, I provide copies of the article to any students who are interested in reading more. The article can be found in Krummel et al. (2003). It is also available at *www.biology.mcgill.ca/faculty/gregory_eaves/articles/KRUMME~1.pdf*.

Web Version

This case and its complete teaching notes and answer key can be found on the website of the National Center for Case Study Teaching in Science at *http://sciencecases.lib.buffalo.edu/cs/collection/detail.asp?case_id=191&id=191*.

References

Krummel, E. M., R. W. Macdonald, L. E. Kimpe, I. Gregory-Eaves, M. J. Demers, J. P. Smol, B. Finney, and M. Blais. 2003. Delivery of pollutants by spawning salmon. *Nature* 425 (6955): 255–256.

Chapter 13
The Great Parking Debate

Jennifer S. Feenstra

The Case

Part I: The Question

At the end of a long day of shopping, Katelyn and Lisa were walking out to Lisa's car, ready to go home. Putting their shopping bags in the trunk, Lisa slipped into the driver's seat while Katelyn took her position as passenger.

Lisa put the key in the ignition and absently said to Katelyn, "Looks like someone's waiting for our spot." Lisa started the car. The vehicle waiting for their spot honked.

"That drives me nuts," Lisa said in response to the horn. "I'm going, I'm going. Be patient," she muttered under her breath to the driver of the other car. Although she had been about to put the car in reverse and pull out of the spot, she dug into her purse for her lipstick.

"Lisa, that other car is waiting. Why aren't you leaving?" Katelyn inquired.

"I will," said Lisa. "I can't put lipstick on while I'm driving. Besides, it's my spot and I can stay in it as long as I want."

"Now you're being rude," Katelyn scolded her friend. "I always try to leave as quickly as I can when another car is waiting."

"You do not," Lisa responded. "I've ridden with you, and you make the other person wait."

Lisa pulled out of the parking spot as Katelyn shot back, "Do not."

"Do too," Lisa continued the argument. "Everyone takes longer to leave a parking spot when someone's waiting. It's an instinctual thing—we're defending our territory."

Katelyn responded with a snort. "Instinct. I don't think so. People are nicer than that. Except for a few rude people like you," she said teasingly, "most of us leave faster when someone's waiting."

Science Stories: Using Case Studies to Teach Critical Thinking   113

Chapter 13

"As a psych major, you could study something like that," Lisa responded.

"Maybe I'll make that my next project for my research methods class," Katelyn replied. "I'll let you know what I find out."

Questions

Help Katelyn find out whether she is right or Lisa is right. In groups of two or three students, address the following:

1. Identify the specific research question(s) implied by their discussion.

2. What hypothesis would your group suggest?

3. What predictions can you make if your hypothesis is correct?

4. How could you test the predictions? In other words, if you were to investigate this issue, how might you do it?

Part II: Finding a Method

A few weeks later, Katelyn met Lisa for lunch. She brought up their parking debate and told Lisa about a study she had found that addressed the question.

In this study, the researchers stood in front of the main entrance to a shopping mall and watched shoppers as they left the mall and walked to their vehicles. They timed the shoppers from the time they opened the vehicle's door to when they had completely left the parking space. The researchers noted how many people were traveling in the vehicle as well as whether or not another vehicle was waiting for the parking space.

Questions

1. What method are the researchers using to investigate the research question (case study, survey, naturalistic observation, or experiment)?

2. How does this method test the hypothesis?

3. Based on the hypothesis, what would you expect they would find?

Part III: And Now, Some Results

The researchers (Ruback and Juieng 1997) designated those drivers who had a vehicle waiting for them to pull out of their parking spot as *intruded upon*. Those who did not have a vehicle waiting for them to leave were *not intruded upon*.

For the 200 drivers observed, these were the average amounts of time to leave the parking space:

Intruded upon	39.03 seconds
Not intruded upon	32.15 seconds

The researchers also looked at whether it took those traveling with others longer to leave than those who were alone:

Traveling alone   30.64 seconds
Traveling with others 37.45 seconds

Questions

1. What do these findings this tell us? Do the data support your hypothesis?
2. What might be a weakness (or some weaknesses) of this method?
3. How might the researchers use another method to explore this research question?
4. Are there additional effects that the researchers should investigate?

Part IV: That's Not All

The researchers were concerned that the type of car waiting or some behavior of the driver of that car might make a difference in the actions of the people in their first study. To deal with this, they designed an experiment. They used a low-status car (1985 Nissan Maxima) and a high-status car (1994 Infinity Q45 or 1993 Lexus SC400). They had three levels of *intrusion*: (1) *no intrusion* (no car was waiting for the spot), (2) *low intrusion* (the other car waited four spaces from and faced the direction of the departing car), and (3) *high intrusion* (the car waited four spaces from the departing car, turned on its turn signal, and honked the horn once after the driver sat behind the wheel). The high-status cars were involved in half of each of the intrusion conditions and the low-status cars in the other half of the intrusion conditions. An observer recorded the number of seconds it took the driver to leave the space after he or she opened the door.

Questions

1. What do you think were the hypotheses of the researchers?
2. What would you predict they would find?
3. What made this study an experiment and not a naturalistic observation?

Part V: More Results

Below are data from the study.

Average number of seconds that vehicles took to leave a parking space in three conditions:

High intrusion	Low intrusion	No intrusion
42.75 seconds	30.80 seconds	26.45 seconds

Chapter 13

The researchers did not find any significant effect for whether the intruding car was high or low status.

Question

What did the researchers find about the effect of intrusion on length of time to leave a space?

Part VI: The Final Word

"I was right!" Lisa crowed when Katelyn told her about the results. "I knew people took longer to leave when someone else was waiting. Everyone knows that."

"Maybe not everyone, Lisa," Katelyn countered. "These researchers did one more study to see if people knew the effect that intruding cars have on people."

Katelyn explained: "The researchers asked participants whether other drivers would vacate their spaces in different amounts of time than they had. Participants made judgments on a scale from 1, 'make the time shorter,' to 7, 'make the time longer,' with 4 in the middle as 'no effect.'

"The researchers asked people to make the judgment for their own behavior and for others' behavior and whether the other car honked or not."

"What did they find?" Lisa asked.

"Here's the table of their findings," Katelyn replied, pulling a sheet of paper from her bag.

Table 13.1. Mean Ratings of Own and Others' Behavior Following Low Intrusion (No Honking) or High Intrusion (Honking) When Leaving a Parking Space

	Low Intrusion	High Intrusion
Own behavior	1.87	4.88
Others' behavior	2.83	4.40

Lisa studied the table for a few seconds. "What does this tell us?" she asked.

Questions

1. What does Table 13.1 tell us about what people think they do and what they think others do in response to intrusion?

2. Do these results and the results of the other studies tell us about anything else besides parking? Do people behave in similar ways in other places besides parking lots? In what other kinds of situations might these results help us understand human behavior?

Teaching Notes

Introduction and Background

An understanding of research methods is essential to understanding material in a number of fields. A day of talking about hypotheses, operational definitions, and the like is not very exciting for most students. "The Great Parking Debate" was developed for use in an introductory psychology course to cover terms and concepts related to research methods in a more engaging and interesting way. The case could be used in other introductory science classes, early in research methods courses, or in upper-level social science courses to remind students of research methods terms and concepts and to provide them practice in evaluating data.

To prepare, students are asked to read about research methods in a textbook (e.g., Myers 2010). The reading should cover the following concepts (if it does not, the instructor should introduce those concepts before the case study begins): hypothesis, research question, naturalistic observation, survey, case study, experiment, and random assignment.

Objectives

- To help students critically evaluate terms and concepts related to research methodology
- To help students evaluate various research methods and understand the positives, negatives, and potential uses of those methods
- To provide students with the opportunity to evaluate data

Common Student Misconceptions

- Hypotheses are simply assumptions or guesses and are not necessarily testable.
- Different research methods (naturalistic observation, experiments, survey method) provide similar types of information.
- Research findings are not applicable beyond the research context.

Classroom Management

The case was designed for a one-hour class period. To finish within an hour, instructors need to keep the class moving at a fairly quick pace, so attention must be paid to the amount of time each part of the case takes. The case could easily be stretched to take up an hour-and-a-half class period.

The case is best done with a class of 40 or fewer students, so the instructor can clear up confusion by wandering among the groups and be timely in presenting each

Chapter 13

new part of the case. With the help of teaching assistants, however, using the case study with larger classes would be possible.

Students should be asked to read the research methods portion of their textbook before coming to class. An instructor could also lecture on research methods and have students complete "The Great Parking Debate" case in a subsequent class period. Students should have a copy of the textbook or the reading on research methods with them for reference in class the day of the case.

Student should be divided into small groups of about two to four students per group.

Part I

Once the students have been divided into groups, they read Part I of the case study. As a group, they discuss the questions at the end of that part.

After the class works on these questions for about 7–9 minutes, the instructor brings the class together to discuss their answers. This may take 5–10 minutes.

Begin the full-class discussion by asking a small number of student groups to share research questions and hypotheses. Write hypotheses on the board. Talk about what a hypothesis is. If a group comes up with a hypothesis that would not be testable, the instructor can discuss with the class the criteria for testable hypotheses.

Next, ask students what their group did in discussing how they might test the predictions. One of the things instructors should point out in the discussion is that to test the predictions, students need to decide what they mean by conditions such as someone waiting for the parking space.

Operational definitions can also be covered in this section as students think about how they would test the predictions they make.

Part II

When the discussion is finished, hand out Part II of the case, letting students know that the method the researchers used may be different from what their groups proposed. The students should discuss in their small groups how the researchers addressed the research questions suggested in the first part, how these methods get at the hypotheses, and what they predict would be the results of the study.

There is no whole-class discussion at this point, just small-group work, which should take about 4–5 minutes. When the groups have finished in their small groups with Part II, hand out Part III.

Part III

Allow students 4–5 minutes to discuss this part of the case in their small groups. Then take about 7–9 minutes to discuss the questions at the end of Part III.

In Part III, students decide whether the data support the hypothesis, discuss some potential weaknesses of the method being used, think about other methods that might be used, and identify other effects that may be of interest.

The data do support the hypothesis. As expected, drivers who were intruded upon (had another vehicle waiting for their spots) took longer to leave their spaces (39.03 seconds) than those not intruded upon (32.15 seconds). Students should be able to use the data to make such a conclusion.

The study is a naturalistic observation because the researchers did not manipulate the situation in any way and simply observed what was happening in a naturalistic situation (Part II).

Naturalistic observation can be a good method to use if you want to look at readily observable behaviors in a public setting. One issue with this method is that the researcher cannot control the variables, so extraneous variables may affect the results (how far away the waiting car waits may make a difference, but the researcher cannot control that) (Part III).

Part IV

Hand out Part IV to the groups and allow 6–8 minutes for discussion in groups. No full-class discussion.

Part V

When the groups have finished discussing Part IV, hand out Part V. After about 3–4 minutes of group discussion time, discuss this section as a class (5 minutes):

- What did the researchers find? (from Part V)
- What made this an experiment and not a naturalistic observation? (from Part IV)

If there is time, terms related to experimentation, in particular *independent and dependent variables* and *random assignment*, might also be discussed.

Part VI

At the end of the whole-class discussion, hand out Part VI. Allow students about 4–5 minutes of discussion in their small groups; then bring the class together for a general discussion.

- This final study introduces the sorts of behaviors that can be studied using a *survey* method; in this case, the researchers asked participants to report on their perception of their own and others' behavior. Once again, students are asked to evaluate data and come to conclusions concerning the meaning of the data.
- Students are also asked to think about what we can learn from research findings. The results of this study go beyond cases of parking in mall parking

Chapter 13

lots and teach us something about what happens when our freedom is being threatened.

Web Version

This case and its complete teaching notes, references, and answer key can be found on the website of the National Center for Case Study Teaching in Science at *http://sciencecases.lib.buffalo.edu/cs/collection/detail.asp?case_id=578&id=578.*

References

Myers, D. G. 2010. Thinking critically with psychological science. In *Psychology*, pp. 15–45. New York: Worth.

Ruback, R. B., and D. Juieng. 1997. Territorial defense in parking lots: Retaliation against waiting drivers. *Journal of Applied Social Psychology* 27 (9): 821–834.

Chapter 14

Poison Ivy: A Rash Decision

Rosemary H. Ford

The Case

Part I: Encounters With Poison Ivy

The year: 1412
Warning among the Onondaga Indians living in the mountains of New York: "Beware of the ko-hoon'ta [stick that makes you sore]!"

The year: 1607
Message from John Smith from the newly founded Jamestown Colony to England: "The poysoned weed is much in shape like our English Ivy, but being touched, causeth rednesse, itching, and lastly blisters, that which howsoever after a while passe away of themselves without further harme, yet because for the time they are somewhat painfull, it hath got itselfe an ill name, although questionlesse of no ill nature."

The year: 2010
Urgent message from Julie, a first-year college student in Florida, to her mother: "Mom! Last Friday I finished writing a long paper and wanted to get away. So on Saturday a group of us went hiking. We followed trails along streambeds and all through the woods. But now it is Tuesday and I am miserable! I wish that I had not gone. My legs are covered in a rash with blisters that are ready to pop. The itching is out of control. I don't understand. What should I do? Help!"

Science Stories: Using Case Studies to Teach Critical Thinking

Chapter 14

Questions

1. If you have seen or touched poison ivy, where was it located?
2. What symptoms are caused by poison ivy?
3. Describe why poison ivy is able to live in very different habitats—in the cold mountains of New York and the coastal forests of Virginia and along streambeds in Florida.
4. Propose a reason why poison ivy produces this reaction in humans even though it was in North America even before the Indians came.

Part II: The Mystery of the Blisters Unfolds

In the phone conversation with her mother, Julie wanted some sympathy and good advice, but her mother merely laughed.

"Julie," her mother explained, "you probably brushed against poison ivy while hiking through the woods. I remember when you were a little girl, you would regularly get into poison ivy while playing outside. A few days later, you would have a rash on your legs and arms and be miserable."

Julie asked, "Do we have it in California? I don't remember getting this awful rash at home!"

"No," her mother replied. "I haven't seen it since we moved here."

Meanwhile, text messages were flying among the students who went on the hike. Who had the most blisters? What can relieve the itching? Meanwhile, Joe, one of the hikers, was not complaining. He was an international exchange student from Spain and had no symptoms. He was pleased. "I must be immune to poison ivy," he bragged to his friends.

Since Julie had lived in California for most of her life, she did not know much about this plant and wanted to know more—what it was, what caused her discomfort, and how long the rash would last.

She decided to do some research on poison ivy using the internet. A lot of information was available: how to recognize it, how to get rid of the plant in your yard, and, most important, what types of cures are best for the itch.

Excited, she made a list of facts about poison ivy to share with her friends who went on the hike. Here are some facts about poison ivy that Julie compiled:

1. An oil called urushiol is found within all plant parts. It is the ultimate cause for the symptoms.
2. When the skin brushes the plant, the leaves bruise and release urushiol. This chemical binds quickly and tightly to the skin's outer cells.
3. These cells now appear foreign to the body.

4. The immune system then launches a chain reaction to create defensive cells that will recognize urushiol the next time it enters the body.

5. At the next exposure, these new defensive cells launch an attack on the urushiol. It takes about two days before the skin shows the effects of the immune system's response.

Julie made a quiz to see how much her fellow hikers knew about poison ivy. You should take the quiz as well (below). Explain your reasoning for each response.

Julie's Quiz

1. The leaves of poison ivy burn the skin on contact. True or false?
2. You can catch poison ivy from someone else. True or false?
3. Merely petting a dog that recently walked through a patch of poison ivy can produce a reaction. True or false?
4. Because Joe did not react to poison ivy, he is immune to the plant's oil. True or false?
5. After scratching the blisters, the rash can spread elsewhere on the body. True or false?
6. Washing the exposed skin with water just after contact will prevent the reaction. True or false?
7. There is a way to prevent the symptoms. True or false?

Part III: Finding a Treatment

Julie and her friends met for dinner later that week. Their rashes were no longer itching uncontrollably and they were laughing about their troubles. All of them had been to the local stores to find lotions to relieve their symptoms. None solved the problem, since the rash and blisters did not go away, but a few treatments dried up the blisters and stopped some of the itching.

One particular treatment that Julie kept seeing on the internet was related to an herbal remedy containing jewelweed, a native plant. A natural and organic approach to control the symptoms seemed appealing. It was also one that her mother had used at nature camp.

"Just rub the leaves of jewelweed where you touched the poison ivy," the camp counselors had recommended. "Then you will not get the rash. It is easy to find since it grows in the same places as poison ivy. This remedy was passed down from the first Americans, the Indians."

Chapter 14

Julie told herself that she had better learn to recognize the plant because she did not plan on staying in her room for the rest of her life!

Your job is to use the scientific method to plan an experiment to determine if jewelweed is an effective treatment.

1. First, develop a hypothesis that includes a justification and prediction.

2. Second, plan the experiment. These questions will help you:
   a. How will poison ivy be applied?
   b. What part of jewelweed should be used for treatment—the flowers, the leaves, the stems, all parts of the plant?
   c. How will jewelweed be applied?
   d. Who will be tested?
   e. How will they be tested?
   f. What are the controls?
   g. What variables must be considered?
   h. What type of data should be collected?
   i. How will those results be analyzed?

Part IV: The Scientific Experiment

Long, Ballentine, and Marks Jr. (1997) decided to find out if the widespread claims that jewelweed can control the poison ivy rash were true. They could not just rub the poison ivy leaves on people and then rub the area with jewelweed. A plan had to be developed so the experiment could be repeated by others.

Controlling for the treatment with poison ivy was easy. To simulate brushing against the plant in nature, they prepared a standard solution of urushiol. But the active ingredient of jewelweed was not known, so a purified chemical was not available. The jewelweed extract was prepared by boiling fresh stems for 30 minutes, diluting the mixture with water, and placing it in small bottles. An identical set of bottles was also filled with water. The participants received a set of two bottles, one of jewelweed extract and one of water, as takeaways so they could administer the treatments themselves. They did not know which bottle contained the jewelweed extract and which the water (or control), although the researchers knew.

Ten participants or subjects were recruited. All were known to develop an allergic reaction based on a preliminary test. Each received the urushiol solution on two areas, on each forearm. The urushiol solution remained on the skin for four hours, then the arms were washed with soap and water. They left with their two bottles and were instructed to apply one of the solutions to the affected area of a single arm four times a day beginning on day 2. The other solution was to be applied to the other arm. On days 2, 3, 7, and 9, the treated areas were evaluated using a scale from 0 (no reaction) to 7 (the most severe reaction) (Table 14.1).

Table 14.1. Descriptions of the Symptoms by Score

Score	Reaction
0	No reaction
1	Erythema (reddish skin) with itching
2	Erythema with edema (swelling) and itching
3	Same as 2 but with 25% or less of the test site with blisters
4	Same as 2 but 25%–50% blistering
5	Same as 2 but 50%–75% blistering
6	Same as 2 but 100% coverage with blisters
7	Same as 2 but with ruptured blisters

On each day the symptoms were observed and scores for the 10 subjects were recorded and analyzed. The reaction range line for each day shows the median (middle) score and the maximum and minimum scores (Figure 14.1). The box on each line is the IQR (interquartile range, or the range in which 50% of the scores reside).

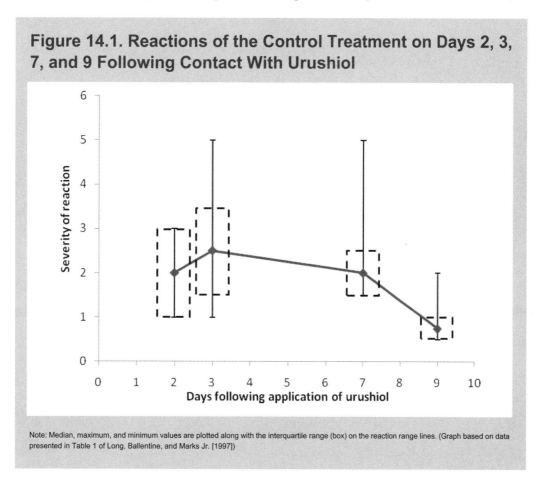

Figure 14.1. Reactions of the Control Treatment on Days 2, 3, 7, and 9 Following Contact With Urushiol

Note: Median, maximum, and minimum values are plotted along with the interquartile range (box) on the reaction range lines. (Graph based on data presented in Table 1 of Long, Ballentine, and Marks Jr. [1997])

Chapter 14

Questions

1. Using the median scores, describe the change in severity of the symptoms throughout the course of treatment.

2. On Figure 14.1, predict new median points on the reaction lines for days 2, 3, 7, and 9 that support the two conditions described below:
   a. The jewelweed extract improved the symptoms. Place a square on the reaction range line (or on a line that extends from the reaction range line) for each day.
   b. The jewelweed extract had no effect on the symptoms. Add an X to the reaction range line (or on a line that extends from the reaction range line) for each day.

Part V: Does the Jewelweed Treatment Work?

The median scores for the jewelweed-treated areas were compared with those of the control (Figure 14.2). The IQRs were not included because they were similar for the two data sets.

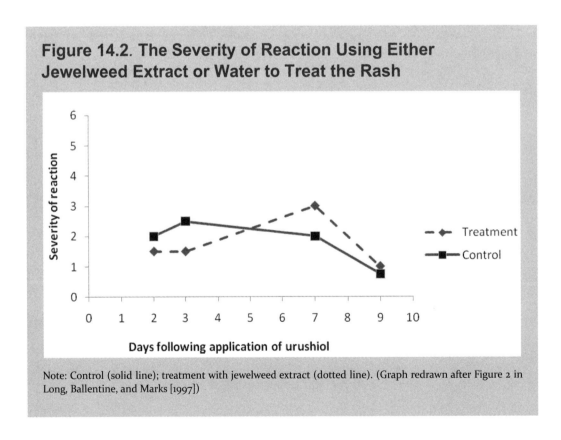

Note: Control (solid line); treatment with jewelweed extract (dotted line). (Graph redrawn after Figure 2 in Long, Ballentine, and Marks [1997])

Questions

1. Do these results support the claim that application of jewelweed is a good treatment for symptoms caused by poison ivy? Explain.

2. If you were walking in the woods and found yourself in a patch of poison ivy, does this experiment let you know if you should rub the affected area with jewelweed? If not, plan an experiment that would let you know what to do in this case.

Teaching Notes

Introduction and Background

The poison ivy plant produces itching, blisters, and swelling in many people today as it did in American Indians. They used herbs to treat medical problems, and the skin symptoms of poison ivy were no exception. This case is based on accounts in the ethnobotanical history from several native Indian tribes who claim that the application of jewelweed alleviates the symptoms of poison ivy.

This case introduces the scientific method in a manner that is appropriate for use in introductory biology classes, either majors or nonmajors. Students with little background knowledge of plants or the immune system can complete the case.

Objectives

The objectives of this case are the following:

- To increase knowledge about the poison ivy plant and the body's response to exposure to the plant

- To analyze information for the purposes of developing and evaluating hypotheses, making predictions, designing experiments, interpreting data, and drawing conclusions

Common Student Misconceptions

- Commonly held beliefs are always true.
- Planning a scientific experiment is an easy task.
- The numbers produced by a scientific experiment can be interpreted without further analysis.
- Product advertisements are based on scientific proof.

Classroom Management

This interrupted case fosters class discussion after each section and can be completed easily within a single class period or in about an hour. For each of the first four parts, small groups of students, depending on the size of the class, discuss their answers and then share them with the class (Table 14.2, p. 128). Groups frequently re-evaluate

Chapter 14

their answers as they listen to the discussions of other groups. Re-evaluation is particularly important in Part III so that students can revise their experimental designs to meet the expectations that the experiment can be repeated by others. Part V can be administered as a written assignment.

Table 14.2. Class Management of the Case

Part	Time (min.)	Outcome
I	10–15	Increase student interest and lay the foundation for the case Introduce the idea that plants produce defensive chemicals to decrease chances of being eaten
II	10	Introduce the connection between contact with urushiol and the immune system
III	20	Formulate a testable hypothesis Develop a simple experimental design
IV	10	Evaluate data based on statistical information Predict an outcome from scientific evidence
V	Written work	Interpret data Evaluate student learning of the scientific method

Web Version

This case and its complete teaching notes, references, and answer key can be found on the website of the National Center for Case Study Teaching in Science at *http://sciencecases.lib.buffalo.edu/cs/collection/detail.asp?case_id=579&id=579*.

Reference

Long, D., N. H. Ballentine, and J. G. Marks Jr. 1997. Treatment of poison ivy/oak allergic contact dermatitis with an extract of jewelweed. *American Journal of Contact Dermatitis* 8 (3): 150–153.

Section IV:
The Scientific Method Meets Unusual Claims

Many of the world's most important discoveries were met initially with skepticism. With repeated testing, some of those ideas were vindicated, but not all. Some initial ideas were intriguing, even sensational, but later shown to be wrong (such as cold fusion). Some were fraudulent from the beginning (such as Piltdown Man and claims of human cloning). Some were based on inappropriate measurements or faulty equipment. And others were based on wishful thinking that led to biased data collection (such as T. D. Lysenko and the Soviet Union agricultural scandal).

There are many ways to go wrong in science without proper experimental design. But how can we find real gold and separate it from the fool's variety? Carl Sagan's famous chapter on "baloney detection" in his book *The Demon-Haunted World: Science as a Candle in the Dark* can help us separate the two. Here are some of the key tools to be used when someone makes an unusual claim, such as "I saw a ghost last night":

- Wherever possible, insist on independent confirmation of the facts.
- Encourage serious debate on the evidence by knowledgeable proponents of all points of view.
- Assume that arguments from authorities carry little weight (there should be no "authorities" in science).
- Propose more than one hypothesis—do not simply accept your first idea.
- Try not to get overly attached to a hypothesis just because it is yours.
- Measure wherever possible.
- Ensure that every link in a chain of argument is strong.

Section IV

- If there are two hypotheses that explain the data equally well, choose the simpler one (also known as Occam's razor).
- Ask whether the hypothesis can be tested—would others duplicating the experiment get the same result?

In this section of the book, we include cases that investigate unusual claims to help students learn to evaluate experimental design and determine whether the evidence supports the conclusion. In approaching these cases, students should be reminded of the famous dictum of skeptic Marcello Truzzi: "An extraordinary claim requires extraordinary proof" (Truzzi 1978, p. 11).

We lead off with "Extrasensory Perception?" This case tells the story of a young girl who is trying to move a spoon with her mind (psychokinesis). This is one category of so-called extrasensory perception (ESP). Thousands of tests have been performed over the years, and the jury is still out on the subject. Yet, as critics are fond of saying, if bending spoons are all we have to show for the millions of dollars that have been spent studying this supposed phenomenon, then even if ESP is real, it is pretty trivial. This case kick-starts a discussion of possible ways to test such claims and the potential for erroneous interpretations.

The case study "A Need for Needles: Does Acupuncture Really Work?" follows. It deals with acupuncture, a therapy based on ancient Asian medical practice. During the case, students research information on both sides of the debate over whether acupuncture really works. One outcome of the case is that students recognize test bias—in this case, the difficulty in testing the claim because a patient knows when he is being tested and what the anticipated outcome is.

The third case in this section is "Love Potion #10: Human Pheromones at Work." This case picks up on a claim made repeatedly by the advertising media: that a particular cologne, perfume, or beauty treatment will do miracles for your love life. In this case, the presumed "love potion" includes a pheromone claimed to make one more attractive to the opposite sex. Students analyze the data through a series of stages to see if the claim is justified.

The fourth case, "The 'Mozart Effect,'" deals with the claim that listening to classical music (especially Mozart) will improve one's concentration and creativity. Students are asked to design an experiment to test this assertion. They then examine the results from two published studies that test these very same claims. Next, they compare and contrast their own work against the published studies and evaluate their results.

As its title makes clear, the case "Prayer Study: Science or Not?" deals with prayer. One might think this would be a difficult topic of study. And it is, especially because many critics contend that religion and science should never mix. Still, scientists have attempted to study the possible effect of prayer on the healing process. The result? A number of studies have concluded that prayer does improve the chances of recovery, if patients know that a prayer is being offered on their behalf. But is this due to the placebo effect? What if we try to eliminate that possibility by ensuring that the patient does not know that prayers are being offered on his or her behalf? What then? A rigorous and skeptical assessment of this research can be instructive for students. But how should one conduct such a study?

The final case, "The Case of the Ivory-Billed Woodpecker," investigates a reported sighting in the wilds of Arkansas of an ivory-billed woodpecker, a bird long believed to be extinct. Students review the evidence and evaluate how they might handle the ensuing interest in the discovery. This is an example of a case that easily could fall into the category of how scientists deal with the media, as the investigators who made this claim endured intense scrutiny from their colleagues and the fascination of the media. Does the claim stand up?

All of these cases deal with issues at the fringe of science. They are wildly different. Some are claims that seem credible on the surface, and others fall into the category of urban legend. Some are clearly testable, while others simply are impossible to measure and assess. If someone claims to have seen a ghost, how do you test it? Even if one can test claims such as the Mozart effect, does anyone in the scientific arena really want to use his or her time to attack this question? And even if one does debunk a commonly held idea, will that change the way people think? And consider the money that might be involved. Do you really want to spend a million dollars looking for the certainly elusive, and possibly extinct, ivory-billed woodpecker?

Chapter 15

Extrasensory Perception?

Sarah G. Stonefoot and Clyde Freeman Herreid

The Case

The spoon's silver surface gleamed in the light coming through the window as it rested in the center of the dining room table. If you looked directly across the table, you could just make out Eliza sitting perfectly still at one end staring at the spoon, her nose resting on the tabletop. Eliza, who usually couldn't sit still for 10 minutes, was now completely frozen. If it weren't for the fact that she was sitting in an incredibly odd position or for her attentive eyes, one might have mistaken her for being asleep. Eliza was far from asleep. She was wide awake and concentrated fully on the silver spoon in the middle of the table.

"Eliza, Eliza, where are you?" Eliza's mother, Jean, came into the dining room. Startled by her daughter's stillness, Jean reached down to touch her.

"Eliza, honey, are you okay?"

"Shhhhh," Eliza whispered, refusing to move from her position.

"Eliza, what on Earth are you doing?" her mother exclaimed, growing tired of her daughter's antics.

Eliza refused to respond. Maybe if she didn't talk to her mother, she would leave Eliza alone and let her get back to work. But Eliza's silence only infuriated her mother.

"Eliza, answer me!" Jean exclaimed as she grabbed the spoon from the table.

"Nooooo," Eliza squealed, leaping up from the table.

"What are you doing?" her mother demanded, holding the spoon out of her daughter's reach.

Eliza jumped up in an attempt to retrieve the spoon.

"I'm going to make it move, or at least I was until you took it away."

"What?" her mother replied in confusion.

Eliza stopped jumping long enough to explain.

"I'm going to make it move," Eliza replied. "Matilda could do it; she moved lots of things. I can do it too. You just have to let me concentrate!"

Chapter 15

Eliza was referring to the title character in the book *Matilda* by Roald Dahl, a little girl who had the ability to move objects with her mind by focusing on them.

After a moment's hesitation, Jean returned the spoon to its position on the table.

"Thanks," was all Eliza managed to get out before resuming her position.

Jean watched her daughter sink back into concentration and then slowly backed out of the dining room. As she did, she bumped into her husband, Ralph.

"Jean, is something wrong?" he asked.

"Oh, Ralph, I didn't see you there!"

Jean reached up to Ralph's shoulders. Laughing slightly, she guided him back into the kitchen. Although she could deal with her daughter's eccentric behavior, she knew Ralph wouldn't allow it. They managed to get into an argument every week when Ralph found Jean reading her horoscope. "Such nonsense," he would say. Jean really wanted to keep him out of Eliza's experiment for now. She knew it wouldn't last long—none of Eliza's phases did—but she could at least let her have some fun. It was good for her.

"What is Eliza doing in there?" Ralph asked, aware of Jean's attempts to hide it from him.

"Oh, nothing, she's just reading."

"But then where's her book?" Ralph asked.

"She's just playing," Jean said in response, trying to get Ralph to lose interest.

"Playing? It looks more like she's being punished. She's just sitting there."

Before Jean could say anything more, Ralph squeezed past her into the dining room.

"Eliza, what are you doing?" Ralph demanded.

"Daaadd, be quiet," Eliza whined.

"What!?" Ralph's voice rose in response to the tone of his daughter's voice.

"Ralph, come here, I'll explain," Jean said, guiding her husband back into the kitchen again. Jean tried to make Eliza's behavior sound as normal as possible, hoping her husband would just see it as a game, something to amuse Eliza for a while before she got bored of it. But Ralph did not take it so lightly. He saw this as "more of the same nonsense" that was overtaking his house.

So, it was beginning, another monumental argument in the McCloud household. Jean was really dreading this one. She knew it would turn into another one of Ralph's condescending speeches, where his professional title—Ralph McCloud, professor of organic chemistry—would loom over the entire discussion. Jean had dubbed them Ralph's "holier than thou" speeches whenever she complained to her friends. She wasn't going to let Ralph immediately put her down with his scientific jargon. She figured if she could at least postpone the discussion long enough to jump on the internet, she could find some information to back up Eliza's little project.

"Ralph, honey, I know what you're going to say, but I don't have the time to argue about this right now. I have to get dinner started. But as soon as we're done with dinner, I promise you that we'll discuss this."

Jean guided Ralph out of the kitchen, being careful not to let him get a word in. She leaned against the door and took a deep breath once he was out of sight. Then she rushed to the computer. They could eat leftovers tonight.

* * * * *

Jean had managed to convince Eliza to interrupt her efforts long enough for dinner. However, right after dinner, Eliza resumed her position at the dining room table, giving Ralph and Jean time to talk.

"Okay, Jean, now I know you want to have a reasonable discussion about this, but I'm telling you now you are not going to be able to convince me that telepathy and clairvoyance exist."

"I know I'm not going to be able to convince you, Ralph, but I would like you to just consider the possibility that people can read minds, that people can move things with their minds. There are a lot of things out there that can't be explained. Who's to deny that one of the reasons for that is that people have extrasensory perception? Moreover, what is the harm in our daughter determining for herself whether or not it exists?"

Ralph eyed his wife. "Jean, there is scientific proof for everything that happens. There are reasonable, logical explanations for everything. The only problem lies in the fact that people aren't educated enough to understand those reasons. And that's when pathetic ideas like ESP take hold, and it is those kinds of ideas we do not want to fill our daughter's mind."

"OK, OK. But what about all of the experiments that have been conducted that show that ESP can exist? Are you just going to ignore all of the work J.B. Rhine did, all of the tests he conducted? What about the Ganzfeld procedure, where 6 of the 10 studies had positive results for telepathy?" Jean said eagerly.

"No, Jean, I am not going to ignore any of those experiments." Ralph was getting incredibly annoyed with the conversation and was not afraid to show it. "Rather, I'm going to acknowledge all of the studies that have been done, including the hundreds that have shown no positive results and have been conveniently ignored. How about you show me an example of an experiment that upholds the crucial scientific principles? Show me an experiment that has been conducted that has positive results and can be repeated. Show me an experiment that has not been influenced by human bias."

Ralph paused for effect, then continued: "Jean, there are hundreds of fundamentals that have to accompany an experiment before you can go around declaring that something exists."

Jean felt flustered. She didn't know enough. She couldn't prove her husband wrong. But she believed that he was wrong. She wasn't going to give up. It would take a lot more than a couple of scientific principles to convince her otherwise. Luckily,

Chapter 15

there was a sound in the background that cut their conversation short for the time being. It was a soft clink of some metal object falling on their dining room table. It sounded just like a spoon.

Questions

1. Define the following terms: (a) ESP, (b) telepathy, (c) clairvoyance, (d) precognition, and (e) psychokinesis.
2. Who is J.B. Rhine? What experiments did he conduct, and what did he bring to the studies of ESP?
3. What is the Ganzfeld procedure? What does it test for? What were the results of the Ganzfeld procedure?
4. What are the general criticisms of ESP? Why do scientists criticize many of the studies that have been conducted on this topic?
5. How do humans obtain information—that is, what sense organs do we have to obtain information?
6. Suppose we have a person who claims to have ESP. How could you possibly test such a claim and be absolutely certain that the person was not using trickery?

Teaching Notes

Introduction and Background

This case is intended for an introductory-level science or psychology class. It can also be used in most human biology classes when the nervous system is explored, taking off from questions about human sensory systems. The case should be presented at a point in the course when the students are dealing with scientific experimentation, data collection, and analysis. This case will help the students understand and evaluate the need for controls and careful experimental design, especially when the claims being investigated are extraordinary. The students do not need to be focused on studies of paranormal activities; rather, they just need a general knowledge of its existence. The case itself is not filled with scientific information; it merely serves to kick-start a discussion on ESP and the need for skepticism in viewing experimental results.

Objectives

The overall purpose of this case is to teach students to be skeptical of "scientific claims," especially those that are sensational and fall outside the boundaries of normal scientific explanation. With ESP, the students are examining a topic at the fringes of "normal science." ESP has yet to be demonstrated to the satisfaction of the scientific community and is often called a pseudoscience. The case serves to illustrate the boundaries and fundamentals of scientific data. Students evaluate information and data to determine whether they believe there is enough scientific evidence to confirm ESP's existence. The questions posed at the end of the case guide the students to consider the following topics:

- The subdivisions of ESP: clairvoyance, telepathy, precognition, and psychokinesis
- J.B. Rhine and his studies at Duke University
- The Ganzfeld procedure
- Criticisms of ESP and the difficulties inherent in performing adequate experiments in ESP testing

Common Student Misconceptions

- ESP is an easy subject to study.
- ESP has clearly been demonstrated.
- It is easy to determine if someone is cheating on ESP tests.
- ESP can be studied in a classroom.
- Scientists are the best ones to study ESP when someone claims that he or she has this ability.

Classroom Management

The case is structured as a directed case study: Students read a story and look up the answers to specific questions, and then the teacher runs a discussion based on the questions and answers. In addition, in this instance, students try out some of the early experimental procedures and criticize the method.

Classroom Case Discussion

The students are given the case ahead of time. All students should hand in a copy of their answers to the questions that follow the case before the discussion and keep another copy for the discussion itself.

If working in teams, students could read the case and then divide up the questions. Each student looks up the answers to his or her questions and then shares the information with the rest of the team at the beginning of the next class period. A quiz could be given after this sharing period. The quiz ensures that the students are prepared for the general discussion, but perhaps, just as important, if the quiz is given to teams of students, they will interact and be warmed up for the discussion to follow. A sample quiz might be as follows:

Chapter 15

1. Define psychokinesis.
2. Describe in general J.B. Rhine's initial experiments at Duke University.
3. What was the Ganzfeld procedure?
4. List two scientific principles that ESP experiments have been criticized for not following.

One way to approach the discussion is to go around the classroom asking the students for their answers to the questions, using this opportunity to develop points that the teacher believes need emphasis. Any questions that the students had about the case can be addressed at this time as well. The discussion can rotate around the topics that were mentioned in the quiz—telepathy, clairvoyance, precognition, pyschokinesis, the Ganzfeld procedure, and J.B. Rhine's experiments—but might not focus at this point on the criticisms of ESP. That topic might be discussed later, after the class has run through their student exercise.

Another approach is less structured and perhaps more exciting. The teacher can start the discussion simply by asking students to guess what made the sound at the end of the story. Do they think that Eliza was able to move the spoon? If her mother went into the room and found the spoon moved, what are the most likely explanations for its movement? The class can be divided into skeptics and believers based on their answers, and these groups could have an informal debate on the issue. This allows the teacher to launch into the student experiment phase of the class, asking how we can possibly test for the presence of ESP with absolute safeguards as to the integrity of the data.

Beginning students might need guidance as to the modeling of the scientific process of hypothesis testing, reproducibility, falsifiability, replication, and control of variables to give them a foundation to base their skepticism for any unusual results.

Student Exercise

After any questions have been answered and the quiz has been fully understood, the experiments conducted by J.B. Rhine are further evaluated. The instructor should prepare before class a deck of 25 cards for each group, similar to those used in the experiments Rhine conducted. Create a deck for each group using 3 × 5 in. cards. Using a pencil or some other writing instrument that will not bleed through the card or be transparent, place one symbol on each card (square, three wavy lines, plus sign, circle, and star), as shown below:

Five cards of each symbol should compose the deck. (Zener cards or ESP cards can be purchased on many internet sites or at magic shops; they have blue patterned backs on them so that the symbols on the other side are not visible.)

Each group of students will conduct an experiment for telepathy. One member of the group will hold the deck of cards and look at one card at a time while other members attempt to guess the symbol that is on the card. The students should record their guesses on a piece of paper numbered from 1 to 25. When all of the cards have been "transmitted," then (and not before then) the student who played the role of the "sender" should look at all the cards and reveal what the order really was. The students should count the number of their "correct" hits. Obviously, by chance they should get about 20% correct. How many such tests should be tried? How can the investigators be absolutely sure that "cheating" has not occurred?

Other tests can be attempted. Students can test for clairvoyance by first shuffling the deck. Then someone takes a card out of the deck and places it facedown without looking at it. The students then record what they think the card is. This should be repeated until all of the cards have been dealt. Only then, and not before, should the symbols on the cards be checked against the students' predictions. Note: If the cards are checked as the test is being conducted (to see how things are going), students will—consciously or not—take mental note of the relative numbers of symbols that have already passed and are likely to end up with a higher number of hits.

Students can test for precognition by writing down the order of 25 cards *before* the deck is shuffled and then dealt face up. As in all tests, it is easy to cheat or to inadvertently give clues. Remember that magicians readily perform ESP effects all the time and have frequently fooled parapsychologists. After performing such tests, it is useful to ask the students if they can design a test that is fail-safe—that is, one where trickery could not be used.

The data that are collected should be placed on a blackboard or overhead. In all tests of this type, you would expect to get an average of 1 out of 5 hits correct; that is, students will guess 5 out of 25 correctly. Obviously, if large numbers of students are doing this many times, you would expect by chance that an occasional value will fall outside the expected probability. This is clearly a chance for the teacher to discuss statistics and what constitutes reasonable evidence for the existence of ESP. In fact, this reveals one of the major problems that plague ESP work: Investigators do multiple testing and only report the top results, discarding all the negative data.

Given the results of the entire class, the groups will then be asked to criticize the procedure of the test. They should readily identify many flaws and can be asked how they might redesign the procedure to eliminate them.

As a final wrap-up to the class, students could be asked to consider the question, "Does ESP exist?" Students should have strong back-up information for their answer. In coming to this conclusion, they should also incorporate the ideas introduced in the case's storyline and the argument that Eliza's parents have. For example, if they

Chapter 15

believe that ESP does not exist, following the beliefs of Eliza's father, even if some of the results were above the probability ratio, they need to support that statement with criticisms of the experiment or the data. After each group has come to a conclusion and written down statements, a discussion of the decisions can follow, using either a discussion or a debate-style presentation.

Web Version

This case and its complete teaching notes, references, and answer key can be found on the website of the National Center for Case Study Teaching in Science at *http://sciencecases.lib.buffalo.edu/cs/collection/detail.asp?case_id=229&id=229*.

Reference

Truzzi, M. 1978. On the extraordinary: An attempt at clarification. *Zetetic Scholar* 1 (1): 11–19.

Chapter 16
A Need for Needles: Does Acupuncture Really Work?

Sarah G. Stonefoot and Clyde Freeman Herreid

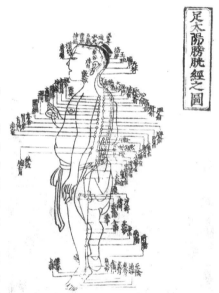

The Case

Janet sat in her car in the driveway of her mother's house and eyed the front yard, which was completely taken over by a vegetable garden. It was possible that somewhere in there her mother was hidden, picking over her prized vegetables. Her mother was a bit eccentric. Actually, the word *crazy* sometimes came to Janet's mind when considering her mother. But she never said it out loud. This was her mother, after all.

Janet took a deep breath and then grabbed the door handle. It was time for another lunch with her mother, a meal that would inevitably turn into an argument.

Audrey greeted her daughter at the door even before Janet had a chance to knock.
"Why, hello, darling."

Janet was carefully unhooking a tomato vine from her foot before her mother noticed. She looked up and greeted her with a sense of apprehension.
"Hello, Mother."
"Oh, Janet, it's so nice to see you. Isn't it just a gorgeous day today?" Her mother was bubbling over with her usual happiness and high spirits.
"It's hot," Janet grumbled, "too hot."
"Well, come on in. I've just put together a delicious salad for lunch."

Salad again, Janet thought. She was glad she had stopped at McDonald's on the way over. She forced a smile and followed her mother into the house.

Lunch went well, until Audrey decided she couldn't hold her idea back any longer. "I was reading that magazine you got me the other day."

Chapter 16

"Oh, really," replied Janet, pleased. She had gotten her mother a subscription to *Time* magazine for Christmas. It was her attempt to get her mother on the same track as the rest of the world, or at least aware of what was happening. She knew her mother used the magazines as coasters on the coffee table more than anything else.

"Yes, and I came across this really interesting article. It was on acupuncture."

Janet sighed. Her hopes evaporated. Of course, the only thing to interest her mother was an article on some sort of nonsense.

"As I was reading it, it began to make a lot of sense. It mentioned that arthritis was one of the things that it helped. And seeing how you're always telling me that arthritis is why my wrists bother me, I thought it might be helpful. My wrists have been acting up a little. I couldn't garden the other day. My poor tomato plants are going to wither away in the sun if I don't get to them soon."

"Mom, you don't want acupuncture. The whole thing is ridiculous. It's a big scam. There is no way that putting needles in your body is going to stop your pain. What you need is to see a doctor." Janet's frustration level was quickly rising.

"I had a feeling you would say that," Audrey said with a sigh. "So I think you should read the article." She handed her daughter the magazine.

"No, Mom," Janet said, pushing it away.

Janet knew she had the final say in this situation. After all, she worked for her mother's insurance agency and had pulled some strings to get her mother covered. She felt strongly that she had a say in what treatment her mother had. Audrey had reluctantly accepted the insurance, knowing that Janet would have some control over her life. She solved the problem simply. She avoided going to doctors.

"Mom, I really think you should go to a doctor about your arthritis. If you don't, it will just get worse. This could become serious. I'm sure there are tons of different medicines you could choose from to help the pain," Janet preached. It was a speech she had given many times before, yet as before it got nowhere with her mother.

"Janet—don't start that again. You know that I'm not putting any drugs into my body, especially when there's no good reason."

"No good reason? Mom, you're in pain," Janet responded, a touch of sympathy entering her voice.

Audrey sighed. She looked down and realized she had been massaging her wrist the whole time. She knew she was being difficult. Her daughter was right, at least about the pain.

"OK—what if we compromise," Audrey began. "What if we talk to a doctor about acupuncture. Will you be convinced to let me try the treatment if you hear from a doctor that it is beneficial?"

"Fine," Janet replied. She knew no sensible doctor would agree to sticking needles in her mother to get rid of her pain. The whole thing was ludicrous.

* * * * *

A Need for Needles: Does Acupuncture Really Work?

It was two weeks later and Janet and her mother were on their way to see a doctor. Audrey had already jumped out of the car and headed into the doctor's office. Janet dragged herself out of the car and followed.

They sat in the waiting room for what seemed like hours. Janet, in her business suit, was hiding behind the *New York Times*, trying not to imagine what people were thinking of her mother. At least her mother had taken off her apron, but she was still in gardening clothes, straw hat resting in her lap. The nurse had seemed amused when she collected the initial data. She now reappeared.

"Audrey Baker, you can come in now."

Janet was glad to escape the eyes of everyone in the waiting room and followed her mother into the doctor's office. Moments later, Dr. Ramirez walked in as they were getting settled.

"Hello, how are you two doing today?" Dr. Ramirez asked.

"Great, thank ...," Janet began.

"Just wonderful," Audrey interrupted. "The reason why we're here today is that my daughter can't seem to grasp the concept of acupuncture. I have some pain in my wrists, and I understand it can help that. So, I was wondering if you could just take a couple minutes to explain it to her."

Janet scowled, but before she could say anything, the doctor asked, "What kind of pain do you have?"

"Oh, nothing serious," Audrey said. "I'm just getting old and my bones aren't what they used to be. When I'm pulling weeds, they tend to get a little sore."

"Oh, you're a gardener," Dr. Ramirez said, pleased. "You know, I have a garden of my own."

"Really!" said Audrey. "Flowers or vegetables?"

"Both, actually."

"OK," Janet interrupted. "I'm sorry to be rude, but I am on a tight schedule. Can you just tell my mother that acupuncture will not work on her pain, so that we can set up a suitable treatment?"

"I see," Dr. Ramirez said.

He could see how this was going. He looked at Audrey's chart and paused as he thought about how to approach this case most effectively.

* * * * *

Your task is to assist Dr. Ramirez in reaching his goal. There is a recent trend in medicine, called evidence-based medicine, in which physicians search the literature to determine effective approaches to treatment rather than just doing what one of their teachers taught them to do in medical school. Your task is to approach the case in this manner: Scientifically investigate the pros and cons of acupuncture treatment,

Chapter 16

consult with Dr. Ramirez on what you find, and offer suggestions on how to best proceed with Audrey and Janet.

You will be divided into groups: Half of your group will search for the pro literature (supporting the use of acupuncture) and the other half for the con literature. Be sure you understand the theory behind acupuncture, the different treatments acupuncturists might use, and the evidence or lack thereof that suggests acupuncture may work, including the argument that any positive results are due to the placebo effect.

When you return to class armed with evidence, your job will be to work out a consensus consultative opinion for Dr. Ramirez in your group and share that opinion with the rest of the class. Part of that sharing will involve the soundness of the evidence. Then you will need to work out among yourselves what you think Dr. Ramirez should do.

A good place to start your research would be the web page "Acupuncture Information and Resources," National Center for Complementary and Alternative Medicines, on the National Institutes of Health website (*http://nccam.nih.gov/health/acupuncture*).

Teaching Notes

Introduction and Background

Acupuncture is at least 2,000 years old. It is a form of therapy in Asian medicine, part of a complete medical system that also includes herbology, physical therapy, dietetics, and special exercises. It is the most widely used healing system on Earth, having first appeared in China and extending from there throughout Asia, Europe, and North America.

Practitioners of acupuncture claim that qi (pronounced chee) is an energy force running through the body. It includes all essential life activities. Qi is composed of two opposites, yin and yang. These opposites must be kept in balance to sustain a healthy life. Yin represents the female, cold, dark, passive, and medial part of the body. Yang is the opposite of the yin, encompassing the male, warm, light, active, and lateral aspects of the body. The qi travels along special pathways of the body known as meridians. The meridians are 12 main pathways on each side of the body. The channels are named after 12 main organs; however, the pathways are not limited strictly to that organ. In acupuncture, the needles are inserted at points along these pathways. The points are where the meridians come to the surface. The aim of acupuncture is to adjust the vital energy of the body so that the correct amount reaches the proper part of the body at the essential time. This improves the body's ability to heal itself.

Acupuncture has been used to treat a variety of ailments, including lower back problems, cervical spondylosis, condylitis, arthritic conditions, headaches, allergic reactions, drug addictions, endocrine disorders, mental disturbances, heart failure, attention deficit hyperactive disorder, immune disorders, carpal tunnel syndrome, cerebral palsy, hay fever, and menopause.

Critics argue there is no solid answer as to how acupuncture might work. The hypothetical energy patterns of qi have not been demonstrated, nor have they been studied to a great extent in Western medicine. This leaves the theoretical foundation of acupuncture very much in question. In tests where acupuncture has been evaluated, it has been difficult to set up proper control groups. People undergoing acupuncture know what is supposed to happen. Thus, results from the procedure may be due to the placebo effect.

This case is intended for an introductory college-level science course such as human biology. It does not require the students to have an in-depth knowledge of the medical procedures of acupuncture or the central nervous system. However, it would be helpful for the students to have a general knowledge of the systems of the body. Students should also be aware of some of the general scientific guidelines used in evaluating scientific data, which will help them evaluate the theories behind the acupuncture procedure.

Objectives

This case is intended to expose students to one of the procedures advocated by therapists of alternative and complementary medicine, acupuncture. Students are asked to take a skeptical look at the procedure in an effort to see where the boundaries may exist between the believers of these alternative therapies and the scientific facts. The overall objective of the case is to expose students to the possibilities of alternative and complementary medical therapies, as well as to encourage them to question their effectiveness.

Students evaluate information on both sides of the issue to determine if there is adequate scientific information to conclude that acupuncture is a helpful method of treatment. During the case, students collect information from internet sources or journal publications with an emphasis on carefully evaluating the credibility of the information they collect.

The major blocks of analysis addressed by the case include

- theories of pain and the body (qi, yin, yang, meridians);
- methods of acupuncture;
- needles, electroacupuncture, and ear acupuncture;
- theories regarding how acupuncture works (gate theory, central nervous system, and endorphins);
- side effects;
- ailments;
- trials that have been conducted on patients; and
- criticisms and evaluations of the believers.

Chapter 16

Common Student Misconceptions

- Acupuncture is an effective treatment that has been accepted by the Western medical establishment.
- We understand how acupuncture works.
- The effectiveness of acupuncture is easy to demonstrate.
- The placebo effect cannot be involved in acupuncture treatments.
- The energy meridians of Chinese medicine are clearly established.
- Qi is a philosophy accepted by the Western medical establishment.
- Evidence-based medicine confirms that acupuncture works.
- Scientists understand how acupuncture affects the nervous system.

Classroom Management

The approach we suggest for teaching this case is described in the instructions to students at the end of the case (pp. 143–144). Working in groups of four, the students are given the case to read in class. They are divided into pairs, with one pair assigned to research the pro side of the argument and the other the con position. The groups come together in a later class (perhaps a week later), discuss the issues within their groups, and try to come up with a consensus recommendation for Dr. Ramirez. The students might recommend a pro-acupuncture position, an anti-acupuncture position, or something in between. They then share their views with the entire class. In this general discussion, which is facilitated by the instructor, it is essential that students cover the theory behind acupuncture, the different treatments that acupuncturists might use, the evidence or lack of it that suggests that acupuncture may work, and the possible complication of the placebo effect. An interesting question to ask the students would be, "Can you design an experiment to eliminate the placebo effect?" (What would happen if you placed the needles in places that are not specified by the Qi meridians?)

Web Version

This case and its complete teaching notes and references can be found on the website of the National Center for Case Study Teaching in Science at *http://sciencecases.lib.buffalo.edu/cs/collection/detail.asp?case_id=370&id=370*.

Chapter 17
Love Potion #10: Human Pheromones at Work?

Susan Holt

The Case

Part I

"Hey, Laurie! Did you notice how the guys at lunch today were flirting with me? I'm sure it's because I'm wearing my new guy-attracting chemical!"

"What's a guy-attracting chemical?" Laurie asked.

Jean explained, "They call it a pheromone. I found an ad for this stuff called Pheromone 10:13 in a magazine. It sounded really great. The ad said that it would increase your sex appeal. You just mix a little bottle of this unscented chemical with your cologne and it gets you more romantic attention. It was expensive, about $100 for one little bottle, but I used it today and obviously it's worth it. You should buy some too."

Laurie really hadn't noticed the guys paying more attention to Jean. She retorted, "If this pheromone stuff really attracts guys, why don't more people use it? Why should I waste my money on a quack product?"

Jean bristled at Laurie's comment. "I should show you the ad! They make two kinds of pheromones—10X for guys to attract girls and 10:13 for girls to attract guys. You should read what people are saying! One guy says that this pheromone stuff 'really has science behind it' and that he notices a big difference when he is wearing it. A woman says that she can almost feel the 'energy of the men's attraction to her.' The ad has testimonials from all sorts of people who say pheromones really work for them."

Laurie couldn't believe that Jean would fall for testimonials. "You see those kinds of things in lots of ads. Diet pill ads have lots of people saying they lost gigantic amounts of weight. But they're a rip-off—they usually don't work, and my mother says that they're not safe."

But Jean had a comeback for that: "A real doctor who has written in medical books and journals discovered these pheromones. She did scientific research that

Chapter 17

proves that pheromones work. These pheromones aren't cheap imitations that fail. She started a company that sells them. And how can it be unsafe if you just put a little on your skin? They even have an internet site that tells you all about this scientific evidence."

Laurie wasn't quite sure now. If these pheromones had been scientifically tested by a real doctor ... well, that made her wonder. "I'll take a look at the internet site tonight. What's the address?"

Jean had the address written down. "It's *www.Athena-Inst.com*. Go see for yourself."

Questions

1. Do you believe that the Athena pheromone really does what the ad claims?

2. What information in the advertisement leads you to believe that the pheromone works?

3. What additional information would help to convince you that the pheromone does what is claimed?

Part II

Jean was annoyed. Laurie had treated her like a gullible sucker. That night she did some research on the website listed in the advertisement. The next day she gave Laurie copies of the information.

"I found more information about pheromones that I think you should read. This stuff should convince you that pheromones really are sexual attractants."

Laurie had decided to prove that Jean was wrong. She had also collected some information on human pheromones.

"The stuff that I read really didn't prove that pheromones work. Some people think it might work, but they really don't have any evidence."

Questions

1. What is a pheromone? How do pheromones work? Do pheromones really affect human behavior?

2. What information in the articles supports the advertising claim?

3. What information in the articles makes the advertising claim questionable?

4. Based on this information, do you believe that the Athena pheromone really does what the ad claims?

Part III

Today at lunch with friends, Laurie continued the argument. "Would you believe that Jean still thinks that pheromone stuff she bought really works? She thinks the pheromones are the reason that she has a date this weekend. How could she be so gullible?"

Love Potion #10: Human Pheromones at Work?

Jean had expected this sort of attack and was ready with a response. "How can you ignore the scientific facts? Dr. Winnifred Cutler, from the Athena Institute, did an experiment that proved 74% of the people who wore pheromones were more attractive to members of the opposite sex. It was too long and boring to read completely, but here's the abstract and data from her experiment. I don't believe you won't even trust the scientific evidence."

This study tested the effect of human male pheromone on the sociosexual behavior of men and, by implication, the sexual responses of the women they encountered, as well as the men's perception of these effects. Thirty-eight heterosexual men, ages 26–42, completed a 2-week baseline period and a 6-week placebo-controlled, double-blind trial testing a pheromone "designed to improve the romance in their lives." Each subject kept daily behavioral records for five sociosexual behaviors— petting/affection/kissing, formal dates, informal dates, sleeping next to a romantic partner, and sexual activity—and faxed them each week. Significantly more pheromone than placebo users increased above baseline in sexual activity and sleeping with a romantic partner. There was a tendency for more pheromone than placebo users to increase above baseline in petting/affection/kissing and informal dates but not in formal dates. A significantly larger proportion of pheromone than placebo users increased in two or three of the five sociosexual behaviors. Thus, there was a significant increase in male sociosexual behaviors in which a woman's sexual interest and cooperation plays a role. These initial data need replication but suggest that human male pheromones affected the sexual attractiveness of men to women. [See Tables 17.1 and 17.2 (p. 150) for data.]

Table 17.1. Average Initial Age, Height, Weight, and Relationship Status for Subjects by Treatment Group

	Pheromone (n = 17)		Placebo (n = 21)	
Age (years)	33.1		33.8	
Height (inches)	69.6		71.7	
Weight (pounds)	189.7		187.0	
Relationship Status	N	%	n	%
Not dating but would like to be	7	41.2	9	42.8
Dating	2	11.8	8	31.8
Keeping steady company	2	11.8	1	4.8
Married	6	35.3	3	14.3

Source: Cutler, W., E. Friedmann, and N. L. McCoy. 1998. Pheromonal influences on sociosexual behavior in men. *Archives of Sexual Behavior* 27 (1): 1–13.

Chapter 17

Table 17.2. Number of Subjects With an Increase Over Baseline for Each of Five Sociosexual Behaviors by Treatment Group

Sociosexual Behavior	Pheromone (n = 17)		Placebo (n = 21)	
	N	%	n	%
Sexual activity	8	47.0	2	9.5
Sleeping next to romantic partner	6	35.3	1	4.8
Petting/affection/kissing	7	41.2	3	14.3
Informal dates	6	35.3	2	9.5
Formal dates	7	41.2	7	33.3

Source: Cutler, W., E. Friedmann, and N. L. McCoy. 1998. Pheromonal influences on sociosexual behavior in men. *Archives of Sexual Behavior* 27 (1): 1–13.

Questions

1. What hypothesis was Dr. Cutler testing?
2. What sort of people did she use as a control group?
3. What sort of people did she use as an experimental group?
4. How many people did she choose for each group?
5. What factors did she keep constant in her experimental and control groups, and how did she do this?
6. What procedure did she follow to make sure the pheromone was tested fairly?
7. What data did she collect? How often did she collect data?
8. How did she decide if the pheromone was effective or ineffective?
9. Do you think that Dr. Cutler's research provides credible evidence that the pheromone Jean bought really does what the advertisement claims? Why might people answer "yes" to this question? Why might people answer "no" to this question?
10. Do you think you would get the same results if you replicated Dr. Cutler's research? Why or why not? Why has her experiment not been replicated by other researchers?

Part IV

Dr. Cutler's research was done with Athena Pheromone 10X (unscented fragrance additive for men). Jean had purchased Athena Pheromone 10:13 (unscented fragrance additive for women). A product should be properly tested to make sure that it does what it is supposed to do. Design an experiment to test Athena Pheromone 10:13 to see if it gets women more romantic attention.

Questions

1. What hypothesis would you be testing?
2. What sort of people would you use as a control group?
3. What sort of people would you use as an experimental group?
4. How many people would you choose for each group?
5. What factors would you have to keep constant in your experimental and control groups, and how would you do it?
6. What procedure would you follow to make sure the pheromone is tested fairly?
7. The data collected need to be "socially appropriate"—that is, the research should involve only behaviors permitted by high school rules. What data would you collect? How often would you collect data?
8. Make a data table that you would use to record data from your experiment.
9. How will you decide if the pheromone is effective or ineffective?

Part V

Scientific literacy means more than being able to design and conduct experiments. You must also be able to understand how scientific research affects your life and your decisions. Consider each of the following mini-cases and indicate what you think should be done in each situation. Be prepared to share your answers with your classmates.

1. A researcher working for an herbal medicine company conducts an experiment on 10 middle-age men. He tells them he is testing an herbal medication that scientists think will make people feel more energetic. He instructs them to take one herbal medicine tablet a day for one week. After a week, he asks them if they feel more energetic. He reports that 9 out of 10 people feel more energetic after taking the medication. *Would you buy this medication? Explain why or why not.*

2. A college professor noticed that the men and women in her class usually formed male-only or female-only groups. She wondered if pheromones would encourage mixed-gender groups. Without explaining what she was doing, she asked the women in the class if they would like to try a free sample of a new perfume. She did not tell them that the perfume contained a pheromone that attracted members of the opposite sex. *Do you think this is appropriate? Explain why or why not.*

Chapter 17

3. Scientists have discovered a medicine they feel has the potential to cure a deadly form of childhood cancer. They need to test the medicine to be sure that it is safe and effective. *Should they test this medicine on animals before they begin tests on humans? Explain why or why not.*

4. Parkinson's disease is a progressive neurological disease that gradually destroys a person's control of voluntary movements. Scientists have tested a treatment for Parkinson's disease in monkeys who showed disease symptoms. They drilled holes in the skulls of two randomly selected groups of monkeys. The experimental group had fetal tissue injected into their brains. The control group had distilled water injected into their brains. The treatment was 80% effective in treating the symptoms of the disease in monkeys. You have recently been diagnosed as having Parkinson's disease. Researchers have asked you to participate in their clinical trials to test the fetal tissue transplant treatment. *Would you agree to become a subject in these clinical trials? Explain why or why not.*

5. Scientists are conducting a long-term (three-year), double-blind, placebo-controlled clinical trial on a chemotherapy they hope will cure prostate cancer. They want to determine if the chemotherapy is safe and effective. They randomly assign patients to be members of the experimental or control group. After six months, they found that early treatment with the chemotherapy results in a 90% cure rate with no dangerous side effects. *Should they offer the medicine to people in the control group now or continue the research as planned? Explain why or why not.*

6. Testing and FDA approval are not required for natural herbal medications such as ginkgo biloba, St. John's wort, and ginseng. *Do you think people should use medicines that don't have FDA approval? Explain why or why not.*

7. During World War II, unethical scientists did research on the effects of radiation on humans. They used people in concentration camps as their experimental group—exposing them to dangerous doses of radiation. The results of these experiments have recently been discovered. *Should this information be released? Explain why or why not.*

Teaching Notes

Introduction and Background

Most nonmajors in biology classes will not actually do scientific research outside a classroom. They will, however, encounter many situations in which they need to consider whether evidence provides adequate support for scientific claims. The "love potion" case was designed to encourage skepticism in evaluating an advertising claim that implies there is scientific evidence to support the claim. The advertisement for pheromones claims these products "get you more romantic attention."

Even relatively unengaged students find this controversial topic an interesting way to discuss scientific methods. Instead of traditional lectures or readings on the "scientific method" and the process of science, small-group and whole-class discussions in this case are used to develop student skills in evaluating and testing a real-life advertising claim.

I have used this case study in advanced, average, and below-average high school (grades 9–12) biology classes. It could also be used in Advanced Placement biology or introductory college biology classes. In my school, the case was used during the first few weeks of school because it served as an introduction to scientific thinking and sent a clear message that we expected students to link biology to the world around them. The science vocabulary in the case is intentionally kept to a minimum because we know that it will be formalized during lab and class activities later in the year.

The case could be used anytime during the school year. The most logical content connection would be during a unit on animal regulation or behavior or as a review of the scientific process at the end of the year. If the case is used later in the year, teachers should use more scientific vocabulary during class discussions.

Objectives

In completing this case study, students should

- understand that biological processes are thought to play a role in human behaviors,
- apply skepticism in evaluating whether scientific evidence supports claims, and
- design an experiment to test an advertising claim fairly.

Common Student Misconceptions

- People should believe advertisements that include recommendations from a scientist, doctor, or celebrity.
- One research study done by a scientist or a doctor is adequate evidence to prove that something works.
- If a friend tried it and it worked, then it must be a good product.

Classroom Management

Because this case study was also used as an introduction to case learning, small-group and class discussions are organized around the answers to the study questions. This allows individual students to prepare for participation and ensures that all students will have written information to share with their group or class. Most of the work is

Chapter 17

done during class so that the instructor can provide support and assistance to students who have little experience in active learning. Done in the manner described (pp. 153–157), this case should require a minimum of four periods (160 minutes) of class time. If used later in the year when students are familiar with case study learning, class time could be reduced.

Consider printing copies of the internet advertisement at *www.Athena-inst.com* and having sample magazines with advertisements for Athena Pheromones for your students.

It is important to emphasize that the case questions have many possible answers. Encourage "science-phobic" students to participate by keeping criticism of student answers to a minimum at the beginning of the year. Student answers that appear trivial or incorrect can often be clarified by asking students, "What does that mean?" or "Why do you think that?" Avoid teacher paraphrasing by encouraging students to summarize in their own words.

Students write their answers to each part of the case study on separate sheets of paper. This allows room for more elaborate answers and for future additions.

Using markers and large sheets of newsprint taped to the wall (or overhead transparencies) instead of writing directly on the blackboard allows you to save and post a record of the class discussion for future reference. You can substitute blackboard for newsprint if you do not need a record of the discussions.

Day 1: Complete Part I and Assign Part II

Try to find a copy of the song "Love Potion No. 9." Playing this at the beginning of the class starts the case with a humorous tone that "hooks" non-science-types. This classic pop hit of Leiber and Stoller has several famous versions by such artists as The Clovers, The Ventures, and even Herb Alpert & the Tijuana Brass. Ask students to begin by working individually for 5 minutes to answer the questions in Part I. Assign students to a small group of 3–5 students. Explain that they have 10 minutes to work with their group to share answers to the questions.

Questions and Teaching Suggestions

1. Do you believe that the Athena pheromone really does what the ad claims?

   Informally survey the class. Put three categories on a newsprint sheet: Works, Can't Tell, and Doesn't Work. Tally the number of students in each category. Save this for later reference.

2. What information in the advertisement leads you to believe that the pheromone works?

   Make a heading on a newsprint sheet—Evidence That Pheromones Work. Ask one member from each group to read one possible statement that answers the question. If time permits, call on volunteers to extend the list.

3. What additional information would help to convince you that the pheromone does what is claimed?

   Make a heading on a newsprint sheet—Additional Information Requested. Ask a different member from each group to read one possible statement that answers the question. If time permits, call on volunteers to extend the list.

Assign completion of Part II for homework. Each student in a group gets the same questions but is provided with different references to read. Do not ask students to find their own internet sites for this case study! Using "human pheromones" for a keyword search will result in sites that are pornographic in content.

Day 2: Discussion of Part II

Students work to share their answers with their small groups for the first 20 minutes of class. For each question, all members of the group read their answers aloud. Other group members add more information to their answers as they listen. Encourage groups to make a list of questions they have about pheromone 10X. Label three sheets of newsprint: Know, Supports, and Doubts. These sheets are used to record the key ideas from class discussion.

Questions and Teaching Suggestions

1. What is a pheromone? How do pheromones work? Do pheromones really affect human behavior?

   Call on one member of each group to help you make a list on newsprint of what students know, or think they know, about pheromones. Ask them to indicate whether they are sure (put an "S") or not sure (put a "?") about their statements.

2. What information in the articles supports the advertising claim?

   Call on another member of each group to help you make a list of the information that supports the claim. Then ask for volunteers to challenge things that are on the list by explaining why they feel that information does not provide support. Put a "?" after statements that are challenged. What should begin to emerge is an understanding that there is weak support and little clear-cut evidence for the advertising claim.

3. What information in the articles makes the advertising claim questionable?

   Call on one or two members of each group to help you make a list of the information that causes them to question or doubt the advertising claim. Examples for this list are more difficult for students to consider. Allow more time for this discussion and try to make this list the same length as the previous one. Often the discussion

Chapter 17

here adds additional "?" marks to the previous column. It should become obvious that there is weak support but little clear-cut evidence for the advertising claim.

4. Based on this information, do you believe that the Athena pheromone really does what the ad claims?

Informally survey the class. Put three categories on a newsprint sheet—Works, Can't Tell, Doesn't Work. Tally the number of students in each category. Compare this with the similar sheet from Part I. Ask students who changed their answers to explain why. Assign Part III, questions 1–8, for homework.

Day 3: Complete and Discuss Part III

Allow 10 minutes for students to share and expand their answers to questions 1–8. Ask them to select one question they would like you to go over in class discussion. Just before the 10 minutes are up, visit each group to ask the number of the question that they found most difficult. Write the problem question numbers on the board. Usually two or three questions are selected by many groups. Call for volunteers from other groups to explain their answers to these two or three questions. Ask students to work in small groups to answer questions 9 and 10. Tell them they have 10 minutes to do this. Each group should make a large, two-column chart on newsprint for question 6—"Yes because" and "No because."

Questions and Teaching Suggestions

1. Do you think that Dr. Cutler's research provides credible evidence that the pheromone Jean bought really does what the advertisement claims? Why might people answer "yes" to this question? Why might people answer "no" to this question?

   Introduce another technique for "class discussion." After 10 minutes, suggest that each group send a "scout" (or two, depending on group size) to visit other groups to get ideas for how to expand their answer to question 9. Allow 5 minutes for "scouting" and 5 minutes for the "scouts" to report back to their groups and add to the newsprint chart. Post these charts prominently. Ask students to look at the charts and identify the five most obvious reasons why Dr. Cutler's research does not provide credible support for the advertising claim.

2. Do you think you would get the same results if you replicated Dr. Cutler's research?

   Why or why not? Why has her experiment not been replicated by other researchers?

   Focus mainly on why people might not get the same results if they repeated Dr. Cutler's experiment. Call on each group to provide one reason their results might be different. List these reasons on newsprint. Call for volunteers to suggest how the experiment could have been changed to overcome these problems.

Day 4: Complete and Discuss Part IV

Set the stage by pointing out that Dr. Cutler's research was done using the male pheromone. It really doesn't address the question of whether the female pheromone that Jean bought really works. Point out that the questions break the task of designing an experiment into "baby steps." In the future, they will learn to design experiments without having "baby steps" provided. Have students work for 15 minutes in small groups to answer these questions. During the remainder of class, call on one group to answer each question and explain why their answer is important in designing a fair test of the pheromone.

Closure

Ask students to work individually for homework (or as a quiz or test) to prepare

- a list of 10 endings for the statement "You should be skeptical if you notice an advertising claim that ..." and
- a list of 10 endings for the statement "When testing a product claim, a good experiment ... ."

Remind students that they can use their notes from the case study as a reference. Take a few minutes at the beginning of the next class to ask each student to share one important thing from each of their lists. Record students' contributions on the blackboard.

Web Version

This case and its complete teaching notes and references can be found on the website of the National Center for Case Study Teaching in Science at *http://sciencecases.lib.buffalo.edu/cs/collection/detail.asp?case_id=173&id=173*.

Chapter 18
The "Mozart Effect"

Lisa D. Hager

The Case

Part I: Enhanced Performance?

"Hey, Bill, what are you listening to?" asked Fred.

"I'm listening to these CDs of classical music that I bought. They're supposed to help me concentrate more and make me become more creative," answered Bill.

Fred frowned. "How can listening to classical music do all of that? Where did you hear about this?"

"Well, I was flipping through this magazine at my girlfriend's house and I saw this ad where you could buy these CDs that are supposed to stimulate the right side of your brain and improve your ability to concentrate and stuff," said Bill.

"And how much did you pay for these CDs?"

"Just $45, and there's a money-back guarantee if they don't work. In the ad, it said that some researchers found that listening to this music made people do better on different mental tests and that it made your brain release these chemicals that made you feel better," Bill said excitedly.

Fred still felt a little skeptical about the power of Bill's new CDs. "So, what else did the ad say?"

"All kinds of cool things. Like, when they played the music for these cows, they gave more milk, and immigrants who were learning English learned faster when they listened to the music, and, this one is really cool, when they played the music next to this yeast, it made better sake," said Bill.

Fred laughed. "So, have you been giving more milk or what?"

"Hey, don't make fun of me. I haven't been doing so great in some of my college classes, so I figured I might as well give it a chance," Bill answered. "Here, you can check it out for yourself on the product website."

Chapter 18

"You know, this might just be the kind of thing I could do for my project in my research methods class. Our professor is encouraging us to be more skeptical about claims just like this one. We've been talking about something called the principle of falsifiability," Fred said.

"The principle of what-ability?" asked Bill.

"The principle of falsifiability. It's a scientific term that basically says that when we study something, like whether these CDs improve concentration and creativity, we have to do it in a way that will allow us to confirm whether the prediction is false. So, if people study this effect using the scientific method and they don't find that the CDs improve concentration and creativity, then we have to accept that there's no truth to the claim being made by the person who produced them," replied Fred.

"Well, that makes sense to me. You know, I think I want to help you with this study. I already have the CDs, so maybe it would be kind of cool to be part of a scientific study. What do we do next?" asked Bill.

Fred and Bill need to figure out how they can determine if listening to classical music really will produce the kind of effects that the product's maker claims. Fred and Bill decide to go visit Fred's psychology professor to see what she thinks about their study idea.

Questions

Answer the following questions based on Bill's description of the advertisement and the information on the website.

1. What claims are made for the product?

2. Is there evidence to support the claims?

3. What suggestions do you have for Fred and Bill?

4. How can Fred and Bill find out if there's any published evidence to substantiate the claims?

5. Evaluate the quality of the information presented on the website. Use Table 18.1, which distinguishes between characteristics of nonscientific and scientific ways of acquiring knowledge, to help you organize your response. For example, decide whether the observations posted on the website for the CDs are based on a handful of anecdotes or rather on systematic, controlled experiments. Next, evaluate whether the reporting of results is biased and subjective, or unbiased and objective. Apply each category in turn and use the dialogue between Bill and Fred to justify your conclusions.

Table 18.1. Characteristics of Nonscientific and Scientific Ways of Acquiring Knowledge

	Characteristics of a Nonscientific, Everyday Approach to Experience	Characteristics of a Scientific Approach to Experience
Observation	Casual, uncontrolled	Systematic, controlled
Reporting	Biased, subjective	Unbiased, objective
Concepts	Ambiguous, with surplus meanings	Clear definitions, operational specificity
Instruments	Inaccurate, imprecise	Accurate, precise
Measurements	Not valid or reliable	Valid, reliable
Hypotheses	Untestable	Testable
Attitude	Uncritical, accepting	Critical, skeptical
General Approach	Intuitive	Empirical

Note: Adapted from Shaughnessy, J. B., E. B. Zechmeister, and J. S. Zechmeister. 2003. *Research methods in psychology*. 6th ed. New York: McGraw-Hill.

Part II: Outlines of an Experiment

Encouraged by Fred's psychology professor, Fred and Bill decide to conduct an experiment to test the effectiveness of the classical music CDs. Outline an experiment that they could conduct. Be sure to address the following questions:

1. What is the research question that Fred and Bill want answered?
2. What would be the scientific hypothesis?
3. What independent variable(s) could be tested?
4. How could the effect of the independent variable(s) be measured? In other words, what should the dependent variable(s) be?
5. What other variables should be controlled, and how should this be done? Describe control procedures that Fred and Bill may want to use, including (but not limited to)
   - holding some variables constant,
   - eliminating the effect of some variables,
   - choosing a random sample, and
   - randomly assigning participants to groups (levels of the independent variable).

Chapter 18

Part III: Research Report Analyses
(A) Rauscher, Shaw, and Ky (1993)
Fred and Bill did a literature review, and one of the research reports they found was from *Nature* (see Rauscher, Shaw, and Ky 1993). Read this report and respond to the following:

1. Identify the independent variable(s).

2. Identify the dependent variable(s).

3. What aspects of the study did the researchers control? What aspects did they fail to control?

4. What were the results of the study?

5. What conclusions do the researchers reach?

6. Based on the design and results of this study, do you believe that the researchers are justified in reaching these conclusions? Why or why not?

7. Now look at the study that you and your group designed. Did you control for any of the problems present in Rauscher, Shaw, and Ky's study? If so, what were your controls?

(B) Steele, Bass, and Crook (1999)
Let's take a look at another research report that Fred and Bill found in the journal *Psychological Science* (see Steele, Bass, and Crook 1999). Read this report and respond to questions 1–6 above as they pertain to this article. In addition, point out ways in which this study is an improvement on the Rauscher, Shaw, and Ky study.

Part IV: Replication? (Optional)
As a class project, we are going to do our own study to see if we can replicate the study done by Rauscher. We will use the 1993 *Nature* article and the Steele article in *Psychological Science*, as well as any additional articles that we come across in a literature review. Your next task is to use the database PsycInfo to conduct a literature review. Spend some time with your group generating some search terms that will be useful for this literature search. Don't forget to include authors' last names as appropriate terms.

Teaching Notes
Introduction and Background
The research methods course in psychology often is taught in the junior or senior year and typically is designed to teach students to apply the scientific method for evaluating research and conducting their own research. In my courses, I show students how the research methods course can help them in their daily lives by teaching them how

to evaluate media reports and advertisements with the skepticism of a scientist. I usually like to start the course with an introduction to the application of the scientific method by having students apply some of the basic concepts related to research design.

This case study combines several approaches, including a directed case approach with the use of journal articles combined with lecture, individual assignments, and small groups. Students are given information regarding an advertisement and a website claiming that listening to the classical music in the advertised CD set will enhance people's cognitive skills and creativity. Students are asked to

- evaluate the claims and the evidence cited to support the claims;
- use the claims as a way of comparing the scientific method to a nonscientific, everyday approach;
- evaluate published research that has been conducted on the claims;
- complete a literature review to obtain further information on the claims; and
- design and conduct a test of the claims.

Objectives
Upon completion of this case, students should be able to

- identify and critically evaluate the claims being made in the advertisement;
- determine how to independently verify the claims;
- compare and contrast a scientific approach with a nonscientific, everyday approach;
- design and evaluate an experiment to test the claims;
- read and analyze the components of scientific reports investigating a phenomenon similar to the one described in the advertisement; and
- determine whether they can rely on advertised claims.

Common Student Misconceptions
- Psychology is just common sense.
- Information on websites can be considered to come from trusted sources.
- There is no need to critically evaluate published scientific claims.
- All researchers come to the same conclusions when evaluating scientific data.

Chapter 18

Classroom Management

Part I

In Part I of the case, students read the exchange that takes place between two students, Bill and Fred. Bill is describing a set of CDs that he purchased. Listening to the CDs is supposed to improve Bill's mental abilities and creativity. Fred expresses skepticism regarding the legitimacy of the product. Students must identify the claims that are being made, evidence to support the claims, and ways in which Fred and Bill could independently verify the claims.

The initial purpose of this first part of the case is to get students into the mindset of thinking like a scientist—being skeptical and critical. In addition, students are asked to use their information literacy skills to identify ways in which they can substantiate claims made in the media or by individuals they know. The goal is for students to realize that they do not have to accept all claims at face value and that they have the resources and knowledge to investigate these claims.

The second purpose of Part I is to get students to consider the differences between everyday approaches to knowledge and the scientific approach to knowledge by using the table adapted from Shaughnessy, Zechmeister, and Zechmeister (2003) presented in Part I. This part of the analysis clarifies the difference between these two approaches and introduces students to the language of science.

I found that it worked well to have students read Part I in class and share their responses in small groups. Small groups then share the consensus of their groups with the rest of the class. If instructors want students to go to the website for the CDs in Part I (*www.springhillmedia.com/b.php?a=DCAMPBELL*), they may want to have students do this outside class.

By using the table in Part I, instructors can conduct a more detailed discussion of the characteristics of the scientific approach by explaining each term and giving examples before students conduct an analysis of the characteristics.

Part II

Students are asked to design an experiment to test the effectiveness of the CDs and given specific design issues they must address. This analysis allows instructors to introduce the following concepts: operational definitions, hypotheses, independent variables, dependent variables, and control procedures.

I allow students to use their textbooks for this part of the case study. Instructors could have students complete this part on their own and then share their responses, or they could use this part as a graded assignment. The information is shared and critiqued as part of the larger class.

Part III (A)

In Part III (A), the instructor provides students with a scientific article from *Nature* by Rauscher, Shaw, and Ky (1993) and asks them to analyze it. As part of this analysis, the students are required to identify basic elements of experimental design. Question 6 requires students to critically evaluate the *Nature* article and the methods used and conclusions reached by the authors. It also provides the opportunity for instructors to discuss important methodological issues through a critique of the study

Instructors could have students retrieve the article on their own. To save time, I provide students with the article. Both this part and Part III (B) below could be used as group or individual assignments, graded or ungraded.

Part III (B)

Students are given a second article (Steele, Bass, and Crook 1999) about research that fails to replicate the results of Rauscher, Shaw, and Ky as reported in *Nature*, and they are asked to apply the same six questions as they did to the Rauscher, Shaw, and Ky article. Using the Steele, Bass, and Crook article allows the instructor to demonstrate good research design to students, as this study provides more control than that provided by Rauscher, Shaw, and Ky.

Reading and reviewing this article gives the student an excellent opportunity to contrast a poorly designed study with a well-designed one. This allows the instructor to point out additional aspects of "good research." It also allows the instructor to discuss operational definitions and clearly written methods sections and how these relate to replication (and its importance) in science.

Web Version

This case and its complete teaching notes, references, and answer key can be found on the website of the National Center for Case Study Teaching in Science at *http://sciencecases.lib.buffalo.edu/cs/collection/detail.asp?case_id=230&id=230*.

References

Rauscher, F. H., G. L. Shaw, and K. N. Ky. 1993. Music and spatial task performance. *Nature* 365 (6447): 611.

Shaughnessy, J. B., E. B. Zechmeister, and J. S. Zechmeister. 2003. *Research methods in psychology*. 6th ed. New York: McGraw-Hill.

Steele, K. M., K. E. Bass, and M. D. Crook. 1999. The mystery of the Mozart effect: Failure to replicate. *Psychological Science 10 (4):* 366–369.

Chapter 19
Prayer Study: Science or Not?

Kathy Gallucci

The Case

Prayer Heals, Scientists Report

Heart patients who had someone unknowingly praying for them suffered fewer complications, according to a study conducted by researchers in Kansas City, Missouri.

William Harris, a heart researcher and the lead author of the study, said in 1999 when the results were published that it's "potentially a super- or other-than-natural mechanism," or a "natural explanation we don't understand yet."

The study by Harris and other researchers involved 990 patients admitted during a single year to the Mid America Heart Institute program of St. Luke's Hospital.

Patients, randomly divided into two groups, either had a community volunteer pray for them each day for four weeks or had no one assigned to pray for them.

That strangers were praying for patients in one group was not revealed to the patients, their families, or their caregivers. They were not even told they were participating in the study.

The volunteers were told to pray daily for the speedy recovery with no complications for patients. They were given only the first names of selected patients.

Patients who were prayed for experienced about 10% fewer complications, ranging from chest pains to cardiac arrest, after four weeks, according to the study.

The research was published on Oct. 25, 1999, in the *Archives of Internal Medicine*. The American Medical Association publishes this internal-medicine journal.

Researchers concluded that prayer may be an effective addition to standard medical care. "Although we cannot know why we obtained the results we did, we can comment on what our data do not show," according to the report. "For example, we have not proven that God answers prayer or that God even exists. It was intercessory prayer, not the existence of God, that we tested here."

Chapter 19

Harris admitted in the study that he could not control all variables. For example, at least 50% of the patients admitted to the hospital said they have a religious preference.

"It is probable that many, if not most, patients in both groups were already receiving intercessory and/or direct prayer from friends, family, and clergy during their hospitalization," according to the report. "Thus, there was an unknowable and uncontrollable (but presumed similar) level of 'background' prayer being offered for patients in both groups; whatever impact that (the) group assignment had on healing was over and above any influence background prayer may have had."

But this study and a similar one in 1988 in San Francisco that involved 393 heart patients had questionable methods, according to an expert not involved in either study.

Both studies used their own scoring systems that tallied complications. Dr. Herbert Benson, a professor of medicine at Harvard Medical School, said that neither scoring system had been proven medically valid. Benson was president at the time of the Mind/Body Medical Institute at Beth Israel Deaconess Medical Center in Boston.

In one study, prayed-for patients suffered more than those not prayed for, and others have found no apparent benefits to being prayed for (Harris et al. 1999).

Benson, however, noted that people who believe in God or in prayer typically fare better than those who don't, according to medical research.

But whether prayer itself makes a difference remains unproven, according to Benson.

Questions
Based on the above, evaluate the science in the study by answering the questions below.

1. What is the hypothesis of the researchers?
2. What predictions did the researchers make?
3. What is the independent variable?
4. What is the dependent variable?
5. What are the controlled variables?
6. Describe the control group.
7. What kind of evidence was collected in the study?
8. What is the conclusion of the researchers?
9. What assumptions did the researchers make?
10. Do you think this study is an example of pseudoscience? Explain.

11. Do you think this study is an example of junk science? Explain.

12. Do you think this study is an example of antiscience? Explain.

13. What questions do you have about details of the research that were not reported in the news article? How would the answers to these questions better help you evaluate the study?

14. What objections to this study do you think people of faith would have?

Teaching Notes
Introduction and Background
The scientific study of prayer is not a new area of research for the scientific community. In 1872, an interesting proposal was made in Great Britain to study the effects of prayer by making one ward or hospital the "object of special prayer by the faithful," over and above the "background" prayer offered by the patients themselves and their families. This proposal set off a flurry of controversy in two prestigious publications of the time, *The Spectator* and *The Contemporary Review*. The objections, from both religious and scientific viewpoints, were published in a book-length collection titled *The Prayer-Gauge Debate* (Tyndall et al. 1876).

Interest in studying the effects of intercessory prayer resurfaced after the publication of a landmark study by Randolph Byrd in 1988, which used rigorous scientific methodology.

This case is in the form of a fictitious news story that was developed based on a study conducted and reported in the *Archives of Internal Medicine* (Harris et al. 1999). The case is intended for use in an introductory biology course for nonmajors at the freshman or sophomore level and can be used to evaluate students' understanding of the scientific method. Students learn the basics of the scientific method (hypothesis testing, collection and analysis of evidence, and making valid conclusions) in experimental as well as observational science.

Before the case is presented, students obtain tools for skeptical thinking from Carl Sagan's "Baloney Detection Kit" (Sagan 1996) and learn to apply these tools to examples of science studies from their textbook and the media. They learn to recognize the fallacies of studies when a part of the scientific method is abused (see Figure 19.1, Scientific Method Concept Map, p. 170). These abuses of the scientific method have each been given their own epithets ("pseudoscience," "junk science," and "antiscience").

Figure 19.1. Scientific Method Concept Map

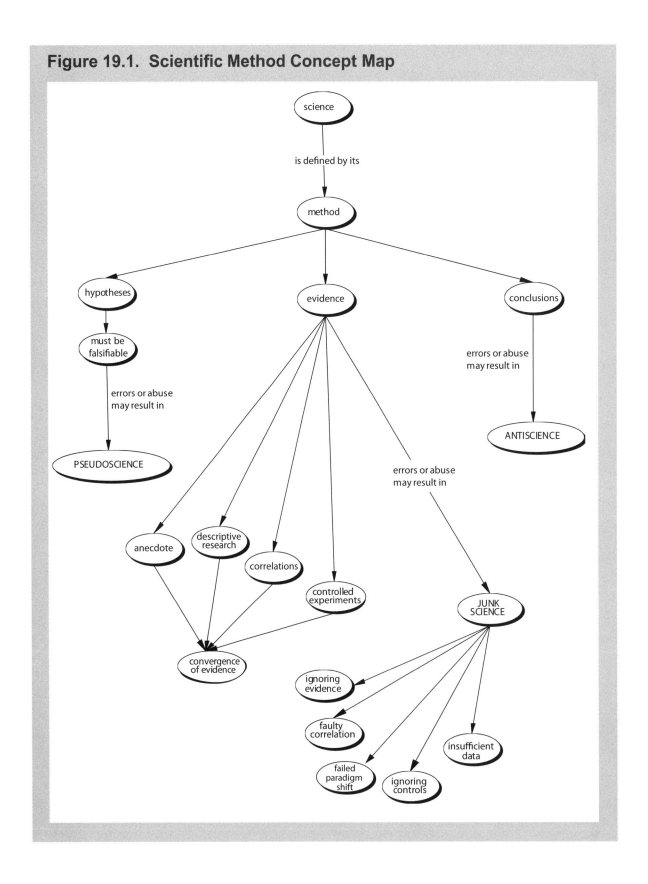

For example, "pseudoscience" proposes hypotheses and makes predictions that are not falsifiable, or cannot be tested. "Junk science," a term coined more recently in the media, uses faulty, insufficient, unreliable, twisted, "dredged," or biased data (see *www.junkscience.com*). "Antiscience" makes conclusions based on cultural norms, popular ideas, or personal biases, rather than objective evidence.

Students also learn to recognize that the popular media typically omit important information necessary to evaluate the validity of a study. If students understand the scientific method, they should be able to ask key questions about the missing information and, when the information is provided, be able to reevaluate the study.

This case may be adapted for courses in the allied health and rehabilitation fields and perhaps for courses in sociology, psychology, and religious studies as well.

Objectives

After completion of the case, the student should be able to do the following:

- Describe why science is distinct from other ways of knowing
- Explain that science is defined by its method and describe the essential parts of the scientific method (hypothesis testing, collecting and analyzing evidence, and making conclusions)
- Identify the independent, dependent, and controlled variables in a study
- Identify the control group and the experimental group in a study
- List the assumptions made by researchers in a study
- Use skeptical thinking to analyze a news story about a study
- Identify missing information that would be essential or helpful in making an objective assessment
- Analyze reports of studies in the news media as "pseudoscience," "junk science," or "antiscience"

Common Student Misconceptions

- Science can study all questions.
- All hypotheses are testable.
- Correlation is the same as causation.
- Close examination of evidence will always lead to the same inferences.
- Science can function independently of social and cultural influences.

Chapter 19

Classroom Management

Before the students evaluate the study for its scientific validity, they must first understand the scientific method. They become familiar with the components of the scientific method in the lab and in class by identifying those components in neutral examples that do not incite emotion. It is important that they learn that the scientific method cannot be applied to every question. The hypothesis must be testable and falsifiable, the method must be objective and controlled, and the conclusions must be made without professional or personal bias.

This case has been used for the lecture portion of a general studies course in biology that fulfills the requirement for a lab science. In an 80-minute class of up to 60 students, I have the students read the news article silently, then allow small groups to discuss the questions or a subset of them. The answers to the questions are then compared in a class discussion. I try to identify weaknesses in their thinking during the class discussion. Throughout the discussion, it is important to promote respect for students' religious faith and the separation of scientific questions from religious ones. I have also used this case study as a homework assignment and as an essay for an hour-long exam.

Web Version

This case and its complete teaching notes, answer key, and references can be found on the website of the National Center for Case Study Teaching in Science at *http://sciencecases.lib.buffalo.edu/cs/collection/detail.asp?case_id=170&id=170*.

References

Byrd, R. C. 1988. Positive therapeutic effects of intercessory prayer in a coronary care unit population. *Southern Medical Journal* 81 (7): 826–829.

Harris, W. S., M. Gowda, J. W. Kolb, C. P. Strychacz, J. L. Vacek, P. G. Jones, A. Forker, J. H. O'Keefe, and B. D. McCallister. 1999. A randomized, controlled trial of the effects of remote, intercessory prayer on outcomes in patients admitted to the coronary care unit. *Archives of Internal Medicine* 159 (19): 2273–2278.

Sagan, C. 1996. *The demon haunted world: Science as a candle in the dark*. New York: Random House.

Tyndall, J., F. Galton et al. 1876. *The Prayer-Gauge Debate*. Boston: Congregational Publishing.

Chapter 20
The Case of the Ivory-Billed Woodpecker

Kathrin Stanger-Hall, Jennifer Merriam, and Ruth Ann Greuling

The Case

Part I: Background

Brad Murky, a graduate student in conservation biology at Cornell University, had been involved for nearly a year in a highly secretive research project taking place in the tupelo swamp of the Cache River National Wildlife Refuge in Arkansas. In the early 20th century, the eastern Arkansas forests were heavily logged to remove most of the large, old trees for timber. Along with the trees went one of the most majestic of all birds found in Arkansas, a bird so impressive that people exclaimed "Lord God" when they saw it fly. Now and again, reports of sightings of the bird were made but dismissed

as rumors. But now, Brad's efforts and those of his mentors had paid off, yielding a huge discovery—a living ivory-billed woodpecker, a bird not documented in North America since 1944. The elusive bird had been captured on video a year earlier, and his team had decided that they now had enough evidence to go public. Brad's elation was unrestrained as he envisioned the history he and his team were making.

Those jubilant feelings, however, had been invaded by a trace of doubt that was increasingly bothering him. Brad stared vacantly at the camera lights for the upcoming press conference at which he and his colleagues would announce their discovery and thought back to the e-mail exchange of the past two days between him and his sister, Mary.

Question

What evidence would convince you that the ivory-billed woodpecker is not extinct?

Chapter 20

Part II: The Main Evidence
Message 1: Brad's First E-mail to Mary
> 4/26/05 12:15 p.m.
> Hey Mary—Things have been crazy here since we decided to go public with our evidence. I wish you could be here—it is so amazing to be part of this huge scientific discovery!!! I am sending you one of our key pieces of evidence, a video clip of "Elvis." This will confirm what we've believed all along—that the "Lord God Bird" is still alive!
>
> The video is pretty short, but it clearly shows that this bird is quite large and has the distinctive white wing patterns of the ivory-billed woodpecker. You can see the extensive white feathers on the trailing edge of the wing as the bird flies, and the white shield on his back after he lands on the tree. Don't you dare tell anybody about this—it is still top secret! Later, Brad.

With hands shaking from excitement, Mary clicked the video link: *www.sciencemag.org/content/vol0/issue2005/images/data/1114103/DC1/1114103S1.mov*.

Assignment
Evaluate the merit of the video as scientific evidence.

Part III: E-mail Exchange
Message 2: Mary's Reply to Brad
> 4/26/05 9 p.m.
> Hi Brad, You've got to be kidding??!! All I can see is a flapping black and white bird!! How can you be so sure it is an ivory-bill? Why not a pileated woodpecker? They have big white patches on the underside of their wings. You guys can't possibly stake your reputation on this!?! I have to go study for my bird class final. Cheers, Mary.

Message 3: Brad's Rebuttal to Mary
> 4/26/05 11 p.m.
> Mary, Come on—it is very hard to get high quality videos in the field. Some of our guys are the top birders in North America and have spent years trying to find an Ivory-bill. And as you can clearly see from their field notes (attached [Figure 20.1]) they saw the white trailing wing patches characteristic of an ivory-billed woodpecker. They know what they are talking about. And OF COURSE the video is not the only evidence. We have more than 17,000 hours of sound recordings that include some "kent" calls typical for ivory-bills. Not to mention that seven people have personally seen an ivory-bill fly by anywhere from 15–150m away! So there! Good luck on your exam, Brad.

The Case of the Ivory-Billed Woodpecker

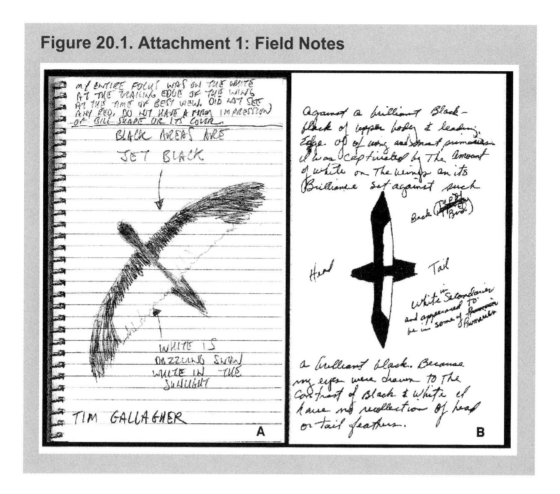

Figure 20.1. Attachment 1: Field Notes

Message 4: Mary's Reply to Brad
4/27/05 2 p.m.
Brad, I sure hope your seven observers have gotten a good look at the bird! Because I am not convinced by your "sound" evidence. Your recording may sound just like an ivory-billed woodpecker, but guess what: blue jays and nuthatches can also make "kent" calls. How do you know it wasn't one of them? I think you should forget about this piece of evidence, until you see an actual ivory-bill making the calls. You guys are crazy! You can't possibly plan a big announcement with that! Mary

Message 5: Brad's Reply to Mary
4/27/05 8 p.m.
Mary, you just don't get it! You have to look at ALL the evidence. We have world-renowned experts on our team: they know their stuff. Sooner or later we will get a perfect picture of the bird. But if we wait too long we may miss our chance of protecting the last surviving ivory-bills! They need huge protected habitats to

stand a chance. We can't wait until we have searched the entire 550,000 acres in the Big Woods area; that would take years! On top of that it gets more difficult to keep this operation secret. We have more and more people snooping around. Stop being such a downer! You should be happy for me. Brad

Message 6: Mary's Final Comment to Brad
4/28/2005 1 a.m.
Brad, are you saying that fame is more important than good science? That playing the odds is more important than the truth? I sure hope you turn out to be right! Not only because I really want to see an ivory-billed woodpecker myself, but because—if you are not—millions of dollars will be spent protecting a phantom bird, millions that will not be available for the protection of other endangered species!! Of course you guys could always say that you saw the last bird before it flew off into the thicket and died, making the species, finally, extinct! But seriously: I really want you to be right AND famous, but if you are wrong, you will be famous too... for all the wrong reasons. Be careful! Mary

As Brad sat waiting for the news conference, he began to feel a sense of dread. He couldn't stop thinking about his exchange with Mary. Suddenly, an excited news reporter approached him. "I got your name from one of the project leaders," she said. "You've been a valued member of the team, and I'd like to get a couple of quotes from you."

Questions

1. What is the major conflict between Brad and Mary in terms of the scientific process? Make a list of Brad's arguments and valid pieces of evidence and Mary's response to each.

2. What do you think about Brad's concern that by waiting with the announcement they could miss their chance to save the birds?

3. Imagine you are the owner of a company that owns the logging rights adjacent to the area of the woodpecker sightings or a biologist trying to protect the habitat of another endangered species in another part of the state. Do you think you would be as satisfied as Brad with the same amount of evidence in this case? Why or why not?

4. What is the right amount of evidence? How can you determine the answer to that question?

5. Give other examples of public discourse, policy decisions, or controversial issues where your insights from this case could be applied.

The Case of the Ivory-Billed Woodpecker

6. Decide how much evidence *you* would need to accept the claim that the ivory-billed woodpecker *is not* extinct.

7. Decide how much evidence *you* would need to accept that the ivory-billed woodpecker *is* extinct.

8. Put yourself in Brad's position—what would you have told the reporter?

9. Does it matter to you who presents the evidence?

10. Who presented the evidence in the real ivory-billed woodpecker case (who was present at the press conference)?

Teaching Notes
Introduction and Background

This case study is fiction. Though based on events that took place in 2004 and 2005, the scenario presented here is the creation of the authors. The main characters in this story and their opinions are fictional, but the presented evidence and the discussion points are authentic and were extracted from published articles in the scientific literature.

The case is based on events in a tupelo swamp of Cache National Wildlife Refuge in Arkansas, which led to a nationally televised press conference on April 28, 2005, and an article published in *Science* on the same day (Fitzpatrick et al. 2005). The focus of this attention was the reported rediscovery of the ivory-billed woodpecker (*Campephilus principalis*). One of the largest woodpeckers that ever lived in North America (the third largest in the world), the ivory-billed woodpecker had not been documented in North America since 1944. In this case study, a hypothetical graduate student who works with the rediscovery team faces a dilemma when he is asked to comment on the exciting news after an exchange with his sister in which he begins to doubt the evidence that the woodpecker still survives.

This case is appropriate for introductory biology classes for both majors and nonmajors. It can also be used in courses on the nature of science; in ornithology, ecology, or environmental studies; in an environmental public policy course; or for teacher training. It works well in both large and small classes. The case is designed as an application of the scientific process. It is best used after the scientific process has been taught in class or after developing the diagram in Figure 20.2 (p. 178) in class together with the students.

Chapter 20

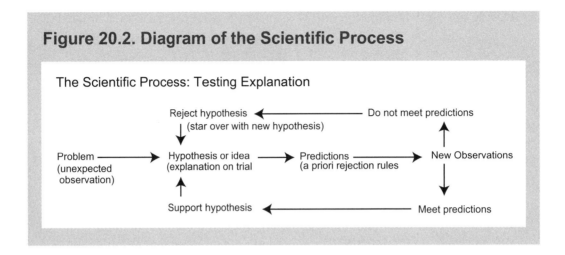

Figure 20.2. Diagram of the Scientific Process

Objectives
Upon completing this case, students should be able to

- apply what they have learned about the scientific process to a real-life situation;
- recognize that science is an ongoing process, not a one-time method to find the final answer;
- explain the importance of the rejection of alternative hypotheses in the scientific process;
- give an example of how science affects decisions in everyday life; and
- be familiar with the story of the ivory-billed woodpecker.

Common Student Misconceptions
- The same amount of evidence is required by all individuals to accept or reject a hypothesis.
- Scientific evidence is always evaluated objectively.
- Personal motivation does not affect how evidence is evaluated.
- All evidence is of equal value for the testing of hypotheses.
- It is equally challenging to document that a species is extinct as it is to document that it is not extinct.

Classroom Management
The case is designed as an interrupted case with three parts of analysis. It can be taught in one 50-minute class period as presented on pages 179–180, or extended over two

class periods if the scientific method is taught as well. Alternatively, an introduction to the scientific method and the case can be presented in one 75-minute class period.

Students work on one part of the case at a time and are asked to answer questions before they receive the next part. Individual parts of the case are handed out to students as the case progresses, or the entire case can be handed out at once with clear directions that students should not look ahead. Another approach is to project individual parts of the case on a screen during class and post the entire case on a class website for students to download after class. Discussion of Parts I and II should be short (less than 10 minutes each) to ensure there is enough time to discuss Part III.

Below, we include suggestions on posing the questions embedded in the case to student groups, but individual instructors should feel free to make adaptations to best serve their individual class sizes and teaching goals. If using student groups, have students form groups of two to four individuals or assign students to groups prior to beginning the case.

Part I: Background (5–7 minutes)
After students read Part I, direct student groups to identify what kind of evidence would convince them that the ivory-billed woodpecker is not extinct. Students should write down their criteria. After a few minutes, ask student groups for their criteria (one per group) and make a list on an overhead.

Part II: The Main Evidence (5–10 minutes)
Give students a minute or so to read through Part II before playing the video (available directly at *www.sciencemag.org/content/vol0/issue2005/images/data/1114103/DC1/1114103S1.mov* as of the time of publication of this book, or find it listed at *www.birds.cornell.edu/ivory/multimedia/videos/index_html/document_view*).

Tell the students that the video was taken from a canoe during a search for the ivory-billed woodpecker. They will see the handle of a canoe paddle and the hand of the videographer's companion. This video is very short and should be played at least twice so students don't miss anything. You may also show the video frame by frame and point out some of the key marks of the ivory-billed woodpecker (e.g., the pattern of the white wing feathers).

The question in this part of the case can be addressed to the entire class. Poll individual students as to their opinion concerning the merit of the video as scientific evidence (i.e., why the video may be good evidence or why it may be weak).

Part III: E-mail Exchange (25–35 minutes)
Students will need about 15 minutes to discuss the scientific method and address the first question in this part. Allow them 3–4 minutes to read the e-mail exchange, then have them work individually to make lists of Brad's arguments and Mary's arguments. Once each group member has a few ideas, have students collaborate with their group members to complete their lists. After 6–7 minutes, bring the class back together and

use an overhead projector to write down answers from each group (make a list with two columns).

Project the scientific process diagram on a screen and take about 5 minutes to discuss the following:

- Where do Brad's arguments fit in the diagram?
- Where do Mary's objections fit in the diagram?
- Why is Mary so intent on testing alternative hypotheses (e.g., pileated woodpecker)?
  * Is supporting a hypothesis the same as proving a hypothesis? Why or why not?
  * What would you have to do to prove a hypothesis?
  * What is the difference between a hypothesis and a scientific theory?

Questions 2–7 will take about 10–20 minutes to address, depending on emphasis and whether the questions are posed to the whole class or discussed in groups. Once the class understands the conflict between Brad and Mary with respect to the scientific process, you can point out that science is rarely done in a vacuum. In many cases, it has real-life implications and consequences. Questions 2–5 bring up several of these issues.

Questions 6–10 are possible follow-up questions and can be used if there is time during class or could be assigned for homework, but they are not essential.

Closure: What Actually Happened?

In this case, the Cornell research team decided to go ahead with the claim that the ivory-billed woodpecker had been rediscovered and was not extinct. They announced it with the secretary of the interior, Gale Norton, in a national press conference on April 28, 2005. The federal government (the Department of the Interior and the Department of Agriculture) pledged more than $10 million to buy land and protect the habitat of the ivory-billed woodpecker. Most of these funds were not new appropriations but were re-allocated from other efforts to protect other endangered species.

The interpretation of the evidence by the Cornell research team has since been challenged by other researchers.

In September 2006, another group of researchers announced ivory-billed woodpecker sightings in a swamp in northwest Florida, but they could not provide any direct physical evidence. They have asked for money and habitat protection (see *www.auburn.edu/academic/science_math/cosam/departments/biology/faculty/webpages/hill/ivorybill/* and *http://news.nationalgeographic.com/news/2006/09/060926-woodpecker.html*).

Web Version

This case and its complete teaching notes, references, and answer key can be found on the website of the National Center for Case Study Teaching in Science at *http://sciencecases.lib.buffalo.edu/cs/collection/detail.asp?case_id=609&id=609.*

References

Fitzpatrick, J. W., M. Lammertink, M. D. Luneau Jr., T. W. Gallagher, B. R. Harrison, G. M. Sperling, K. V. Rosenberg, R. W. Rohrbaugh, E. C. H. Swarthout, P. H. Wrege, S. Barker Swarthout, M. S. Dantzker, R. A. Charif, T. R. Barksdale, J. V. Remsen Jr., S. D. Simon, and D. Zollner. 2005. Ivory-billed woodpecker *(Campephilus principalis)* persists in continental North America. *Science* 308 (5727): 1460–1462.

Section V:
Science and Society

The case studies in this section focus on the various roles that society plays in the scientific enterprise. Science is not done in a vacuum. Even a legendary lone investigator may have millions of people to thank for making his or her discoveries possible. Teachers, parents, neighbors, friends, and spouses all have been there to nurture and care for our intrepid discoverer before his or her eureka moment. Nor can our genius claim credit for the tools of the trade. Newton did not discover mathematics. Einstein did not invent theoretical physics. Nor did Darwin create biology.

Matt Ridley, writing in the *Wall Street Journal*, explains:

[T]he sophistication of the modern world lies not in individual intelligence or imagination. It is a collective enterprise. Nobody—literally nobody—knows how to make the pencil on my desk (as the economist Leonard Read once pointed out), let alone the computer on which I am writing. The knowledge of how to design, mine, fell, extract, synthesize, combine, manufacture and market these things is fragmented among thousands, sometimes millions of heads. Once human progress started, it was no longer limited by the size of human brains. Intelligence became collective and cumulative.

In the modern world, innovation is a collective enterprise that relies on exchange. As Brian Arthur argues in his book, *The Nature of Technology*, nearly all technologies are combinations of other technologies and new ideas come from swapping things and thoughts. (My favorite example is the camera pill—invented after a conversation between a gastroenterologist and a guided missile designer.) We tend to forget that trade and urbanization are the grand stimuli to invention, far more important than governments, money or individual genius. It is no coincidence that trade-obsessed cities—Tyre, Athens, Alexandria, Baghdad,

Section V

Pisa, Amsterdam, London, Hong Kong, New York, Tokyo, San Francisco—are the places where invention and discovery happened. Think of them as well-endowed collective brains. (*Wall Street Journal*, May 22, 2010)

Any discovery, if it is to be known at all, must find an audience within the scientific community and in the world at large. Scientists consult with one another. They write papers, give talks, apply for grants, and publish their findings. Flaws in data and faulty logic are exposed—not necessarily immediately, but eventually, and reputations are made and destroyed in the process. Promotions and tenure decisions come and go. All of this is part of science. And the public at large reaps the benefits as well as the fallout of any discovery. When the wheel, the printing press, the internal combustion engine, and computers were developed, the world was changed forever.

In the first case study in this section, "Moon to Mars: To Boldly Go ... or Not," we begin with a subject that has been debated off and on for decades: Should the United States spend the necessary resources to go to Mars? If the answer is yes, then how should we do it? The public's interest in such questions sweeps across the spectrum, ranging from the financial implications to the fascinating possibility that life might be found there. The case is interesting too because it is presented in the form of a public hearing, with the characters in the story presenting arguments, both pro and con.

In the second case, "And Now What, Ms. Ranger? The Search for the Intelligent Designer," we encounter another controversial topic, one based on real-world litigation: Should the theory of intelligent design be taught in a science classroom? The narrative is presented as a discussion by a school board debating the matter. From a legal perspective, the verdict is in: Intelligent design conflicts with the U.S Constitution by promoting certain religious doctrines in that it posits the "designer" as a supernatural being (i.e., a god). From a scientific perspective, the verdict is in: The hypothesis of intelligent design has long been dismissed because it cannot be tested (and thus falls into the category of pseudoscience). Moreover, the process of evolution is understood enough so that we do not have to evoke a supernatural explanation to explain the diversity of life on the planet. Still, a social debate simmers: Evolution is in conflict with some religious faiths. The case allows a skillful teacher to delve deeply into the results of evolution and the surrounding social turmoil. Nonetheless, caution is the watchword here—local communities have a large stake in this issue.

The case of "The Case of the Tainted Taco Shells" deals with the question of genetically modified corn. This is an issue of great importance. Our technology has advanced so far that we can use genetic engineering to modify not only food but also animals and humans. How science can accomplish these technological feats is one aspect of this case, but the question of whether we should be doing them at all is a question of profound importance. The teaching method used in this case is notable because the students in the class are divided into different interest groups. As part

Science and Society

of the assignment, they must read different primary sources to prepare for a general discussion. In addition, the case author combines the case with relevant lab work.

Like it or not, students are interested in drugs. Some people see marijuana as a harmless recreational drug; others believe it to be a gateway substance leading to more serious addictions. In the past several years, more than 20 states have passed laws permitting the sale, growth, and use of marijuana for medical reasons. Accumulated data show marijuana to be effective in treating pain and some illnesses. This has opened the door to new abuses and has promoted a new cottage industry in the sale and distribution of the drug, especially in California. The case study "Medicinal Use of Marijuana" explores this topic using a teaching method called intimate debate. Pairs of students square off, first arguing one side of the question and then the other, before finally (we hope) reaching a consensus. The ability to analyze an argument effectively is a hallmark of critical thinking, and that skill is developed here in its clearest form, dealing with one of the current problems facing society.

The Food and Drug Administration (FDA) is charged with guarding the health of our nation. Promising drugs must pass stringent tests to be judged safe for the American public. The prescription drug Vioxx successfully passed through that process and was touted as an effective palliative for pain associated with joint-related afflictions, including arthritis. Unfortunately, after several years of use, data began to accumulate suggesting Vioxx carried with it a significant risk of cardiovascular complications. The manufacturer, Merck, withdrew the drug from the marketplace despite its proven effectiveness in treating pain. In the case study "Amanda's Absence: Should Vioxx Be Kept Off the Market?" students must sift through information about the risks and benefits of the drug, weigh the consequences of returning it to the market, and deliberate over whether Merck should be fined for its role in the inadequate product testing.

The final case in this section, "Sex and Vaccination," involves the human papillomavirus (HPV), a causative agent in uterine cancer. In 2007, by executive order, Texas governor Rick Perry directed that all young girls attending public schools be vaccinated for this virus. A political firestorm ensued, with a rapid retraction of the mandate. What is noteworthy about the case is the strong role the public played in overturning Perry's directive. The case also touches on the role government plays in the lives of its citizens, explores the economic concerns at issue, and delves into the complex question of vaccinating people against their will. This case exemplifies the complex role of science as it interacts with society, the topic of this entire section. Discoveries are not made in a vacuum; they can have profound economic and political consequences.

Chapter 21
Moon to Mars: To Boldly Go ... or Not

Erik Zavrel

The Case

Part I: Prelude to Space

The following case study presents a transcript of a fictionalized conference held at the Jet Propulsion Laboratory in Pasadena, California, in 2007 and attended by NASA officials, space scientists, aerospace engineers, space enthusiasts, ardent NASA supporters, and strident NASA critics. The panel of experts consists of the following fictional cast: James

Everett, administrator of NASA; Susan Bowman, the founder of the Ares Society, a space advocacy group with the goal of sending astronauts to Mars; and Richard Greene, executive director of the Space Exploration Society, a space advocacy group interested in exploring the entire solar system. Apart from NASA, any resemblance to real persons or societies with these names is strictly coincidental.

The topic of the conference was then-president George W. Bush's Vision for Space Exploration (VSE), commonly referred to as the Moon-to-Mars Plan. Those in attendance were able to listen to experts debate whether the VSE was what NASA should be committing itself to and also voice their own opinions. The Vision for Space Exploration, announced by Bush in January 2004, outlined an ambitious plan to return to the Moon before voyaging to Mars. Moreover, the Space Shuttle was to be phased out upon completion of America's obligations in the construction of the International Space Station around 2010 (Sietzen 2004). The Space Shuttle was to be replaced by the Orion Crew Exploration Vehicle, an enlarged, modernized Apollo capsule that would have used proven technology developed four decades earlier. The plan called for NASA to return humans to the Moon by 2017 at the earliest and 2020 at the latest (Sietzen 2004)—a full half century after the first landing on the Moon in 1969.

Chapter 21

The Bush administration did not allocate any additional funding for NASA to meet these goals, yet the NASA administrator at the time, James Everett, was adamant that NASA could make VSE a reality on its existing annual budget of about $17 billion. Everett earned the ire of many space scientists who found their pet projects either indefinitely delayed or canceled outright as money was reallocated to meeting the VSE timetable (Stover 2004).

Questions

1. Should Americans return to the Moon? If so, why? If not, why not? What reasons could justify the great expenditure of funds, time, and national will?

2. Is this return long overdue, or should NASA focus on other goals? Should America skip the Moon and head to Mars? Why might it make sense to return to the Moon before voyaging to Mars? Consider such factors as the length of the journey and hazards involved.

3. Is the cost of human space flight justifiable with the numerous problems confronting the nation today, or could robotic exploration return comparable results at a fraction of the cost?

4. What factors were involved in President John F. Kennedy's mandate of "landing a man on the Moon and returning him safely to the Earth"? Are any of these factors present at the current time?

5. Do you think there is wide public support for a return to the Moon, or is the American public indifferent or even hostile to the idea?

Part II: Public Hearing

Note: The members on the expert panel portrayed below are fictitious and the transcript contrived; however, the views expressed in the audience comments and the technical responses correspond to actual views held by leading space advocacy groups, such as the Mars Society and the Planetary Society, as well as NASA. The fictionalized dialogue should not be attributed to any representative of these organizations.

Moderator: Now that we have outlined the basics of President Bush's Vision for Space Exploration, we will open up the floor to members of the audience who may put forth a comment, opinion, or question. Members of the panel will be given a chance to respond, as will others in the audience; there will be ample time for rebuttals, so please refrain from interrupting the speakers.

Armchair Space Enthusiast: I'm not a scientist or engineer, but I have closely followed the space program since its inception half a century ago. The president's plan is exactly what NASA has been in desperate need of for the past 35 years—a mission. NASA has been without direction or a mandate for far too long. John F. Kennedy challenged NASA to land a man on the Moon and return him safely to the Earth by 1970. NASA rose to the challenge and got there half a year ahead of time. NASA has not had a major goal since Eugene Cernan and Harrison Schmitt left in 1972. Yes, there were some impressive feats such as *Skylab, Pioneer, Voyager,* and the Martian rovers, but no major goal to strive for. The Shuttle never lived up to its promise of cheap, reliable access to low-Earth orbit, as evidenced by the tragic disintegration of *Challenger* in 1986 and *Columbia* in 2003, as well as the half-a-billion-dollar price tag per launch. Returning to the Moon will help channel NASA's attention in the coming years and will culminate in an achievement that will captivate the public in a way that has not been done in nearly four decades.

Aerospace Engineer: I've been employed here at JPL for nearly 40 years; I worked on the *Voyager* mission in the '70s and '80s and the Martian rovers from the '90s up to the present time. These robotic emissaries returned vast amounts of knowledge. Robotic probes cost a fraction of comparable manned missions and don't jeopardize human life; no special consideration has to be paid to radiation shielding or artificial gravity. How can you possibly justify the financial cost and the imperilment of human life when we have the ability to construct competent robotic probes to go in our stead?

James Everett (NASA administrator): We are not trying to supplant robotic craft; it will play an important role in the Moon-to-Mars Plan. However, astronauts offer versatility and flexibility; they won't get stuck on a rock or have a power connection fail. They also have the potential to conduct novel experiments beyond automated analysis. As just one example, David Scott, commander of *Apollo 15,* in a simple yet powerful demonstration, dropped a hammer and feather in front of a camera while on the Moon's surface, illustrating Galileo's discovery that objects fall at the same rate regardless of their mass. Also, on many occasions during Apollo missions, humans proved their worth, while the onboard AI proved to be the weak link. During the descent of the *Apollo 11* LEM, Neil Armstrong took manual control of lander guidance, overriding the descent computer, when he noticed that the computer was guiding the lander down into a patch strewn with boulders. Also, this time we are returning to the Moon for good; we fully intend to stay. The astronauts will construct a permanent lunar base that will be manned on a continuous basis, with one batch of astronauts being relieved every few months.

Astronomer: Hello. I am an astronomer and have worked as part of Project Spaceguard for several years, helping to discover and track near-Earth asteroids that might at some time cross paths with Earth. Many of my colleagues and I are concerned with the precarious state humans find themselves in. We have placed all our eggs in one basket, so to speak. We must establish an off-world colony as insurance. If something untoward were to happen to Earth, at least humanity would survive in some form. Space exploration should thus involve humans, as its most critical concern is the survival of the human race. A lunar colony would be the first step in that direction.

Chapter 21

Female College Student: Hello. I attend UCLA and have worked with international relief agencies the past two summers. I can't see how you can justify such a waste of money when we have so many pressing problems right here on Earth: the war on terrorism, the national debt, a failing public school system, AIDS, and global warming. All these things deserve higher priority than planting footprints and a flag on some dusty alien world.

James Everett: If Christopher Columbus had waited for every social problem of his day to be remedied, the timbers of the *Santa Maria* would be rotting in a Spanish harbor to this very day. In the 1960s, America was involved in a Cold War, entangled in a war in the jungles of Southeast Asia, and facing massive social unrest at home, yet still managed to pull off Apollo. America must meet its goals in a parallel manner—not serially, not sequentially.

Retired Person: A generation of baby boomers is set to retire in just a few years. They will place an unprecedented burden on Social Security and Medicare. Many experts claim these social service programs will go bankrupt. Millions will need money to offset medical bills, prescription drugs, home care, and heating bills. I agree with the last audience member. It is unjustifiable at the present time to spend billions on the Moon. Why does it have to be now? I mean, it's been 35 years since we last landed on the Moon. Why can't we wait a bit longer?

James Everett: There are several reasons why sooner is better. In the early 15th century, China was arguably the world's greatest power. The Ming Dynasty sent grand fleets with tens of thousands of mariners as far as eastern Africa. But the emperor decided to recall the fleet and isolate China. As a result, China grew withdrawn and introverted, and the world scene became dominated by Portugal, Spain, and England.

Heckler #1: Any other reason than an obscure Chinese emperor?

James Everett: Yes. If you had not interrupted me, I would have gotten to it. The American historian Frederick Jackson Turner, in his book *The Frontier in American History*, argued that the presence of the frontier, a region where independence and self-reliance and inventiveness were fostered and nurtured, was instrumental to maintaining the vitality of American democracy.

Richard Greene: Also, there is an innate drive within humans to explore, to go where no one has ever been before, whether that be the jungles of central Africa, the Mariana Trench, or the surface of the Moon. It is what compelled Edmund Hillary to scale Mount Everest—simply because it was there. This drive is an integral part of what makes us human.

Female College Student: But what about all the social ailments?

James Everett: Man does not live by bread alone. Where there is no vision, the people perish. America needs an inspiring goal. Simply meeting basic requirements is not enough. Returning to the Moon will encourage more American students to study science and engineering, for one thing.

Male College Student: Hi. I'm a student at UCLA and was wondering why it is deemed necessary to return to the Moon before going to Mars. The Moon was the destination of my parents' generation. Why not skip it and go directly to Mars?

Susan Bowman: As founder of the Ares Society, an organization of dedicated amateurs and professionals with the collective goal of sending a manned mission to Mars in the immediate future, I completely agree with the young man. We have already been to the Moon. Six lunar modules descended to its surface, and 12 men walked around, collected rocks, set up experiments, and even played a round of golf. The Moon has been done. Mars should be our objective, not the Moon.

James Everett: I must disagree. As the president outlined, America will be going to Mars, but only after returning to the Moon. There are several reasons for this sequence. The Moon is the ideal testing ground for the equipment that astronauts will use on the Red Planet. Divers test their scuba tanks in a swimming pool before descending to great depths. The Moon is only three days away; if anything goes wrong, there is a good chance we could send a rescue mission. But Mars is nearly a year out; there will be no chance of earthly assistance.

Susan Bowman: But we don't need to use the Moon as a testing bed. The Ares Society runs several Martian analog camps around the world. Equipment can be tested in Chile's Atacama Desert, the Utah desert, or Devon Island in Canada. All these environments are startlingly similar to Mars.

Heckler #2: So what good is the Moon? A source for more Moon rocks?

James Everett: Aside from its use as a testing bed, the Moon has many other appealing features. The Moon offers a unique window from which to observe the cosmos. It is geologically dead; there are no "moonquakes." This offers the ability to achieve something called optical spectrum long-baseline interferometry, which involves linking many telescopes together so that the effective telescope is equal to the distance between the telescopes. With such a telescope array, astronomers could see farther into the cosmos than ever before. Radio telescopes on the lunar far side would be isolated from the radio noise emanating from Earth. This would block out annoying bogies that plague the SETI [Search for Extraterrestrial Intelligence] program. No turbulent atmosphere means access to the full EM [electromagnetic] spectrum; Earth's atmosphere blocks the infrared, microwave, x-ray, and gamma ray portions of the spectrum. Deep craters at the poles may serve as natural cold-traps to establish IR [infrared] telescopes. The low lunar gravity means very large telescopes can be constructed without the optics sagging under their own weight.

Businessman: Hello. I am a small-business owner and have always prided myself on my ability to deliver a desired product or service to the customer at a reasonable price. Aside from its scientific returns, which can only be appreciated by a tiny minority of people in the rarefied heights of academia, what practical, tangible returns can the public expect? After all, it is the public who are footing the bill, and they should be able to expect some return on their investment.

Chapter 21

James Everett: The space program has always been one of America's wisest investments. The space program has spawned entire industries and innumerable technology spin-offs.

Heckler #3: Like Tang and Teflon! How could we live without those breakthroughs?

James Everett: Actually, those two products existed well before the Apollo program. But how about communication satellites? How about XM radio, GPS, and weather and climate satellites to assist weather forecasting and crop monitoring? How about the miniaturization of the computer? There are personal computers today because the onboard navigation computer of the lunar module had to be shrunk down. Fuel cells were developed from a chemical novelty into a practical technology to provide the astronauts with potable water and electricity. Were it not for the space program, and specifically the Apollo program, the contemporary world would be very different indeed.

Businessman: But does the Moon offer anything directly worthwhile to the public?

Richard Greene: As a matter of fact, it does. There exists on the Moon a rare isotope of helium called helium-3, which arrives via the solar wind—the stream of ionized particles traveling at hundreds of kilometers per second emitted from the Sun. However, helium-3 particles carried by the solar wind cannot penetrate the Earth's magnetosphere. Consequently, they circle the Earth until striking the Moon (which has no blocking magnetosphere) and embed in the lunar regolith—the finely granulated soil pulverized by billions of years of asteroid and meteoroid bombardment. This isotope is the ideal fusion fuel because all of the products of the fusion of helium-3 are easily contained and nonradioactive. There is enough He-3 available on the Moon to provide for the entire energy needs of humanity at current consumption rates for 10,000 years. If the Space Shuttle's cargo bay were loaded with He-3 (25 tons), it could power the entire United States for a year.

Heckler #4: And all we have to do to make use of it is develop nuclear fusion!

Astrobiologist at JPL: The Moon-to-Mars program is a total waste. To fund it, NASA has robbed worthwhile space science programs like the *Terrestrial Planet Finder*, an array of space-bound telescopes designed to seek out Earth-like planets in orbit around other stars. It has also led to the cancellation of the *Jupiter Icy Moons Orbiter*, a craft that would have probed the Galilean moons for signs of subsurface water, a potential indicator of alien life. These are projects that my colleagues and I have been involved in planning for years. I can't adequately convey how devastating it is to suddenly have your life's ambition terminated on a political whim.

Susan Bowman: I agree completely. From a biological point of view, the Moon holds no interest. It is a dead world. On the other hand, Mars, Europa, an icy satellite of Jupiter, and Enceladus, an icy satellite of Saturn, offer tantalizing hints of past or present existence of water. And where there is water, there may be life. The search for life, no matter how basic, is much more compelling than establishing a lunar base. If we discovered life elsewhere in our solar system, we could determine whether it was based on DNA or whether there was a Second Genesis. If the latter were the case, and life arose independently in two different locations within one stellar system, we could feel comfortable in assuming life must abound throughout the cosmos.

Richard Greene: The Space Exploration Society has launched a petition-based campaign to keep NASA from pilfering its space science budget to pay for the Moon-to-Mars program. We still support the president's Vision for Space Exploration, but not at the expense of space science. So far, we have been successful in getting Congress to provide some extra funding to certain space science missions that otherwise would have ended up on the chopping block.

Moderator: I'm afraid we are out of time. Please give our panel of experts a round of applause for sharing their views on the Vision for Space Exploration. And thank you for coming out.

Questions

1. After reading the above, do you think America should be returning to the Moon? Are there more worthy destinations? Are the potential returns worth the investment?

2. Is the human space program unnecessary? Can robotic probes accomplish just as much? What are the advantages and drawbacks of human spaceflight?

3. What factors could set back or altogether prevent a return to the Moon?

Teaching Notes

Introduction and Background
This fictional public hearing case study is based on the decision by then-president George W. Bush to set NASA's primary goal as a return to the Moon, followed by a mission to Mars.

Objectives
Upon completing the case, students should

- recognize that governmental agencies (as well as scientific societies) are not monolithic entities with each worker solidly in line with the agency's proclaimed mission statement, but rather an assemblage of people with very different interests and views;

- realize that many citizens expect scientific programs funded by public tax dollars to return material benefits;

- be able to draw historical parallels to present situations and discern whether historical allusions and precedents cited by others are accurate or misleading;

Chapter 21

- have an appreciation for carefully analyzing proposed large-scale science projects and programs for nonscientific motives; and
- learn the reasons for and against a space mission to the Moon.

Common Student Misconceptions

- Large-scale science undertakings and projects are always initiated for purely scientific reasons.
- Technology is infallible.
- Understanding history and precedent is not important in determining a scientific agency's objectives, goals, and overall mission.

Classroom Management

This is a trigger case that is designed to raise issues rather than focus on factual information. For this reason, the case is not overly technical and is appropriate for use in an introductory engineering course, the purpose of which is to expose students to a variety of problems that engineers face. It similarly can be used in a planetary geology or astronomy course, an honors seminar, a public policy class, or an economics class. In such classes, additional reading and further follow-up assignments might be given to emphasize the particular concerns of each course.

There are two parts to the case: a prelude and the public hearing. Each is followed by a series of questions. For both parts, I recommend that students work in teams of four to five students.

Part I and its questions should be handed out to the students prior to the class in which the public hearing is held so that they may do some library research. The students can be asked to bring written responses or just their research notes to the next class. When class meets, the students should gather in their groups to share their answers to the questions in Part I with their teammates. Provide at least 15 minutes for this exchange. At the end of 15 minutes, the instructor asks the groups to share their thoughts with the class before assigning roles for the public forum. As the students answer each question, the instructor can make notes on the blackboard or overhead as he or she probes the logic and justification for the answer, eliciting contrary views where appropriate.

Part II (the public hearing portion of the case) features 16 different characters. The script should be given to all students and individual roles read aloud by members of the class. (If the number of roles exceeds the number of available students, then multiple roles must be assigned.) Following this reading, the student groups should reconvene and answer the questions that follow the script. After the students have finished, the instructor should call on a member of each group to share their group's thoughts with the entire class. The instructor again should provide guidance and emphasis where needed.

This case is versatile and lends itself to modification to suit an instructor's needs. As described above, the case could be completed within a single class session. However, the instructor could extend the case by having the student groups research in greater detail the various views outlined in the case, summarize the salient points, and give a presentation to the class in an effort to convince the other students of the merits of their viewpoint.

Web Version

This case and its complete teaching notes, references, and answer key can be found on the website of the National Center for Case Study Teaching in Science at *http://sciencecases.lib.buffalo.edu/cs/collection/detail.asp?case_id=175&id=175*.

References

Sietzen, F. 2004. A new vision for space. *Astronomy* 32 (5): 48–51.
Stover, D. 2004. Are we really going to the Moon again? *Popular Science* 264 (4): 60–67.

Chapter 22
And Now What, Ms. Ranger? The Search for the Intelligent Designer

Clyde Freeman Herreid

The Case

"Ms. Ranger, please see me immediately." Janet Ranger, a high school biology teacher, fingered the note apprehensively, reading it again. What was this all about? Which one of her freshmen was in trouble? Or was it the school board on a tear again? Moments later, the petite brunette was striding hurriedly to the principal's office.

"Good morning, Janet. Have a seat; this will only take a moment. You'll be back in time to meet your first class." Gerald Talley leaned back in his chair and motioned the teacher to be seated. He could see that she was tense.

"Yesterday, several parents called me up to complain about your science class. They said that you told the students that they came from apes. Is that true? Give me a little background on this. As you know, this is a conservative state and the sentiments on this issue are running high, especially now that the school board is pressing to have intelligent design taught."

Janet sighed and then patiently outlined her evolution lesson. Talley nodded appreciatively as she concluded. "That sounds reasonable to me, Janet. But we do have a major problem on our hands. Robert Bagley, the president of the school board, is dead set on introducing intelligent design into the biology classes. And there isn't much I can do about it, although I have certainly tried. I appeared in front of the board a couple of times with no success. It looks like Bagley has the votes to pull this off. He is insisting that we use the creationist book *Of Pandas and People* for your students."

"But that is outrageous! We already have picked out our biology book for next year. It deals fairly with the issue of evolution. It is a terrific text. All of the biology

Chapter 22

teachers agree that it is up-to-date on the paleontological evidence on missing links, showing how whales evolved from land ancestors. It has the most recent data on human fossils and the DNA results on Neanderthals showing that they were a different species from us and ..."

"Look, I'm sure you are right, Janet. But this isn't about science as much as it is about politics and religion."

"Gerry, come on. We have been through this ID issue before. The creationists have lost every court case in the country, in Arkansas and Louisiana and even the Supreme Court. It is against the Constitution to inject religion into the science classroom."

"I know. I know. But the school board thinks they have a way around this. Tomorrow night they are having a public hearing on their proposal and will then take a vote. Frankly, it looks like window dressing to me; they have already made up their minds. The reason that I called you is to let you know that that meeting will be the last time you and the other teachers will have a chance to influence the vote. So make your plans accordingly."

* * * * *

The auditorium was packed. People were standing. Here and there Janet could see her students and their parents. There were lots of angry faces. The nine members of the school board, with stony looks, sat on the stage behind a long table, much like members of a jury. In a way they were, she thought. Subdued conversations sputtered on and off throughout the audience, but they ceased immediately when Robert Bagley pounded the gavel to call the meeting to order.

"Ladies and gentlemen, it is good to see all of you here tonight. We will dispense with the usual business to focus on a crucial question that has come before us—the teaching of evolution to our biology students.

"As most of you know, ninth-grade biology is a required class by the state. The curriculum is essentially set. But there are many families in our community that have serious questions about the topic of evolution. Evolution is only a theory and a controversial one at that. All scientists agree that there are many gaps in our knowledge, such as how did life originate, or what is the origin of human consciousness and morality and the soul. Evolution answers none of these things.

"Because of the inadequacies of Darwin's theory and because there are alternative theories, the school board has decided that a balanced approach is best. We believe we have a solution to the problem. We propose that all ninth-grade biology teachers read an announcement to their classes next month. Our secretary, Mrs. Katherine Simler, will read the announcement."

The Pennsylvania Academic Standards require students to learn about Darwin's theory of evolution and eventually take a standardized test, of which evolution is a part.

And Now What, Ms. Ranger? The Search for the Intelligent Designer

Because Darwin's theory is a theory, it continues to be tested as new evidence is discovered. The theory is not a fact. Gaps in the theory exist for which there is no evidence. A theory is defined as a well-tested explanation that unifies a broad range of observations.

Intelligent design is an explanation of the origin of life that differs from Darwin's view. The reference book Of Pandas and People *is available for students who might be interested in gaining an understanding of what intelligent design actually involves.*

With respect to any theory, students are encouraged to keep an open mind. The school leaves the discussion of the origins of life to individual students and their families. As a standards-driven district, our class instruction focuses on preparing students to achieve proficiency on standards-based assessments.

Bagley continued, "Now the floor is open for discussion. Please keep your comments short, to the point, and by all means, be civil. Mr. Curtis, president of our PTA, has asked to speak first."

"Good evening. Thank you for this opportunity, Mr. Bagley and members of the board. I am Tom Curtis, and I represent a large number of our parents who have concerns about this issue. Seems to us only fair that all sides of any question be considered. If there are differences of opinion, let them be aired. Our children have the right to know when a serious controversy exists about a subject. The scientific process demands that any weaknesses in a theory should be exposed.

"The idea of intelligent design has been around a long time. Thomas Aquinas talked about it in the 13th century. It should be plain to everyone that the world cannot be due to chance. There must have been a designer. As I understand it, that is what the theory of intelligent design is all about. If the book *Of Pandas and People* points this out, what is the harm in that? Let's have the debate in the classroom and let the chips fall where they may. Let the children decide. It is only fair. Surely, everyone must agree that this debate is good for developing our children's critical-thinking skills. I thank you."

The auditorium erupted with strong applause as Curtis sat down. Scattered voices were heard calling out "Amen," and a tall man in the center of the audience yelled, "But who is the designer?"

A voice answered, "God, of course."

Bagley called the room to order and remarked, sternly, "Ladies and gentlemen, please respect one another. I believe that Reverend Daly of our board has a comment on this point."

"Yes, I do, Bob. Our attorney has advised us that this is not an issue of God. This is an issue of free speech and the right to teach our children about a current controversy

Chapter 22

in science. We don't know who or what the designer is. And we may never know, but the evidence is clear, there was a designer. He may be a time-traveling cell biologist, he may be an extraterrestrial, he may be a space alien, as Nobel Prize winner Francis Crick has suggested. He may be something that some of us might call a God, but it is important that we not get tangled up with the First Amendment. So, let us agree to set the identity of the designer aside. Thank you."

Among the smattering of applause, once again a voice called out, "Who designed the designer?"

Bagley ignored the disturbance and said, "I think it would be helpful at this point to have Ms. Tilley Wilford outline some of the problems with Darwin's theory so that we can all appreciate the difficulties. Ms. Wilford."

"Thank you, Mr. Bagley. I don't claim to understand all of this, but as a layman, I have read about a lot of problems. First: If evolution did occur over millions of years there must be billions of fossils out there. There ought to be a lot of them that show the changes between organisms. You know, missing links. But where are they? Do we have fossils that show how a mouse became a bat, or a dinosaur became a bird, or a chimpanzee became a person? No. Where are the fossils?

"Second, the biochemist Michael Behe has written in his book *Darwin's Black Box* that many of our bodily systems are irreducibly complex. That means that things like the immune system or blood clotting can't function if any piece is missing. If that is true, then they can't have slowly evolved one step at a time like Darwin believed, can they? They had to come into existence all at once or they wouldn't work. My son, Jimmy, is a good case. He has hemophilia. You know, bleeder's disease. His blood doesn't clot right. That is because he is missing one out of a whole bunch of chemicals so his blood doesn't clot. He would be dead if it were not for the fact of modern medicine. Just one piece is missing and the whole system fails."

Sounds of sympathy swept through the listeners. Ms. Wilford concluded, "Biologists simply don't have the answers to a lot of fundamental questions. We are all made of cells, but where did the first cell come from? It seems like a miracle, doesn't it?" She sat down.

"Now the teacher's union representative, Mr. Juan Martinez, has asked to be heard. Mr. Martinez, it is your turn. You can use one of the microphones in the aisles."

A man rose from the middle of the audience. "I can talk from here, Mr. President. Ladies and gentlemen, the union objects to these proceedings. We object because the board has tried to coerce the teachers into using the creationists' book *Of Pandas and People* instead of the biology book they wanted. In fact, Mr. Bagley explicitly said that he would block the use of the biology book unless *Pandas* could be used as supplemental material. Also, it is clear that the designer that the board is talking about is the fundamentalist Christian's version of God. As a Catholic, I object to this on a personal level, and as an American, I object because this is a violation of the

And Now What, Ms. Ranger? The Search for the Intelligent Designer

First Amendment to the U.S. Constitution. There must be a separation of church and state. Even though the board has forgotten it, the teachers haven't."

Mr. Martinez smacked his fist forcefully against his palm. "And last, we object to being forced to read an announcement to our classes that we do not believe in. This is an abridgement of our rights as teachers and a gross violation of academic freedom. If we are threatened with reprisals, we will seriously have to consider a strike action."

Amidst a few boos, a contingent of teachers burst into applause. A red-faced Bagley narrowed his eyes and angrily replied, "Martinez, remember that it is against the law for you to strike. The board will not be bullied by your grandstanding. Now I believe the teachers themselves would like to speak on their own behalf. Ms. Ranger."

"Thank you, Mr. Bagley. We know this is a difficult topic for you and the board and for many parents. This is not about whether or not there is a God. Many evolutionist biologists believe in God. In fact, Dr. Kenneth Miller, the author of the biology textbook we want to use, is a Catholic. Even the pope has accepted that evolution has occurred. We teachers object to the board's proposal because it is our obligation to teach the most honest version of the subject of biology. The theory of evolution has been accepted for 150 years, while the theory of intelligent design has been rejected. This is not because we are atheists. We aren't. It is because we are scientists and teachers.

"Look, you know this yourselves: Not all explanations of the world are equal in worth. If someone wanted us to teach that the Earth is the center of the universe, we could not do that. If someone wanted us to teach that thunder and lightning are caused by the gods battling in the sky, we couldn't do that. We can't teach intelligent design either, and for the same reason. It isn't what scientists believe. Intelligent design isn't science, Mr. Bagley, it is religion."

Someone in the audience called out, "Why isn't it science?"

Bagley looked sharply at the speaker, "No outbursts, please."

Ms. Ranger continued. "I'm glad you asked that. It is pretty simple, actually. Science tries to understand the world in terms of natural laws. We don't try to explain thunder by talking about poltergeists or sprits or gods in the sky. Anything like that is off limits—no supernatural explanations, please. Biologists have a perfectly good answer for who the designer is. It is natural selection working on genetic variation. We don't need to turn to extraterrestrials for help. There are thousands of papers explaining how evolution works. There is not one single scientific paper explaining how intelligent design works. Mr. Bagley and members of the board, I ask your indulgence. I would like one of my students to speak."

Bagley glanced at the board members and then nodded.

A student walked hesitantly to a microphone. "Hi, my name is Daniel Epstein. Here is what I think. I think we ought to let Ms. Ranger teach us what the scientists believe. She showed us how all of the animals have the same basic bone structure in their arms—no matter how they move. I mean that bats and birds and whales and

Chapter 22

even fish have the same bones. And they have the same DNA. Their embryos are alike too. I mean, how do you explain that except by evolution? And the designs aren't perfect either. Look at all of you. You are wearing glasses. That isn't intelligent design, is it? Your eyes have been designed poorly. None of us are even close to perfect. No intelligent designer would do this." Daniel walked rapidly back to his seat.

"OK, Danny. That's fine. And now ... Excuse me, what is it? ... OK ... Dr. Dermet, a member of our board, would like to comment."

"Mr. Bagley, with due deference to Danny, the similarities that he talks about among bones, embryos, and DNA are not hard to explain. After all, the same designer is involved, isn't he? Why would the designer abandon a good design? And I believe that Becky Conner, also a student of Ms. Ranger's, might wish to speak."

"Yes, sir, I would. Danny is a friend of mine, but we don't agree on this. I mean, why can't God be involved in science? If God is the one that created the universe and everything else in science, why can't we talk about it? Why can't scientists use Him as an explanation?"

"Because, Becky, they can't!" Danny called out, practically leaping from his seat. "We don't have a clue about God, for God's sake. If we start claiming that little green men are here causing our lights to go on and off, what good is that? We don't get anywhere talking about things that we can't see and measure. Science stops!"

Bagley was at it again, banging on the table. "Danny. Danny. Please!"

Almost immediately, a cluster of students sitting near Danny erupted in song. "Give me that old time religion. Give me that old time religion. Give me that old time religion. It's good enough for me ...."

Calls from the audience and commotion on the stage brought Bagley to his feet. "Ms. Ranger! Stop this immediately." After a few confusing moments, order was restored.

"Mr. Bagley, I am sorry that the students' enthusiasm got the better of their judgment."

"Is this the way that you teach your class, Ms. Ranger? You ought to be ashamed," Emma Cromwell said from the stage.

Bagley asked, "Ms. Ranger, is there anything further that you would like to add?"

Janet Ranger sat there exhausted and embarrassed, wondering what else she could say. If this proposal passed, neighbors would stop speaking. The classroom would never be the same. It would be a long time before teaching was fun again. If ever. And would there be a strike? That would damage the community for years.

And now what, Ms. Ranger?

Teaching Notes

Introduction and Background

Intelligent design (ID) is a hot political and educational topic in some sections of the country. ID conjectures that living organisms are so complex that they must have been created by some kind of higher force; nature must have had an intelligent designer. ID does not resonate within the scientific community as a serious topic for study. Scientists no longer question that evolution is the designer; that debate was settled 150 years ago. There are many issues left for discussion, however—issues that involve just how evolution happens.

The issue of intelligent design spills over into all fields of science; it is not confined to biology, as one might think at first glance. Chemistry and physics become involved when questions of the age of the solar system are considered. Astronomy becomes engaged when the age and origin of the universe are discussed. Geology gets involved on questions about fossils. And even mathematics and statistics are enmeshed when issues of probability are debated. Moreover, as I will emphasize later, science itself is under siege, for the IDers want to broaden the definition of science to include supernatural causes.

This case is presented as a classical dilemma case. A school teacher is thrust into the middle of a school board controversy. Although the details are fictitious, the case is clearly based on the Dover, Pennsylvania, decision, which instructors can read about at *www.pamd.uscourts.gov/kitzmiller/kitzmiller_342.pdf*. It is suitable to be taught in any class where the issue of intelligent design is relevant.

Objectives

- To learn the basic arguments made for and against the teaching of intelligent design in the science classroom
- To better understand how social and political forces may get involved in scientific matters
- To learn how to evaluate arguments and marshal evidence for or against a position
- To learn how to discuss a controversial topic civilly, looking at both sides of the question

Common Student Misconceptions

- Intelligent design should be taught in the science classroom.
- Scientists know how evolution works.

Chapter 22

Classroom Management

This is a discussion case, with the instructor asking probing questions that focus on the parts of the case that are most important. In a biology class, the science behind the discussion would be emphasized—that is, the major evidence for evolution would receive top billing. All major biology books will treat this adequately as they consider anatomical and embryological homologies, DNA and biochemical resemblances, fossil data, and so on. But the case is more than that, as there is a strong focus on what makes science, science and why intelligent design fails to make the grade. Furthermore, it takes on the legal questions as well as the political and religious issues. Because the case has several themes, it can easily be extended over several days. If it is to be covered in a single class, the focus must be seriously narrowed.

Depending where the case is taught, the case can be highly emotionally charged. I strongly recommend that the instructor cover the scientific data first. Much of this can be done in the classical lecture mode, leaving the case to act as a capstone piece. But obviously there is another option: to use this case as a launching pad that initiates discussion of the topic of evolution. Used this way, the instructor would lead a discussion that would identify the evolutionary issues that need to be explored. Following the case, the students should recognize the importance of the evidence that makes the rationale for evolution so compelling and the lack of evidence for ID.

Whichever way the case is used, the opening question the instructor poses to the class at the start of the discussion is particularly important because this question sets the tone and the entry point into the material. If the teacher wants to talk about the science, he or she should shy away from starting with a question such as "What do you think Janet Ranger should do now?" Such questions are better left until later; otherwise the science is harder to introduce.

There are several major issues that should be discussed:

The Legal Issues

The complete teaching notes for this case (see "Web Version," p. 206) cover the essential information, including the decision that was made in Dover, Pennsylvania, when a similar case was decided. All teachers interested in the ID controversy should read the judge's opinion, as it covers the essential arguments in the ID debate: *www.pamd.uscourts.gov/kitzmiller/kitzmiller_342.pdf*.

The Scientific Issues

In the Dover case, Judge John E. Jones III concluded that ID is not science. It violates three ground rules of science: (a) It invokes supernatural causation; (b) It produces a false dichotomy pitting ID against evolution (evidence against evolution is not evidence for intelligent design); and (c) "ID's negative attacks against evolution have been refuted by the scientific community."

The National Academy of Sciences, the most prestigious scientific organization in the United States, concluded, "Creationism, intelligent design, and other claims of supernatural intervention in the origin of life or of species are not science because they are not testable by the methods of science" (National Academy of Sciences 1999, p. 25).

There are no peer-reviewed articles by anyone advocating for intelligent design. There are no peer-reviewed papers supporting claims that the complex biochemical molecular systems—such as the bacterial flagellum, the blood-clotting cascade, and the immune system—were intelligently designed or "irreducibly complex."

The Social/Political Issues

What are the responsibilities of teachers in dealing with controversial topics (i.e., responsibilities to parents, society, and students)? What is academic freedom, and what does it mean in this situation? What rights and responsibilities do school boards have? What rights and responsibilities do teachers and teacher unions have?

Is the fairness issue relevant? That is, must teachers present all sides of an issue, even if the scientific community believes an issue is settled? Not all arguments have equal weight. The idea of a supreme designer producing the world and universe by fiat in its present form has been supplanted by the notion of evolution—change through time. The evidence is in on that point and has been for 150 years. (This does not mean that a God or gods are or are not involved.) But the details are not all known and the mechanisms are certainly not clear. It is useful to relate some of this during class, but to place the designer argument on equal footing with evolution would be dishonest.

Summary

This is not a debate about if there is or isn't a God. It is a question of whether the scientific evidence is adequate to explain the origin of the universe, Earth, and its organisms—recognizing that we do not know everything. Do we need a supernatural or an extraterrestrial designer to explain the gaps in our knowledge?

The basic strategy of the ID proponents is to criticize evolutionary theory and reiterate the gaps in our knowledge and thus claim they win by default as their explanation must be correct. They commit a well-known fallacy, "false dichotomy." Neither evolution nor the ID explanation may be correct—and it is the duty of the teachers to point this out. But it is also the responsibility for teachers to point out that there is no evidence for ID and that virtually all scientists working and publishing in peer-reviewed journals accept evolution. As teachers we are obliged to instruct students based on our current understanding of the facts.

Janet's Options

I would save this topic for the end of discussion after the scientific, political, and religious issues have been vetted. Janet's options include (1) resignation, (2) acquiescing to the board's decision, (3) joining in a lawsuit against the board, or (4) refusing to read the announcement to the students. All of these options have both short- and long-term consequences, which need to be explored. In the Dover case, the teachers chose option 4. As a result, the administrators had to read the announcement to the students themselves. A group of parents mounted the lawsuit that led to the trial and later decision.

Web Version

This case and its complete teaching notes and references can be found on the website of the National Center for Case Study Teaching in Science at *http://sciencecases.lib. buffalo.edu/cs/collection/detail.asp?case_id=332&id=332*.

Reference

National Academy of Sciences. 1999. *Science and creationism: A view from the National Academy of Sciences.* 2nd ed. Washington, DC: National Academies Press.

Chapter 23
The Case of the Tainted Taco Shells

Ann Taylor

The Case

"Hi, Dad, I'm home! Can Chris join us for supper?" Mark asked as he walked into the kitchen. "Practice ran over and he's got to go straight to the fields and help his dad, but there's always time for taco night, right?"

Dad lifted his head up from the newspaper. "Only if he likes tacos without the shells. Hi, Chris. Got a date for the prom yet?"

"Nope. Still looking, Mr. Schumer."

Dad continued, "I went to Kroger today, and all of the Taco Bell shells were gone. The cashier said there was some sort of recall. And you know your sister can't have the flour ones because of her wheat allergy."

"Well, we've got some hot dog buns, so let's improvise!" Mark said with a flourish. "What's up with the recall? I thought that only happened with bad hamburger."

Dad folded the paper over and pointed to a headline. "It's all in the genes. This article says the shells accidentally contained some genetically modified corn called StarLink. StarLink is a feed corn, but it isn't approved for human consumption."

"So the cows can eat it, but we can't? That doesn't make sense," said Mark.

Chris jumped in. "StarLink hasn't been proven to be safe for humans to eat—you know, not toxic or allergenic. That's important, especially to your sister."

"What's important to me?" asked Michelle as she walked in and sat at the table. "Hi, Chris."

"Oh ... Hey, Michelle." Chris was blushing. "We were just talking about, um, taco shells."

Mark rolled his eyes.

Dad turned to Chris. "Chris, your family farms corn, right?"

"Right."

"Well, why would your dad want to grow genetically modified corn?"

"Do you remember the year Mark helped us detassle?" Chris asked.

Chapter 23

Mark winced. "That was the worst job ever. My hands felt like they were at a paper cut convention!"

"That was the easy part," Chris said. "What if you had to walk the fields twice a year and pull all the weeds? That's what my dad had to do when he was a kid. Now he uses Roundup spray and Roundup Ready seeds. He sprays the field once and all the weeds die. But the crop is okay because the seeds have been genetically modified so the plants won't be affected. The only bad thing is that the seeds are a little more expensive and you have to buy the spray, but it is still a lot cheaper than weeding by hand."

"And my hands thank you," Mark said, waving his hands and bowing.

Chris ignored Mark and continued. "My dad uses another seed that is resistant to insects. When I was 10, we almost lost the entire crop to corn borers, so Dad was pretty happy when this seed came out."

"So how does it work?" Michelle asked, smiling.

"Well," said Chris, his face still a little pink, "all I know is from talking to the seed salesman. He said the corn produces a bacterial protein that is toxic to insects. So when a bug chews on the plant, it dies."

Michelle frowned. "Bacteria? You mean they're loading up my food with extra chemicals and toxic proteins just to make life easier for farmers?"

"Wait a minute. Farming isn't easy, Michelle. My dad had to get a second job just to keep the farm that's been in our family for three generations!"

"Well, what about the butterflies?" Michelle asked, setting her jaw. "They're related to corn borers. Are they killing them, too?"

"I don't know, Michelle." Chris was getting defensive. "Besides, what does that matter? Butterflies don't eat corn!"

"But I do!" Michelle sputtered. "And I'm not going to eat any of this Frankenfood if I can help it!"

Mark stuck his arms forward and wobbled around. "Igor, it's alive! ALIVE!"

Dad looked up again. "Mark, grow up. Michelle, calm down. This could be a good thing. There's another article in here about a company that is genetically modifying rice to include vitamin A. Scientists think it will greatly reduce childhood blindness in Third World countries."

"That's great," Mark said, "but with all this modification and insect resistance, I wonder what happens if the weeds become resistant to the Roundup?"

"Good question," said Dad. "Let's continue this over dinner. Mark, sit over here. Chris, why don't you sit by Michelle?"

Questions

1. Is genetically modified corn safe?
2. Can farmers make a living without genetically modified corn?
3. Will Chris ask Michelle to the prom?

Project Design

"Interest groups" have been defined for this case study: entomologists, farmers, and immunologists. Each student in the class will be assigned to an interest group. Each group will hand in written answers to the common questions and their group questions. Each group will also prepare a 15-minute oral presentation to be given in lab. Your written work will be due at the time of your oral presentation. Following the presentation, there will be an opportunity for questions from the other groups and a general discussion after all of the presentations are complete. Your grade will be 50% for the group written work, 40% for the group presentation, and 10% for your participation in the discussions of the other interest groups.

Common Questions (All groups answer these in their written answers.)

1. Read "Are you ready for [a] Roundup?" *J Chem Ed* 78 (June 2001): 752–756, "What is a transgenic plant?" *www.colostate.edu/programs/lifesciences/TransgenicCrops/what.html* and "How do you make a transgenic plant?" *www.colostate.edu/programs/lifesciences/TransgenicCrops/how.html*; watch the animations on this page.

2. Besides the gene of interest, what other DNA sequences must be inserted into the plant to make it transgenic?

3. Describe two different methods that can be used to generate a transgenic plant.

4. Explain the three most common genetic modifications of plants and why each modification has been made.

5. Compare and contrast the general ELISA (enzyme-linked immunosorbent assay) and PCR (polymerase chain reaction) methods for detecting genetically modified foods. Which method is most commonly used, and why?

Interest Group Questions (Only answer the questions for your group.)
Entomologists

1. Read "Transgenic pollen harms monarch larvae," *Nature* 399 (May 20, 1999): 214; "Monarch Larvae Sensitivity to *Bacillus thuringiensis*-purified Proteins and Pollen," *Proceedings of the National Academy of Science* 98 (21): 11925–11930; and "Impact of *Bt* Corn Pollen on Monarch Butterfly Populations: A Risk Assessment," *Proceedings of the National Academy of Science* 98 (21): 11937–11942.

2. Describe and compare the experimental procedures used in the *Nature* paper and the first *Proceedings of the National Academy of Science* paper.

3. What effect did exposure to Bt corn pollen have on larval survival, leaf consumption, and larval weight?

Chapter 23

4. An insect is called an instar when it is between two molts. A newly hatched insect is called a first-instar or larva. An adult is a final instar. Most caterpillars (butterfly and moth larva) have five or six instars. (To see the different instar stages, go to *www.gpnc.org/monarch.htm*.) Does pollen from Bt corn affect all instars equally?

5. In addition to the effects of Bt corn pollen on monarchs, what other factors should be considered in evaluating the risk of such transgenic crops?

Farmers

1. Read "Farm-level Effects of Adopting Genetically Engineered Crops," *Economic Issues in Agricultural Biotechnology*, Economic Research Service/USDA, Bulletin AIB-762, 10–15; "Widely Used Crop Herbicide Is Losing Weed Resistance," *New York Times*, Jan 14, 2003: C1; and the abstract of "Glyphosate-resistant Goosegrass: Identification of a Mutation in the Target Enzyme 5-enolpyruvylshikimate-3-phosphate synthase," *Plant Physiol.* 129 (3): 1265–1275.

2. What factors have encouraged farmers to use genetically modified plants?

3. Under what conditions do farmers reap an economic benefit from using GM crops?

4. The enzyme affected by Roundup (glyphosate) is 5-enolpyruvylshikimate-3-phosphate synthase, or EPSP synthase for short. EPSP synthase is involved in the synthesis of aromatic amino acids, such as phenylalanine and tyrosine. Why is this pathway crucial for plants but not for animals?

5. Describe how the sequence of the resistant goosegrass EPSP synthase enzyme differs from the wildtype (normal) goosegrass EPSP synthase enzyme.

6. What can farmers do to prevent resistance in weeds?

Immunologists

1. Read "What's Hiding in Transgenic Foods?" *Chemical and Engineering News*, Jan. 7, 2002, 20–22; "Digestibility of Food Allergens and Nonallergenic Proteins in a Simulated Gastric Fluid and Simulated Intestinal Fluid—A Comparative Study," *Journal of Agricultural and Food Chemistry* 50 (24): 7154–7160; and "Screening of Transgenic Proteins Expressed in Transgenic Food Crops for the Presence of Short Amino Acid Sequences Identical to Potential, IgE-binding Linear Epitopes of Allergens," *BMC Struct Biol.* 2 (1): 8 (*www.biomedcentral.com/1472-6807/2/8*).

2. What methods are used to predict whether a protein may be an allergen? What are the advantages and disadvantages of each of these models?

3. Describe in detail the conditions used to test the digestibility of proteins and how digestion was evaluated. What criteria should be used in "establishing a globally

used standardized assay condition"? Based on the results of this study, what should those criteria be?

4. Computational methods may also help screen for potential allergens. Describe the possible algorithms that could be used for such a screening.

5. What is the probability that a protein would contain a given six-amino-acid sequence? Seven-amino-acid sequence? Eight-amino-acid sequence? What are the advantages and disadvantages of long and short reference frames?

6. What would be the advantage of discontinuous epitope searches, and why aren't they currently used?

Teaching Notes
Introduction and Background
The biggest challenge facing farmers is controlling weeds and diseases while maintaining crop yield and quality. This case focuses on some of the issues associated with the use of genetically modified (GM) plants, including ecological risks, resistance, and allergenicity. The case was originally developed for use in a biochemistry course taught in our chemistry department, but it is also appropriate for use in nonmajors general, organic, and biochemistry courses and in a general biology course as well.

The case emphasizes the basic biochemistry and scientific ethics issues associated with genetically modified foods. The case story is used as a launching point to read and discuss primary and secondary literature articles. The papers and questions are appropriate for use with nonscience majors. The articles require little, if any, prior experience with scientific literature, and use very basic methods. It is appropriate for liberal arts schools, as it touches on many of the ethical issues associated with science, including making decisions in the absence of complete information, balancing the needs of the few versus the many, and economics versus health issues. It is useful if the students have a prior knowledge of the central dogma of molecular biology, gel electrophoresis, and enzymes.

Objectives
After completing this case study, students should

- understand how a transgenic plant is made;
- feel comfortable reading a primary literature article (at an appropriate level for the class);
- understand the issues surrounding the use of GM foods, including environmental effects and concerns about resistance;
- see how scientists have to deal with uncertainty and risk-assessment;

Chapter 23

- understand why Roundup is toxic to plants but not to animals; and
- understand the methods used to predict allergenicity.

Common Student Misconceptions
- All genetically modified plants are the same.
- Predicting allergenicity is straightforward.
- Plants don't change from generation to generation.
- All insects are equally affected by insecticide proteins.

Classroom Management
I use this case study during the DNA unit at the end of the semester in conjunction with a laboratory experiment where we test cornmeal samples for the presence of genetic modifications (Taylor and Sajan 2005). The pairing of the case study with a laboratory activity has worked well, as it provides a context for the laboratory activity and allows students to productively use the "waiting" times in the laboratory. This approach is also an easy way to add a case to a course without having to take time out of the traditional lecture period.

During the discussion period, each group presents its results; then, as a way to wrap up the case, the entire group discusses questions such as the following:

- Why do different scientists obtain different answers to questions such as "Are genetically modified foods safe?"
- What is an acceptable level of risk?
- People sometimes complain that scientists tell them one thing this week and the opposite the next week. Why does this happen? How should consumers deal with this ambiguity?

Web Version
The case and its complete teaching notes, references, and answer key can be found on the website of the National Center for Case Study Teaching in Science at *http://sciencecases.lib.buffalo.edu/cs/collection/detail.asp?case_id=610&id=610.*

Reference
Taylor, A., and S. Sajan. 2005. Testing for genetically modified foods using PCR. *Journal of Chemical Education* 82 (4): 597–598.

Chapter 24

Medicinal Use of Marijuana

Clyde Freeman Herreid and Kristie DuRei

The Case

Introduction

Marijuana is classified as a Schedule 1 drug under the Controlled Substances Act (1970). Accordingly, it is currently legislated as having "no accepted medical use in treatment in the United States." At the federal level, marijuana can be used for research purposes only and is illegal to possess otherwise. However, it is commonly argued that marijuana can be used to manage a host of medical problems. Over the past few decades, more than a dozen states have legalized medical use at the state level. The resulting tension between federal and state authority has led to a number of high-profile confrontations between the federal Drug Enforcement Administration (DEA) and state and local agencies that allow the open use of marijuana for medicinal purposes, especially in hospices.

Should the use of marijuana be legalized for medical purposes? Before coming to an informed opinion regarding this controversial issue, you will examine two positions during a classroom debate. To prepare for this, your assignment is to collect all of the information that you can both for and against the medical use of marijuana. Your instructor will designate the date of the debate. At the time of the debate, you must be prepared to debate both sides of the issue. Your instructor will provide you with ground rules for the debate. First, read statements by two individuals who have widely different views.

"Terminal Cancer" by Anonymous

"In October '05, my mother, a 74-year-old elegant lady full of beauty, died due to the effects of liver, lung, and breast cancer. ... The world came crashing down in February

Chapter 24

'05 when she sat me down and gave me the worst possible news. There was nothing the docs could do.

"On October 4th, my 40th birthday, my mom was unable to attend my birthday dinner. On October 7th, the family was convened for a meeting with a social worker from the Hospice. I had no idea what 'Hospice' was and was shocked to my foundations when this fellow started talking funeral homes and cremation services for my mom, basically preparing us for her death. I decided I had heard enough and went upstairs to where my mom was lying. She was once this beautiful creature that never seemed to age, and here she was in bed lying in a fetal position, reduced to a wisp, looking miserable and frightened. There's no feeling more helpless than watching your mom starve to death in front of your eyes knowing there's nothing 'modern medicine' can do for her except give her morphine via a convenient IV, providing the zoned-out opiated comfort of lying in a faux warm pool of sensory deprivation as it accelerates and fogs her remaining days on Earth.

"October 7th was a Friday and she looked like she would not make it through the weekend. I told her, 'I have something in the truck that might help you. …'

"Soon afterward I let her take her first tiny, gentle breaths of marijuana. Only seconds later, she began rubbing her stomach. I asked her how she felt. 'I feel a little woozy but my stomach feels better!' After eating her first food in a while, she regained some strength and, instead of needing help to get to the bathroom, got up under her own power and began walking herself. Her voice started sounding better; the change in her was nothing short of a miracle. With just a few breaths of MJ [marijuana] vapor every four or six hours, she eliminated her nausea, increased her appetite, was able to keep the food she ate down, and turned back on her will to live. MJ made whatever pain she was feeling go away for awhile naturally, with not a single contra-indication or harmful side effect to the body. Most importantly of all, every bit of anxiety she was feeling about this whole cancer ordeal was gone. Gone. What pill can Glaxo possibly create in a lab that effectively treats so many things at once, within seconds of taking, while being so gentle to the body? When her friends came by to visit she would say, 'I'm on pot and it's great!'

"Mom got every piece of information about the world from Fox News and Rush Limbaugh, much to the chagrin of most of the family, but she didn't need convincing to come to the conclusion that everything she'd ever been told about marijuana during the course of her life had been a long succession of smokescreens designed to veil its true worth to humanity. … She was talking coherently up until the final day. … For the family it was a great relief to have that extra peace and comfort knowing that she died an elegant, beautiful, graceful death instead of a morphine drip, 'out of it,' dirty death."

Note: Adapted from Marijuana: The Forbidden Medicine, RxMarijuana.com, *www.rxmarijuana.com/shared. htm*.

Medicinal Use of Marijuana

"The Story of the Lotus Eaters" by Anonymous

"I am a marijuana addict because, when using pot, it was the most important thing in my life. More important than anybody or anything. It helped to suppress all the inadequacies I felt. It helped me not to feel the pain of not living up to expectations. It enabled me not to worry about anything. It helped me to not care about the things I really cared about. It enabled me to stay in my own little world and not deal with emotional feelings that would continually come up when I wasn't smoking. It would drive the fear away, but after a while, the fear would return.

"Pot helped me not worry about not having a relationship with women, even though I wanted this to happen. Because of negative feelings about myself, I always thought deep down that I was worthless and didn't deserve to be happy. Instead of dealing with these issues I would smoke pot and the feelings would go away. Therefore, I never learned very many social skills or problem-solving skills.

"Problems would come up and they would seem too huge to deal with. I would smoke pot and look for the answers after smoking, because then the problems seemed smaller. In reality, they were only day-to-day issues that could be resolved if dealt with, instead of running away from them. I would smoke and not deal with the problems and let them fester inside until I thought, 'I just can't handle it.' I would try not to think about them, or go somewhere I could start all over, escape, and hope that would teach me how to deal with them the next time. But the next time, they would continue and I would do the same thing, over and over, until it was killing me.

"Later, I started to turn to other things (alcohol, cocaine, gambling) in the hope that these things would give me pleasure, or at least let me not care about the problems that followed me wherever I went, and that these feelings I carried around would go away. They didn't."

Note: Adapted from Marijuana Anonymous World Services, *www.marijuana-anonymous.org*

Teaching Notes

Introduction and Background

In this case study, we deal with the controversial issue concerning whether marijuana should be legalized for medicinal purposes. The particular case is suitable for any class where medical issues and social issues of science are discussed. I have used it in honors seminar classes for freshmen. Depending on the sophistication that the instructor expects, it can be used in practically any classroom setting from high school to professional schools such as nursing and pharmacy.

We teach the case via the Intimate Debate Method. This method is a powerful technique for dealing with case topics that involve controversy. Basically, pairs of students face off across a small table. One pair is assigned the pro side of the argument and the other pair is the con side. The pro side speaks for 4 minutes and then the con side speaks. They then switch roles for 2 minutes each. At the end of this exercise, the group of four students must abandon their artificial positions and try to come to a consensus as to a reasonable solution to the problem being debated. (The classroom will have several such tables of students presenting their arguments at the same time; that is, all students are participants.)

The advantages of such an approach, compared to formal debate, include time efficiency, because multiple intimate debates take place simultaneously; dispassionate scrutiny, because switching positions tends to reduce initial "buy-in" or commitment to a given side of the issue; greater participation by the reticent, because there is no audience; and increased realism, because the call for consensus opinion mimics real-world policymaking, in which decisions must be made within given time frames and with (possibly) insufficient information.

Objectives

- To learn the medical evidence for and against the use of marijuana
- To better understand how social, political, and societal forces are involved in scientific and medical decisions
- To learn how to evaluate arguments and marshal evidence for or against a position
- To learn how to discuss a controversial topic civilly by looking at both sides of the question

Common Student Misconceptions

- Marijuana does not have any beneficial uses.
- People with differing viewpoints on drugs cannot come to any agreement.
- Laws dealing with drugs cannot be altered.
- The person with the best scientific argument always wins any public discussion.
- The neurological effects of marijuana are understood.
- Information gleaned from the internet is reliable.

Classroom Management

The Intimate Debate method is ideally designed for a small-class setting where students can move their seats freely. Here is the sequence I use.

1. First, have two students read to the entire class the two different anonymous stories about marijuana. This sets the stage for the debate between students who will take the pro and con sides on the question of whether to legalize marijuana.

2. Arrange groups and assign roles. Suppose there are 24 students in the class. Take 2 groups of 6 students (12 students altogether) and assign them the pro side of the argument. Assign the other 12 students (in 2 groups of 6 students) the con side.

3. Tell the groups to share their information and organize it around the major talking points (arguments) that they can make from their respective sides. Tell everyone they are individually responsible for keeping notes because their current teams are going to be split up. If you fail to inform them of this point, they will not be adequately prepared. This general discussion takes a minimum of 15 minutes.

4. Take 2 students from a pro group and seat them across from 2 students from a con group. This is repeated for all students. In my hypothetical example, I started with 24 students, so I would have 6 debates about to start. The opponents should be either seated across small tables or seated in chairs facing one another.

5. To start the debate, tell the students on the pro side of the argument to speak for 4 minutes to their con opponents. The students representing the con side may not interrupt. Tell all students that the con side must take good notes because they will soon be arguing the pro side and will not have the pro information sheet to guide them; thus, they must listen carefully to the other side's remarks. At the end of 4 minutes, the instructor calls a halt.

6. Tell the students for the con side they may speak for 4 minutes to their pro opponents without interruption. The pro side must listen carefully and take good notes. The instructor calls a halt when the time is up.

7. Allow all groups 3 minutes for a caucus. The teacher explains to the class that now the roles will be reversed: Pro speakers must shift to the con side, and vice versa. Because groups will have only their notes to guide them in their new roles, they should use the caucus time to briefly consult about their best arguments and get ready for a new debate.

8. To start the next round of the debate, the new pro groups will have only 2 minutes to make their best arguments while the new cons listen.

9. Then new con groups have 2 minutes to make their arguments while the new pros listen.

10. Inform all students to abandon their formal positions. In their groups of 4, their job is now to come up with what they believe to be a reasonable solution to the controversy. If possible, they should try to reach a consensus. This usually takes the students about 5 minutes.

11. As the instructor, you now ask the groups to report the results of their deliberations. This can take as little as 5 minutes or considerably longer, depending on how

much discussion the instructor wishes to instigate. The entire intimate debate process lasts about 45 minutes.

There are legal, health, and social issues involved in this case. The essence of the debate boils down to these points, below.

Summary—Pro Arguments

The basic arguments for the pro side will be as follows: (1) Marijuana relieves unbearable pain that cannot be alleviated by medical agents (morphine) without incapacitating the patient. Moreover, it is useful for even more modest medical complaints such as migraine headaches. (2) The claimed medical risks of marijuana are nonexistent or mild. Clearly, the benefits outweigh the risks. (3) The current drug policy is a failure. Society will benefit because there will be a reduction in the illegal drug trade. (4) Once marijuana is legalized, it will be easier to regulate; its dosage and purity can be controlled and taxed.

Summary—Con Arguments

The basic con arguments will include the following: (1) There are various purported health risks. The fear factor will weigh heavily in this—who knows what long-term effects may show up? (2) There are other well-understood drugs that alleviate pain—drugs that have been well studied and given FDA approval, and that have a regulated manufacturing process with studied ingredients. In contrast, marijuana has not been medically approved, its ingredients are multiple and not well studied, its manufacture is not regulated, and there is no guarantee of its purity. (3) There will be an argument that marijuana is a gateway drug to other more dangerous and addictive substances such as cocaine. (4) There will be the strong argument that if it is legalized, there will be no way to prevent marijuana from getting into the hands of people who do not have a medical reason for having it. California's difficulties will be cited as clear evidence of the impossibility of control.

Web Version

This case and its complete teaching notes and references can be found on the website of the National Center for Case Study Teaching in Science at *http://sciencecases.lib.buffalo.edu/cs/collection/detail.asp?case_id=289&id=289.*

Chapter 25

Amanda's Absence: Should Vioxx Be Kept Off the Market?

Dan Johnson

The Case

Part I: Withdrawals

"Amanda missed bio lab again this week."

This was not what Dr. Sharpe had expected to hear when his teaching assistant knocked on his door late one Tuesday afternoon in March. Amanda was an exceptional student. She had been born with Type III osteogenesis imperfecta, a painful and debilitating condition in which bones break and deform easily. Despite her condition, Amanda had earned top academic honors during high school and came to college on a merit scholarship. Before a month ago, she had never missed a single lecture or lab, but thinking back, Dr. Sharpe could not remember her attending his class for at least a week. Was there something seriously wrong?

The student health center had no information, but a phone call to Dr. Rutter, the dean of students, cleared up the mystery. "Amanda has asked to withdraw for this semester for medical reasons. It seems she can't control her pain anymore now that Vioxx is off the market."

Dr. Sharpe remembered that the Food and Drug Administration (FDA) had announced that two pharmaceutical companies were withdrawing Vioxx and Bextra from the market because they caused heart attacks. Both were in the same drug class, the Cox-2 inhibitors. He also vaguely remembered commercials pitching Vioxx to senior citizens for arthritis pain.

Questions

1. Some prescription drugs may remain on the market for 20 years or more. Others are removed shortly after being introduced. Give three or four reasons why a drug might be removed from the market, either by the FDA or its maker. Is every reason true for every drug?

Science Stories: Using Case Studies to Teach Critical Thinking

2. How might the manufacturers have determined that Vioxx increases the risk of heart attack? Based on your answers, when would they have learned this information?

3. What other facts or information might Dr. Sharpe (or you) want to know about Vioxx? About Amanda's condition?

Part II: Press Release

Dr. Sharpe was stunned to hear that Amanda would be forced to leave school just because one medication was not available. He went to the Food and Drug Administration's website, where he found their initial press announcement.

> Sept. 30, 2004*
> FDA Issues Public Health Advisory on Vioxx; Manufacturer Voluntarily Withdraws Product
>
> The Food and Drug Administration (FDA) today acknowledged the voluntary withdrawal from the market of Vioxx (chemical name rofecoxib), a nonsteroidal anti-inflammatory drug (NSAID) manufactured by Merck. The FDA today also issued a public health advisory to inform patients of this action and to advise them to consult with a physician about alternative medications.
>
> Merck is withdrawing Vioxx from the market after the data safety monitoring board overseeing a long-term study of the drug recommended that the study be halted because of an increased risk of serious cardiovascular events, including heart attacks and strokes, among study patients taking Vioxx compared to patients receiving placebo. The study was being done in patients at risk of developing recurrent colon polyps.
>
> "Merck did the right thing by promptly reporting these findings to the FDA and voluntarily withdrawing the product," said Acting FDA Commissioner Dr. Lester M. Crawford. "The risk that an individual patient would have a heart attack or stroke is very small. Yet Merck's study does suggest that patients taking Vioxx chronically face twice the risk of a heart attack compared to patients receiving a placebo."
>
> In June 2000, Merck had submitted to the FDA a separate safety study that showed an increased risk of heart attacks and strokes in patients taking Vioxx chronically for arthritis, compared to patients taking Aleve. After reviewing the results of the earlier study, the FDA required additional label and prescribing information on Vioxx, but did not recommend withdrawal.
>
> The FDA approved Vioxx in 1999 for the reduction of pain and inflammation caused by osteoarthritis, rheumatoid arthritis, and acute pain in adults. It is a Cox-2 selective NSAID; other NSAIDs target both Cox-1 and -2. When Vioxx was approved, it was hoped that it would have a lower incidence of gastrointestinal ulcers and bleeding than other NSAIDs like Motrin (ibuprofen) and Aleve (naproxyn).
>
> * Based on the original FDA press release, which has been modified for the purposes of this case study.

Questions

1. What are the advantages and disadvantages of Vioxx versus other pain-relieving medicines? Why are these significant?

2. According to the data provided to the FDA, are all patients taking Vioxx at greater risk of a heart attack or stroke? Why or why not?

3. What are two other questions you have about the Vioxx withdrawal that were not addressed by the press release?

Part III: Prepared Testimony

The more he read, the more Dr. Sharpe realized that the Vioxx withdrawal was a major event. The FDA had posted pages of reports, press briefings, and letters to physicians and the public. There even was a formal hearing before the Senate Committee on Finance on the matter. Dr. Sharpe read the opening statements from the testimony of Sandra Kweder, M.D., deputy director of the FDA's Office of New Drugs, given on November 18, 2004.*

> Members of the Committee, we appreciate this opportunity to discuss drug safety and the worldwide withdrawal by Merck & Co. of Vioxx. Modern drugs provide significant health benefits. We believe the FDA maintains the highest worldwide standards for drug approval.
>
> All drugs pose some level of risk. Unless a new drug's demonstrated benefits outweigh its known risks for the intended population, the FDA will not approve the drug. The FDA only grants approval once a sponsor demonstrates through clinical trials that a drug is safe and effective. However, our experience has shown we cannot anticipate all adverse effects of a drug before approval, because not every adverse drug reaction occurs during pre-approval trials.
>
> Occasionally, serious adverse effects are identified after approval, in post-marketing clinical trials or through spontaneous reporting of adverse events. Adverse effects also result from errors in drug prescribing, dispensing, or use. The FDA has a strong postmarket drug safety program designed to uncover adverse events that happen after initial approval. Drug safety staff evaluate and respond to adverse events identified by ongoing clinical trials or as reported by physicians or patients. Our recent actions concerning the drug Vioxx illustrate the importance of continuing to assess the safety of a product once it is in widespread use.
>
> Detecting and limiting adverse reactions can be challenging. How do we weigh the impact of adverse drug reactions against the benefits of a product to individual patients and the public health? The question is multifaceted and complex, involving scientific as well as public policy issues.

* Based on the original testimony, which has been modified for the purposes of this case study.

Chapter 25

Questions

1. What are two strengths and two weaknesses in the current system of drug approval? Why did you choose these particular strengths and weaknesses?

2. Based on your answers to question 1, did the FDA approve Vioxx too soon? Why or why not? Could the approval system be changed in a way that prevented the heart attack deaths attributed to Vioxx?

3. Should Merck be fined for putting an unsafe drug on the market? Why or why not?

Part IV: Review Panel

Merck has petitioned the FDA for permission to return Vioxx to the market, with additional warning labels and dispensing information. Both you and Dr. Sharpe have been appointed to a review panel that must make a written recommendation to the assistant director of the FDA, either for or against allowing Vioxx back on the market. What would you recommend, and why?

In crafting your response, you may discuss the general issues with other staff members (members of your work group or class), and you are free to include research findings or data from outside primary resources. However, you must write your own individual recommendation and clearly explain the rationale for your recommendation. Your recommendation and rationale for it is limited to one typed page.

Teaching Notes

Introduction and Background

When chronic pain forces a top student to withdraw from college, biology instructor Dr. Sharpe learns that medications may be removed from the market for many reasons, including safety concerns. As the case unfolds, students learn how the FDA balances drug safety against medical needs.

The storyline of the case is based on an actual student's experience, but all names and identifying details have been changed. The texts of the original press release and testimony in Parts II and III are in the public domain and were obtained from the U.S. Food and Drug Administration's public website (*www.fda.gov*). Both were edited heavily for brevity, but no relevant facts have been altered.

As written, the case is appropriate for a nonmajors course. I have found that most students can participate in a relevant discussion without any prior background.

Objectives

Upon completing this case, introductory-level students should be able to

- describe in general terms the purpose and structure of a clinical trial,
- explain how the decision is made to bring pharmaceutical drugs on and off the market,

- describe or define in general terms the concept of risk/benefit analysis, and
- distinguish between relative risk and absolute risk.

If the instructor modifies Parts I and II to fit a particular topic, additional learning objectives can be achieved as well.

Common Student Misconceptions

- When the FDA approves a drug for use by doctors, it has been proven safe as long as it is used as directed.
- If a scientific experiment is designed and performed correctly, it does not need to be repeated again.
- If a drug is pulled from the market, it is because the company that makes it did something illegal or irresponsible.
- All people respond to drugs in the same way.
- If the outcome of an experiment is proven to be statistically significant, it must be true.
- If a drug is linked to some bad outcome (in this case, heart attack or stroke), that means almost everyone taking the drug will experience the same outcome.

Classroom Management

This interrupted case can be run over the course of one or two class periods. If introductory-level students work in groups and consider just the first two questions for each part, the case can be completed in 50 minutes. If students discuss all of the questions, the case can be extended to 75 minutes. Rather than having students read just the one-page excerpt from Dr. Kweder's congressional testimony, the instructor may choose to have them read the entire transcript. If so, the case should be conducted over two days, giving students time to read the transcript as homework. The fourth part of this case is a possible homework writing assignment; if it is not used, another assignment should be given that requires students to summarize their position and thoughts on the case.

Students read and discuss the questions for each part of the case in small groups of four to seven. During the ensuing classroom discussions in which the entire class participates, the instructor asks each group to report their answers.

In Part I, students learn that while the decision to recall or withdraw a drug usually is a slow, methodical process, in some instances clinical trial data are so convincing that the withdrawal occurs suddenly, as was the case with Vioxx. However, withdrawal rarely is a simple issue. Inevitably there is a long-term toll on quality of life for individuals who had relied on the drug, a factor often overlooked by the popular press as it moves on to other news. Part I attaches these issues to a character (a student) with whom case readers can empathize.

Students read Part I when they first come to class, then discuss it in their small groups and then as an entire class before proceeding to Part II. Alternatively, Part I can be a prelecture reading assignment. When using this case in an introductory-level class, students should be allowed to ask for definitions of concepts or terms. If the students have some relevant background, it would be appropriate to require students to explain concepts to one another (i.e., team learning) rather than ask the instructor.

Parts II and III are read in class, and the questions at the end of each part are discussed following the same procedure as for Part I described above. Part II introduces some of the factual data for the case. The abridged press release that appears in this part of the case summarizes the regulatory history of the drug and why it was approved originally. Briefly, two separate studies, one in 2000 and a second one conducted in 2004, showed a two-fold increased risk of heart attacks in patients taking Vioxx.

Part III is an abridged and edited excerpt of testimony Dr. Sandra Kweder, Deputy Director of the Office of New Drugs, presented on November 18, 2004, to the U.S. Senate's Finance Committee regarding Vioxx's withdrawal. Dr. Kweder outlines the drug approval strategy of the FDA concisely and the agency's policy of continued monitoring. She underscores the fact that no drug is safe in all situations. Finally, she explains that the decision to withdraw a drug depends on many factors, not just its safety profile.

When teaching this case, it is recommended that the class time be split so that students spend approximately 25% of the time discussing Part I, 35% of the time discussing Part II, and 40% discussing Part III. If the instructor wants to encourage more in-depth discussion, students can be assigned a larger excerpt of the testimony transcript to read as homework. In this situation, Part III would be discussed at a second class meeting.

Web Version

This case and its complete teaching notes, references, and answer key can be found on the website of the National Center for Case Study Teaching in Science at *http://libweb.lib.buffalo.edu/cs/collection/detail.asp?id=180&case_id=180*.

References

U.S. Food and Drug Administration (FDA). 2004. FDA issues public health advisory on Vioxx as its manufacturer voluntarily withdraws product. *www.fda.gov/NewsEvents/Newsroom/PressAnnouncements/2004/ucm108361.htm*.

U.S. Food and Drug Administration (FDA). 2004. Statement of Sandra Kweder, M.D., before Committee on Finance, United States Senate. *www.fda.gov/NewsEvents/Testimony/ucm113235.htm*.

Chapter 26
Sex and Vaccination

Erik Zavrel and Clyde Freeman Herreid

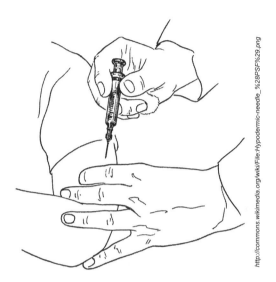

The Case

Part I: A Texas Tempest

The Republican governor of Texas, Rick Perry, caused a whirlwind of controversy on February 2, 2007, when he issued an executive order mandating that all girls in the state's public school system be vaccinated against the human papillomavirus (HPV) prior to entering sixth grade. This virus is strongly implicated as the causative agent of cervical uterine cancer. "The HPV vaccine provides us with an incredible opportunity to effectively target and prevent cervical cancer," said Perry. "Requiring young girls to get vaccinated before they come into contact with HPV is responsible health and fiscal policy that has the potential to significantly reduce cases of cervical cancer and mitigate future medical costs" (Office of the Governor 2007). Governor Perry believed it was his obligation to safeguard the public's health and safety, while many parents in this deeply conservative state were outraged by what they perceived as a governmental intrusion into a private family matter.

Here was the health situation at the time of Governor Perry's pronouncement. The human papillomavirus (HPV) is the collective name for a group of more than 100 viruses, 30 of which are sexually transmitted. HPV is spread during sexual activity by skin-to-skin contact and not by the exchange of bodily fluids. HPV is responsible for genital warts. Most HPV infections occur without any symptoms and go away without any treatment. But a strong link has recently been established between HPV and cervical cancer. The American Cancer Society (ACS) estimated that in 2007, more than 11,000 women in the United States would be diagnosed with cervical cancer and more than 3,600 would die from this malignancy (ACS 2006). Hispanic women develop cervical cancer twice as often as Caucasians; African American women get the disease about 50% more often than non-Hispanic white women (ACS 2006). The Centers for

Disease Control and Prevention (CDC) estimates that at least 50% of sexually active men and women will contract HPV during their lifetimes (ACS 2006). In the United States each year, 6.2 million people are newly infected with HPV, and as many as half of them are 15–24 years old (CDC 2004).

The pharmaceutical giant Merck recently developed a vaccine for HPV. The new vaccine, Gardasil, was approved by the Food and Drug Administration on June 8, 2006, for girls and women age 9–26. It protects against two HPV strains believed to be responsible for about 70% of cervical cancer cases and against two other strains that cause 90% of genital wart cases. The vaccine is given by intramuscular injection in three doses over a six-month period and costs about $360 for the full series (CDC 2006). The HPV vaccine is recommended for 11- to 12-year-old girls, but can be given to girls as young as 9 years of age. There is no U.S. federal law requiring HPV immunization. State laws regulate immunizations for school and child care facilities (CDC 2006). In October 2007, the British government announced that all girls 12 years of age and older would be vaccinated free of charge (NeLM 2007). Similar programs are planned for several Canadian provinces (O'Brien 2007).

Questions

1. If you were a state representative, would you favor mandating the vaccination against sexually transmitted diseases (STDs) such as HPV if vaccines were available? Should the prevention and treatment of STDs factor into the safeguarding of public health?

2. Why might some parents object to having their daughters vaccinated against HPV?

3. Would you consider getting the vaccine yourself or recommending it to a family member? Why or why not? What are the questions that you would like to have answered before you decide?

Part II: The Governor's Case

In the public debate that followed, a number of important arguments in favor of the vaccination order were made. Most statements below are derived from the CDC (2006).

- Though the recommended age for girls to be vaccinated may seem young, it is due to the fact that it is best for girls to be vaccinated before becoming sexually active.

- Children in the public school system are required to be vaccinated for other diseases such as diphtheria, polio, tetanus, and hepatitis. Head-lice checks are still common. The HPV vaccination requirement is no different.

- HPV is the most common sexually transmitted disease in the United States. Studies have found the vaccine to be nearly 100% effective in preventing diseases caused by the four HPV types covered by the vaccine.

- The FDA has approved the HPV vaccine as safe and effective. This vaccine has been tested in more than 11,000 females (age 9–26) around the world. These studies have shown no significant adverse effects.

- At least 50% of sexually active people will get HPV at some time in their lives. Every year in the United States, 6.2 million people get HPV. HPV is most common in young women and men who are in their late teens and early twenties. Approximately 20 million people are currently infected with HPV.

- The American Cancer Society estimates that in the United States each year more than 9,700 women will be diagnosed with cervical cancer and 3,700 women will die from cervical cancer.

- The use of prophylactics during intercourse does not guarantee that HPV will not be transmitted. Condoms do not cover the entire genital area. They leave parts of the sensitive anatomy uncovered, and contact between these areas can transmit HPV (National HPV and Cervical Cancer Prevention Resource Center 2008).

- HPV can be contracted from one partner, remain dormant with no evident indications of infection, and then later be unknowingly transmitted to another sexual partner, including a spouse. This can lead to accusations of infidelity and destroy relationships and marriages (National HPV and Cervical Cancer Prevention Resource Center 2008).

Questions

1. With the information provided, do you agree with Governor Perry? Do you think he should be commended for his initiative and concern for the public welfare, or is he overstepping his bounds?

2. How would you respond to parents with moral reservations about having their young daughters vaccinated against STDs such as HPV?

Part III: Arguments Against the HPV Mandatory Vaccination Policy

Once the executive order became public, the Texas legislators were besieged by phone calls, letters, e-mails, and personal confrontations from the public that were widely reported in the press. Many argued against the order on the basis of personal freedom, religious objection, and quoting "the well-known fact" that occasionally vaccinations were known to kill some people. Additional arguments included the following:

Chapter 26

- HPV is a sexually transmitted disease; it can only be spread by intimate contact. In this regard, it is unlike other diseases for which children in the public school system are required to be vaccinated (Irvine 2007).

- Gardasil would give young people a false sense of security and undermine abstinence-only education and the push to use prophylactics during intercourse. It also serves to challenge parental autonomy (Irvine 2007).

- Some members of the medical establishment have expressed concern with the lobbying efforts from Merck. The company that spent millions developing the vaccine would stand to reap a fortune if it were mandated for every girl in the U.S. public school system (Irvine 2007).

- The vaccine is not cheap. The series of three shots costs $360. This prompts the concern that only the affluent could afford it. Also, if the vaccination were publicly funded, some taxpayers may object on moral grounds (Irvine 2007).

- More research is needed to determine the long-term effects of the HPV vaccine (Irvine 2007).

- Governor Perry is not an unbiased politician acting in the best interests of his constituents. Merck's lobbyist in Austin, Texas, Mike Toomey, was chief of staff for Governor Perry from 2002–2004, as well as for a Republican predecessor, Governor William P. Clements. Merck also contributed to Perry's election campaign (AP 2007).

- A recent medical study is evidence for prudence. Just 2.2% of women were carrying one of the two HPV strains most likely to lead to cervical cancer, about half the rate found in earlier surveys. And just 3.4% of the women studied were infected with one of the four HPV strains that the new vaccine protects against (*USA Today* 2007).

- The vaccine was approved only very recently. It could have adverse effects that will not manifest themselves until millions have been inoculated and until many years have passed. Researchers don't even know how long the vaccine offers protection (*USA Today* 2007).

Questions

1. With the information provided, do you think Governor Perry did the right thing by mandating that all girls in the Texas public school system receive the HPV vaccination? Or is it too early to be mandating Gardasil? Should it be offered to students but not required? How long should tests be conducted before the vaccine is deemed safe?

2. The series of shots costs $360. Should the cost of the immunization be borne by the state or by the individuals receiving the shots?

3. Do you think Governor Perry's decision stemmed solely from his concern for the health of his constituents, or did politics play a factor?

4. Is it ethical for the company that creates a vaccine to lobby for its mandatory use in the public schools?

Teaching Notes
Introduction and Background
This case study is centered on the debate concerning the decision by Texas governor Rick Perry to mandate the compulsory vaccination of girls in the Texas public school system against the human papillomavirus (HPV) prior to entering sixth grade. The interrupted case method is particularly appropriate for this subject, with the successive case sections providing a general overview of the disease, the reasons for such a mandatory vaccination program, the reasons against such a program, and finally a disclosure of what ultimately transpired in Texas.

The case as presented here is designed to be used in an ethics or public policy course. Ethical and public policy issues are emphasized and readings are not assigned. The case could easily be used in courses that emphasize biological and medical topics if additional information is added or additional research on the part of the students is required. As it now stands, the case would serve primarily as a stimulus for students in these classes to look up additional information. If the biological aspects of the case are stressed, before beginning the case students should read the chapters in their course textbook on the cell cycle and how it is affected by oncogenes and anti-oncogenes.

Objectives
- To probe the contested boundary between elected officials' duty to protect the public health and the right of families to decide issues with moral or religious overtones on their own without governmental interference
- To consider a controversial issue objectively, weighing pros and cons without simply appealing to personal convictions
- To develop the ability to discern potential bias and partiality in scientific issues
- To reach a compromise that makes concessions or allowances for dissenting or minority viewpoints while still making significant strides or inroads in addressing a major public health issue

Common Student Misconceptions
- The best time to vaccinate people against sexually transmitted infections (STIs) is after they become sexually active.

Chapter 26

- Only a very small percentage of people will contract an STI in the course of their lives.
- Public vaccination is a simple and straightforward process with the only relevant issue being the science behind the vaccine. The process is insulated from such factors as economics, parental control, and lobbying by pharmaceutical companies.

Classroom Management

The case is designed to be used in a single 50-minute class period. We recommend that the class be broken up into discussion groups of four or five students. Because of the nature of the subject matter, groups might be drawn along gender lines (i.e., the instructor may wish to create single-sex groups). Doing so will ensure a more frank and open discussion among group members.

Within the groups, have students read each section silently or select a person to read a section aloud. Then have the students discuss in their groups the questions that follow each section. This should be followed by a whole-class discussion moderated by the teacher.

To cover all sections of this case in one 50-minute class would only allow 10 to 15 minutes per section for reading and discussion. This is adequate if the goal of the instructor is just to raise student awareness of the issues. Using the case over several class periods would allow for more discussion and would give students time to do their own research on the topics.

There are several key issues that are explored in this case. First, there are the public health issues: What are the health risks involved? Does safeguarding the public health and welfare include the prevention and treatment of STDs? Is the public school system the proper venue for dealing with the treatment and prevention of STDs? Second, there are political, economic, and ethical considerations encompassing issues of lobbying and the ethical obligations of pharmaceutical companies. There are economic questions and the rights and concerns of the parents. Third, there are biological principles. If the instructor wishes to emphasize these principles, special assignments, readings, or additional sections can be added on topics such as immunology, vaccination, the cell cycle, and mitosis. General biology textbooks deal with the essentials, and the web version of the case has a section on the cervical cancer of the uterus with questions for students to research.

Web Version

This case and its complete teaching notes and references can be found on the website of the National Center for Case Study Teaching in Science at *http://sciencecases.lib.buffalo.edu/cs/collection/detail.asp?case_id=238&id=238*.

References

American Cancer Society (ACS). 2006. Overview: Cervical cancer. *www.cancer.org/docroot/CRI/content/CRI_2_2_1X_How_many_women_get_cancer_of_the_cervix_8.asp?sitearea.*

Associated Press (AP). January 30, 2007. Merck lobbies for HPV vaccine to become law. *www.foxnews.com/story/0, 2933,248781,00.html.*

Centers for Disease Control and Prevention (CDC). 2004. Sexually transmitted diseases, surveillance and statistics. *www.cdc.gov/nchstp/dstd/Stats_Trends/Stats_and_Trends.htm.*

Centers for Disease Control and Prevention (CDC). 2006. HPV and HPV vaccine: Information for healthcare providers. *www.cdc.gov/std/hpv/stdFact-hpv-vaccine-hcp.htm.*

Irvine, M. 2007. Mandates complicate HPV vaccine debate. *Washington Post. www.washingtonpost.com/wp-dyn/content/article/2007/03/16/AR2007031601133.html.*

National Electronic Library for Medicines (NeLM). 2007. HPV vaccine recommended for NHS immunisation programme. *www.nelm.nhs.uk/en/NeLM-Area/News/492225/492500/492511.*

National HPV and Cervical Cancer Prevention Resource Center. 2008. Learn about HPV: Myths and misconceptions. *www.ashastd.org/hpv/hpv_learn_myths.cfm.*

O'Brien, E. 2007. Two Canadian provinces to offer HPV vaccination to grade-school girls. *www.lifesite.net/ldn/2007/aug/07089704.htm.*

Office of the Governor. 2007. Gov. Perry establishes HPV vaccination program for young women. *www.governor.state.tx.us/divisions/press/pressreleases/PressRelease.2007-02-02.0949.*

USA Today. March 2, 2007. Worth a shot? *http://blogs.usatoday.com/oped/2007/03/post_6.html.*

Section VI:
Science and the Media

In an essay in the *Wall Street Journal*, Matt Ridley (2010) argues that effective communication among *Homo sapiens* is at the heart of their eventual triumph over the larger-brained and physically stronger Neanderthals. Perhaps.

Human communication has improved dramatically since the era of human origins, moving from gestures and calls to speech and writing, amplified by the printing press, telephone, telegraph, radio, and internet. For millennia, news of discoveries and inventions in one kingdom took years to reach distant lands, if at all; now, such news is transmitted instantly across the world.

The invention of the printing press in 1440 changed everything about the process of science. But it was not until scientific societies began publishing their own journals in the 1660s that the scientific enterprise began to coalesce into the structured enterprise we know today with its accepted rules of conduct. Even back then, however, scientists were wary of being scooped—of having their ideas attributed to someone else. Isaac Newton, ever the neurotic, kept notes about his experiments and mathematics in a personal code so that no one could decipher his musings. Today, priority of discovery plays a vital role in the dynamics of science. Credit for discoveries usually goes to those who get their work published first.

The rush to publish the first draft of the human genome was chronicled closely by the press and culminated in simultaneous publication by two groups, one headed by the maverick scientist Craig Venter and the other by the director of the National Institutes of Health (NIH), Francis Collins. The race ended in a tie. The competition to be the first to determine the structure of DNA was recounted in a book, *The Double Helix*, written by one of the winners, James Watson. Again, first publication of Watson and Francis Crick's short paper in the journal *Nature* determined who won. Publication has become an integral part of the social fabric of the scientific enterprise and the world as a whole. It not only constitutes priority, but it advances careers, determines promotions and tenure, and is essential for winning grants.

Section VI

In the past, scientists waited to announce their findings in scientific publications reviewed by their peers. This time-honored process takes time, however, and more and more scientists are turning to the press in a bid to capture both the public's attention and, they hope, the attention of supporters and funding sources as well. Those who do not, ignore the press at their peril. The 24-hour news cycle has produced an insatiable demand for anything and everything. This has led scientists to make public statements about their research that have not been vetted by the traditional peer-review process, occasionally with disastrous consequences. The cold fusion affair is a notorious example, with two chemists who thought they had discovered a cheap way to generate energy announcing their breakthrough "discovery" at a press conference. Had they followed a rigorous review process, their errors would have been discovered, and the public furor and embarrassment they and their institutions suffered could have been avoided.

Scientists have an uneasy relationship with the press; they depend on it but are suspicious of it, especially the popular press, with its tendency to sensationalize. Some scientists have taken it upon themselves to try to educate the general public about their business in the hope that they can improve awareness and understanding of their work. Books, articles, television, radio, and movies all play an important role in shaping the public's perception of science and its ethical canons. Astronomer Carl Sagan was a master at communicating with the public through his books, both nonfiction and fiction, and through his television series *Cosmos*. He, like many others, was chastised by his colleagues, who sniffed that communicating with people directly via public media signaled that he was not a serious scientist. Only recently has this stigma begun to fade. Stephen Hawking, arguably the most famous theoretical scientist alive, has written two well-received nonfiction books for lay readers without falling from grace.

The role of the media in the scientific enterprise is both complex and enormous, and it figures to some degree in all of the cases in this book. We follow some of its specific threads in the case studies in this section.

In the first, "Tragic Choices: Autism, Measles, and the MMR Vaccine," we deal with a sensational case that chronicles how a scientist manipulated data to claim that vaccinations cause autism. His publication, since withdrawn because of fraud, was of enormous popular interest and received massive television coverage. The reports prompted many parents to forgo vaccinating their children and resulted in the rise of measles and other childhood infectious diseases.

The second case in this section, "Ah-choo! Climate Change and Allergies," asks students to step into the shoes of scientists who work for a public relations firm. They must evaluate data on the role of climate change in increased pollen counts and the resulting increase in people suffering from allergies. Then students must design a PR campaign, including developing a brochure, to broadcast their views. The use of teams and peer evaluation is a vital part of the exercise, which is spread over several days.

The third case, "Rising Temperatures: The Politics of Information," also takes climate change as its theme, but this time students see how news reporters and editors can alter the way a story is pitched. We see conflicts everywhere: within the scientific community, between scientists and politicians, and among people in society, all with differing political agendas. Is "truth" anywhere in the mix?

The final case in the section, "Eating PCBs from Lake Ontario," deals with the pollutant PCBs and how their presence in the fish of Lake Erie may be altering women's menstrual cycles and pregnancies. Students must analyze a news release reporting on an actual research study. The case emphasizes experimental design and data analysis, but the media are clearly involved in the story, because there are missing pieces that the reporters did not provide. Serious citizens reading such news stories often are hard pressed to find the truth of the matter based on the information provided.

The media play a role in all science. They both reflect and mold society's values and beliefs. Scientists from childhood on learn about the world through the media, including books, television, and the internet. And as professors and experimentalists, their understanding is shaped by articles and journals, and it is to those very journals that they turn to publicize their discoveries. No matter how much they may occasionally rail at the shortcomings of TV columnists, bloggers, and reporters as to how they distort or mangle the facts, scientists are beholden to the media.

Reference

Ridley, M. *Wall Street Journal.* 2010. Humans: Why They Triumphed. May 22.

Chapter 27
Tragic Choices: Autism, Measles, and the MMR Vaccine

Matthew Rowe

The Case

Part I: The Choice

Kristen typically loved the monthly Sunday brunches with her mother, Anne, and older sister, Carly. Kristen adores them both. Her mother was, in Kristen's eyes, a saint, having worked two jobs as a single mom to not only raise both girls but also put them through college. And Carly was that great older sister, always supportive, helping Kristen with her math homework when Mom was working the night shift at a local restaurant. Carly graduated from East Texas State University four years ago; she's now married and has a four-year-old son named Ian and is a successful graphic artist in Dallas. Kristen completed a BA in marketing last year, works as a sales manager for Time-Warner Cable in Austin, is also happily married, and has a 14-month-old daughter named Alissa. The monthly "girls day out" is an excuse for the three women to celebrate the love and commitment that helped them through the difficult times when Anne was working 80 hours a week and Carly was helping care for Kristen. The monthly brunches are also an opportunity for Carly's and Kristen's husbands to bond with their young children, who (although Anne loves them dearly) are not invited to the brunches.

The last gathering, however, had been strained. Following Kristen's graduation, Anne courageously tackled her lifelong dream and enrolled in a nursing program at East Texas State. After the first year of course work, Anne was near the top of her class. "Not bad," thought Kristen, "for a 50-year-old ex-waitress from Fort Worth." Anne's increasing expertise in health issues made her, once again, a resource for the entire clan, as Ian had been diagnosed with autism spectrum disorder (ASD) just four months after his first birthday. Autism can take a heavy toll on a family, especially those with children exhibiting the most extreme form of the disease. Ian unfortunately was in this group—regressing from a happy and interactive one-year-old to a toddler who was completely unresponsive to his parents and whose long bouts of

Chapter 27

repetitive rocking were interspersed with brief but intense periods of uncontrollable aggression and self-abuse. Anne's love for her daughters and grandchildren, coupled with her access to the medical school's library, makes her a knowledgeable and compassionate counsel for Carly.

Two seemingly random events led to an uncharacteristic and heated exchange at their last gathering. Anne was taking a course in microbiology, and Carly had stumbled onto an episode of *Larry King Live* while channel surfing.

Kristen was running late. When she was directed to a patio table to join her mother and sister, she found them engaged in the same battle left unresolved a month ago.

"Mom, Jenny McCarthy is just one of many celebrities speaking out against the measles vaccine—and she has the right to. Her own son is autistic, and Jenny is convinced that the combined measles, mumps, and rubella (MMR) vaccine is the cause. She and actor Jim Carrey made a very convincing case on Larry King's show, while the scientists looked to be in bed with big pharmaceutical companies, Big Pharma as they call it, to make money from selling vaccines. It's a huge business. And there has to be something behind this epidemic of autism. Did you know that the MMR vaccine replaced the old simple measles vaccine in 1988? And did you know that during the 20 years following its introduction, autism rates increased by almost 600%? And the rates are still increasing—one in every 110 kids today develops some form of ASD." (See Figure 27.1.)

Anne tried not to sound smug or motherly in her response, but a recent topic in her microbiology course dealt with the amazing benefits of vaccines, including the shot (or "jab," as it was known in Great Britain) against measles.

"Carly, both the original and the more effective MMR vaccines have nearly eliminated a serious and deadly disease. The measles virus kills. A paper we had to read for class showed that during the 20 years following its release in 1963, the measles shot prevented 50 million cases of measles and saved at least 5,200 lives in the United States alone. And that's not all. A common complication from measles is encephalitis, a massive swelling of the brain, which can lead to brain damage and mental retardation. During that same 20-year period, at least 17,000 American kids were spared that tragedy. And I don't believe the link between MMR and autism has been firmly established."

Carly interrupted before Anne could finish. "I know the link has been made," she said with more than a hint of anger in her voice. "Just look at Ian. He was a happy and interactive one-year-old. I take him in for his MMR shot at 15 months, just like my pediatrician said to, and a week later he stops smiling, he stops talking, and he stopped looking me in my eyes—he just slipped away. Jenny McCarthy has spoken with hundreds of parents of autistic children, all caused by the vaccine, and a highly respected scientist in Great Britain proved the connection."

"But Carly, epidemiological studies show that if less than 90% of a population receives the measles vaccine, the disease will return, killing ...."

"Mom, stop, just stop right there! I refuse to sacrifice my children for the greater good. I won't do it, I can't do it anymore. I've lost Ian. And I'm responsible. I thought the vaccine would help, I was told it would help, and I held his tiny little hand while he took it. He didn't even cry. Now he is gone. And I'm scared. We're running out of time. Kristen has to choose. I don't want her to carry around the guilt that I will carry forever. I don't want us to lose Alissa, too."

Part II: The Connection

Carefully examine the evidence presented in Figure 27.1 and Table 27.1 (p. 240); then answer the questions following Table 27.1.

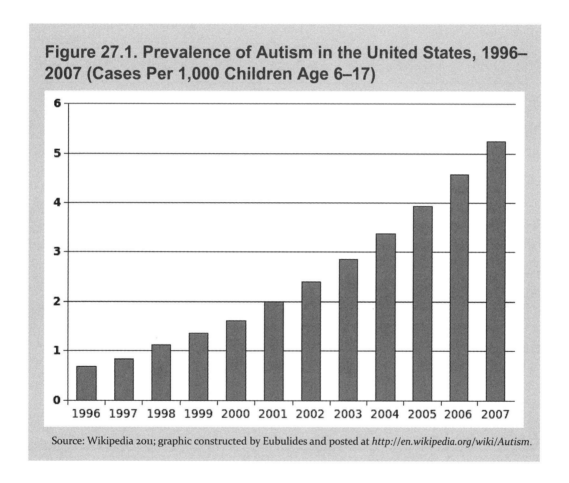

Figure 27.1. Prevalence of Autism in the United States, 1996–2007 (Cases Per 1,000 Children Age 6–17)

Source: Wikipedia 2011; graphic constructed by Eubulides and posted at *http://en.wikipedia.org/wiki/Autism*.

Table 27.1 has been taken from the scientific investigation referred to by Carly when she snapped at her mother that "a highly respected scientist in Great Britain proved the connection" between autism and the MMR vaccine.

Chapter 27

Table 27.1. Neuropsychiatric Diagnosis and Interval Post-MMR Vaccination*

Child	Behavioral Diagnosis	Causal Agent (Identified by Parents or Pediatrician)	Interval From Exposure to Symptoms
1	Autism	MMR	1 week
2	Autism	MMR	2 weeks
3	Autism	MMR	48 hours
4	Autism? Disintegrative disorder?	MMR	Dramatic deterioration immediately after MMR booster at 4.5 years
5	Autism	None, but MMR at 16 months	Self-injurious behavior started at 18 months
6	Autism	MMR	1 week
7	Autism	MMR	24 hours
8	Post-vaccinial encephalitis?	MMR	2 weeks
9	Autistic spectrum disorder	Recurrent inner-ear infections	1 week (MMR 2 months previously)
10	Post-viral encephalitis?	Measles (previously vaccinated with MMR)	24 hours
11	Autism	MMR	1 week
12	Autism	None, but MMR at 15 months	Developmental deterioration noted at 16 months

Note: Adapted from Table 2 in Wakefield et al. (1998), p. 639.

Questions

1. What should Kristen do?
2. How long does she have to decide?

Part III: The Conference

On February 28, 1998, the same day that his article was to be published in *The Lancet*, Dr. Andrew Wakefield called a press conference. He announced that he and his team believed they had found a link between a routine childhood vaccine and autism. Of the 12 children in their study, 8 had developed this neurological disorder within a few days of receiving the MMR "jab." Somehow the vaccine was causing intestinal inflammation, allowing harmful proteins to enter the child's bloodstream and eventually travel to the brain, causing the disorder. To prevent this, parents simply needed to request that their pediatrician separate the MMR vaccine into three separate shots, one each for measles, mumps, and rubella, the same way these immunizations had been delivered prior to the development of the combined vaccine.

Question

How do you think the public reacted to Wakefield et al.'s findings?

Part IV: The Consequences

Figure 27.2 shows what happened in Great Britain. The picture in the United States is similar. After examining these patterns, answer the questions on page 242.

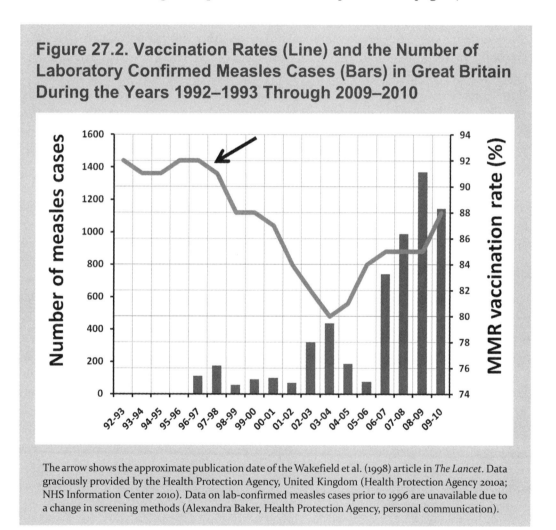

Figure 27.2. Vaccination Rates (Line) and the Number of Laboratory Confirmed Measles Cases (Bars) in Great Britain During the Years 1992–1993 Through 2009–2010

The arrow shows the approximate publication date of the Wakefield et al. (1998) article in *The Lancet*. Data graciously provided by the Health Protection Agency, United Kingdom (Health Protection Agency 2010a; NHS Information Center 2010). Data on lab-confirmed measles cases prior to 1996 are unavailable due to a change in screening methods (Alexandra Baker, Health Protection Agency, personal communication).

Chapter 27

Questions

1. Did Wakefield et al.'s data and interpretations merit the resulting public fear of the MMR vaccine? Why or why not?

2. Design a better study to examine the purported cause-and-effect relationship between the MMR vaccine and autism. Be prepared to discuss the details of your study design.

Teaching Notes

Introduction and Background

This case was developed for and has been used in an introductory science course for nonscience majors. The primary goal of the course, and of this case, is to help undergraduates better understand the nature of science. Explicitly, we want the science-timid college student to leave the university with an appreciation of the power and limitations of science as a way of knowing. This can best be accomplished by helping students understand the distinctions between good science, not-so-good science, and pseudoscience. Similar to the cliché about real estate ("location, location, location"), the three most important rules in science should be "evidence, evidence, evidence"—coupled, of course, with a healthy dose of critical thinking.

In this case, we tackle a hot topic that many students are aware of, namely, the purported connection between the measles, mumps, and rubella vaccine (MMR) and autism. While Anne, Carly, and Kristen are fictitious, the underlying details of this case are real, starting with Wakefield et al.'s (1998) provocative publication in *The Lancet* and culminating, just recently, with *The Lancet*'s retraction of that article (Editors of the *The Lancet* 2010).

The hard-hitting investigations of London *Sunday Times* reporter Brian Deer revealed the shocking conflicts of interest that certainly biased Wakefield's pronouncements. Meanwhile, celebrity Jenny McCarthy and Generation Rescue continue their campaign to scare the public into believing vaccines cause autism, including her recent appearance on CNN (*Larry King Live* 2009). As a result, many people are choosing not to vaccinate their children, leading to a resurgence of childhood diseases, including outbreaks of measles in Great Britain, the United States, and Canada.

The entire exercise is designed to work in a single 50-minute class period, but it can easily be expanded to an 80-minute block or two 50-minute blocks. The case presumes no advanced knowledge on the part of the student regarding autism, immunology, or experimental design.

Objectives

After completion of the case, students should be able to do the following:

- Critically analyze and evaluate a scientific claim

- Identify the control group and the experimental group in a study
- Understand the distinctions between clinical studies and epidemiological studies
- Understand the importance of sample size in scientific investigations
- Understand the importance of replication when evaluating scientific claims
- Understand how undeclared conflicts of interest can lead to bad science
- Appreciate that peer review is an important fire wall against bias and bad science but is not infallible
- Design their own experiments to test a causal relationship between vaccines and autism
- Appreciate the emotional, medical, and economic difficulties faced by families touched by autism, especially severe forms of the disorder
- Understand the importance of herd immunity in protecting communities against disease
- Evaluate the ethical implications of choosing whether or not to immunize one's children
- Employ critical thinking when analyzing media reports about a scientific claim
- Appreciate that thinking scientifically is an important (and learnable) skill that will help them make intelligent choices for themselves, their families, and society

Common Student Misconceptions

- Vaccines cause autism.
- Correlation proves causation.
- Talk shows (e.g., *Oprah*, *Larry King Live*) are good sources for health advice.
- Every article published in a highly respected scientific journal (e.g., *Science*, *Nature*, *JAMA*, *NEJM*, *The Lancet*) must be true.
- Thinking scientifically is only important for scientists.

Classroom Management

Many, but not all, students will be familiar, some personally, with the continuum of developmental and neurological disorders referred to collectively as autism spectrum disorder (ASD). We have found it useful to begin the discussion by asking students to share what they know about ASD, which ranges from Asperger's syndrome at one end to severe autism at the other. It is not important that students understand subtle differences among the various ASD classifications, but they should appreciate the difficulties facing parents of severely autistic children. To this end, we often show a short YouTube clip. The important point is that students realize why parents of severely autistic children are desperate not only for a cure for the disorder but also an understanding of its cause. Following this brief (5–10 minute) introduction, we move to Part I.

Part I: The Choice
Although we provide each student with a copy of the dialogue between Kristen, Carly, and Anne, feedback from previous classes indicates that many undergraduates enjoy having the story read aloud while they follow along; also, when doing this, changing a few phrases here and there during the recitation helps hold the students' attention. After reading the story, which takes about 5 minutes, proceed to Part II.

Part II: The Connection
We start this block by providing each student with his or her own copy of two different data sets. The first, Figure 27.1, shows the dramatic rise of autism in the United States (Wikipedia 2011). It is better to reserve a discussion of whether the apparent increase is real or artificial for Part IV of the case study. Here, it is important that students simply understand Carly's reference in the dialogue to an "autism epidemic." The second handout, Table 27.1, shows the actual data presented by Wakefield et al. (1998) in their now infamous *Lancet* article. It was this paper, along with the press conference called by Andrew Wakefield the day the article was published, that ignited (or at least significantly fueled) the antivaccine movement in the United States and Great Britain. While distributing the table, we let the students know that *The Lancet* is a highly reputable, peer-reviewed scientific journal.

We allow the students, working in groups, approximately 5 minutes to discuss the two questions that end this section (i.e., What should Kristen do, and how long does she have to decide?); we request that the members of each group try to reach consensus. We then have the groups share their answers, tallying responses on the chalkboard.

Part III: The Conference
This brief summary of Wakefield's conference statements can be read to the students or distributed as a handout. We direct the question (How do you think the public reacted to Wakefield's findings?) to the entire class—they need only a minute or two to offer their suggestions.

We transition between Part III and Part IV by providing each student with a copy of Figure 27.2 (which illustrates the declining uptake rate of MMR and resulting increase in measles cases in Great Britain).

Part IV: The Consequences

To appreciate the public health consequences resulting from Wakefield's press conference and *Lancet* article, it is important that students understand the concept of herd immunity. While the students are examining Figure 27.2, we give them a short and informal introduction to the topic. We make the following points: (1) Immunizations are not safe for everyone, including children with suppressed immune systems. (2) Herd immunity occurs when a sufficient fraction of a host population is immune to infection, either because individuals have been vaccinated or because they were previously exposed. (3) Epidemiologists estimate that if 92–95% of a human population were vaccinated against measles, the disease would die out.

Prior to Wakefield's press conference, immunization rates in Great Britain averaged 91%; following the media frenzy, uptake rates dropped below 80% and, in selected metropolitan areas, approached 50%. There were only 56 confirmed cases of measles in Great Britain in 1998, the year of Wakefield's pronouncements; in 2008, there were 1,348. Similar outbreaks (of both measles and mumps) have occurred in the United States and Canada.

The two questions/exercises that close Part IV (i.e., "Did Wakefield's results merit the public panic?" and "Design a better study.") are the pedagogical cornerstones of this case. We give the groups approximately 10 minutes to discuss both questions together, then have them "report out," dealing with each question in turn. The discussion that ensues can easily take 10–20 minutes or longer.

Web Version

The case and its complete teaching notes, references, and answer key can be found on the website of the National Center for Case Study Teaching in Science at *http:// sciencecases.lib.buffalo.edu/cs/collection/detail.asp?case_id=576&id=576*.

References

Editors of *The Lancet*. 2010. Retraction—Ileal-lymphoid-nodular hyperplasia, non-specific colitis, and pervasive developmental disorder in children. *The Lancet* 375: 445.

Larry King Live. 2009. Jenny McCarthy and Jim Carrey discuss autism; medical experts weigh in. Interview by Larry King. April 11. *http://transcripts.cnn.com/TRANSCRIPTS/0904/11/lkl.01.html*.

Wakefield, A. J., S. H. Murch, A. Anthony, J. Linnell, D. M. Casson, M. Malik, M. Berelowitz, A. P. Dhillon, M. A. Thomson, P. Harvey, A. Valentine, S. E. Davies, and J. A. Walker-Smith. 1998. Ileal-lymphoid-nodular hyperplasia, non-specific colitis, and pervasive developmental disorder in children. *Lancet* 351: 637–41. [retracted 6 February 2010]

Wikipedia. 2011. Autism. Wikipedia. *http://en.wikipedia.org/wiki/Autism*.

Videos

Autism Every Day (7 min. version): *www.youtube.com/watch?v=FDMMwG7RrFQ*

Dominic's autistic meltdowns: *www.youtube.com/watch?v=h-tGimFAhMY*

Chapter 28
Ah-choo! Climate Change and Allergies

Juanita Constible, Luke Sandro, and Richard E. Lee Jr.

The Case

Scenario

You work for ScienceSpeak, a public relations firm that educates the public about scientific issues. Your company has won a contract with the World Health Organization (WHO) to supply materials for their new multimedia public health campaign about climate change. The WHO is specifically interested in the relationship between climate change and the increasing prevalence of allergies and asthma worldwide.

Your boss calls a meeting to discuss the contract. She gives you a set of data tables (see "Web Version," p. 252) prepared by two expert scientists to summarize recent evidence on the effects of increasing carbon dioxide and temperature on allergenic plants. Your job is to design and produce a communication product such as a brochure, poster, web page, or television program that informs the public about potential links between climate change and allergies.

Procedure

Note: The meeting descriptions below are intended as guides only. Be prepared to spend time on this project outside class and to have additional meetings if necessary.

Meeting 1: Evaluate Data

1. Examine the data prepared by your company's experts. Write a summary statement (one or two sentences) to describe the main trend for each table in the Planning Worksheet (see "Web Version"). Remember that the public is interested in the big picture.

2. Choose your target audience. For example, are you trying to reach children or adults? People who live in the city or in the country?

Science Stories: Using Case Studies to Teach Critical Thinking

3. Use the Planning Worksheet to rank your summary statements in order of importance (1 = most important, 6 = least important). For each data table, consider the following:
   a. The strength of the scientific evidence. For example, how were the data collected? Are there sufficient data to support a strong claim? How consistent is a response across plant species? Across geographical areas?
   b. The relevance of the data to your target audience. For example, where does your audience live? What plant species might they encounter? In what types of professional or recreational activities are your audience members engaged?
4. Explain why you ranked the top two statements the way you did.
5. On the back of the Planning Worksheet, brainstorm possible messages for your communication product.

Meeting 2: Generate Strategies

1. Use the results of your brainstorming session at the end of Meeting 1 (see "Web Version") to decide what your main message will be. Remember, a respected international organization will use your product to inform the public about an important issue. Make sure your message is clear, factual, scientifically accurate, and catchy.
2. Decide how you will communicate your message to your target audience. You may choose a product from this list: brochure, poster, web page, or television program (you may provide the script or a video). (Get your instructor's approval if you have another idea.)
3. Assess your knowledge by answering these questions:
   a. What do we know?
   b. What do we need to know?
   c. How can we find out what we need to know?
4. Before your next meeting, you will have to collect the information you need to complement the data provided by the experts. Decide what tasks need to be completed before the next meeting. Make sure everyone in the group gets an equal amount of work.

Meeting 3: Produce Draft of Product

1. Review what each team member has learned since Meeting 2.
2. Outline your product. At a minimum, your product will need to include the following:

- An introduction that catches the reader's or viewer's interest and gets your message across
- A summary of your top-ranked data that includes at least two graphics. Even within a given table, not all the information is equally important. You will need to emphasize the critical points. You should include at least one graph, but you also may use maps, flow charts, diagrams, or other graphics.
- A summary of your less important data
- A conclusion that reinforces your message and ties loose ends together

3. Produce a draft of your product based on the outline you made in step 2. You will have time to revise your draft later.

Meeting 4: Conduct Peer Review

Scientists (and many other professionals) use a process called peer review to improve their work. Before a scientific document can be published, it must be approved by a panel of fellow scientists. The procedure you are about to follow incorporates elements of a scientific peer review:

Your group will present its work to another team to be evaluated. Give one copy of the Peer Review Form (see "Web Version") to each person on the other team. Make sure your group name and intended audience are filled in on the form you give to the other team.

Another group will present its work to you. Each member of your team will provide an individual review of the product by

- providing positive feedback on what worked well in the product;
- suggesting changes that will help the product, and pointing out any errors or overlooked or misinterpreted data; and
- thoroughly evaluating the product according to the Peer Review Form.

Reviewers should be strict but fair in their assignment of points to each category.

Meeting 5: Revise Product

1. Incorporate the peer-review suggestions you think would improve your communication product. One member of your team should keep a record of each suggestion you actually use. Effective implementation of appropriate suggestions will mean a higher score for your product.

2. Make final adjustments to your product.

Chapter 28

Meeting 6: Present Final Product

1. Give the Peer Review Forms from Meeting 4 back to the respective reviewers.

2. Present your product to the class. When the team you evaluated in Meeting 4 presents its product, repeat the peer review process.

3. Submit the Planning Worksheet, the Peer Review Forms, and your final product to your instructor.

Teaching Notes

Introduction and Background

Pollen is a major risk factor for respiratory allergies and asthma, which together affect 10–50% of people worldwide (Bousquet, Dahl, and Khaltaev 2007; ARIA 2007). The prevalence of chronic respiratory disease is difficult to measure. Allergies and asthma are underreported, and reporting methods are inconsistent across health agencies.

Several national and international agencies report, however, that allergies and asthma are becoming more prevalent (Bousquet, Dahl, and Khaltaev 2007). Although declines in indoor and outdoor air quality are largely to blame, a growing body of evidence suggests that the upswing is also related to climate change. In today's warmer, more CO_2-rich atmosphere, allergenic plants seem to be producing increasing quantities of highly allergenic pollen throughout more of the year (Beggs and Bambrick 2005).

When the pollen from an allergenic plant contacts mucus membranes in the nose, mouth, or lungs, it releases water-soluble allergens. In an allergic person, these allergens trigger an inflammatory response called allergic rhinitis (hay fever) that can include sneezing and itching. Allergic rhinitis is serious business in the United States, costing the health-care system more than $4.5 billion and resulting in 3.8 million missed work and school days per year. Allergic rhinitis can also be associated with other diseases, most notably asthma (AAAAI 2000).

In this problem-based learning activity designed for nonscience majors, students assume the role of scientists working for a public relations firm. Their assignment is to work in teams to design a communication product linking climate change to pollen allergies. The students are given data to work with for four allergenic groups of plants (oak, birch, ragweed, and grass). The data are presented in a series of tables that deal with changes in the start of pollen season in Europe and experimental data on how pollen production changes with CO_2 levels and temperature conditions.

Objectives

In completing the activity, students develop a variety of process skills critical to scientists, including

- working collaboratively,
- interpreting and prioritizing data,
- telling a story through graphics and text, and
- defending an argument.

Common Student Misconceptions

- The things I learn in science class are only useful if I want to become a scientist.
- Climate change doesn't affect me personally.

This case would be most suitable for a lower-level course in biology, human health, or environmental science.

Classroom Management

Before the activity starts, students should understand that our climate is warming as a result of an increase in carbon dioxide (CO_2) and other greenhouse gases. Instructors may wish to show students the clip from the documentary *An Inconvenient Truth*, in which Al Gore discusses the relationship between CO_2 and global temperatures. Alternatively, instructors could show side-by-side graphs of CO_2 concentrations and temperatures and ask students to describe the relationship between the two lines (e.g., *www.ipcc.ch/graphics/graphics/2001syr/large/05.16.jpg* or *http://rst.gsfc.nasa.gov/Sect16/Web-20vost04k.jpg*).

As a guideline for students, we split this activity into six "meetings." At a minimum, Meetings 1, 4, and 6 should be completed in class. Use of additional class time is at the discretion of individual instructors.

Assessment

There are multiple ways to prioritize the data, but students should be prepared to defend their rankings. When they make their final presentations, instructors should look for reasoning about the strength of the scientific evidence and relevance to each group's target audience. For example, the data suggest that birch trees are producing more allergenic pollen in warmer microhabitats, but that does not mean we can rule out the effect of other environmental factors (e.g., soil type, slope, and aspect). Although the data might be relevant to some audiences, the study results are not that convincing. Therefore, the table should not be top ranked.

Students and instructors can use the same Peer Review Form for assessing the group projects that is included in the case for student use. Instructors should fill in the Points column before duplicating the form for the class. We recommend that students receive a grade for this project based on a combination of a group score and an individual score. The individual score can be drawn from the quality of an individual's Peer Review Form, self-evaluation, peer evaluations of the group process, or instructor evaluations of research notes collected by each student.

Web Version

This case—including the data tables, planning worksheet, and peer review form—as well as complete teaching notes, resources, and references, can be found on the website of the National Center for Case Study Teaching in Science at *http://sciencecases.lib.buffalo.edu/cs/collection/detail.asp?case_id=465&id=465*.

References

Allergic Rhinitis and Its Impact on Asthma (ARIA). 2007. ARIA teaching slides (PowerPoint download). *www.whiar.org/docs.html*.

American Academy of Allergy, Asthma and Immunology (AAAAI). 2000. *The allergy report*. Milwaukee, WI: AAAAI.

Beggs, P. J., and H. J. Bambrick. 2005. Is the global rise of asthma an early impact of anthropogenic climate change? *Environmental Health Perspectives* 113 (8): 915–919.

Bousquet, J., R. Dahl, and N. Khaltaev. 2007. Global alliance against chronic respiratory diseases. *Allergy* 62 (3): 216–223.

Chapter 29
Rising Temperatures: The Politics of Information

Christopher V. Hollister

The Case

In their *New York Times* article "Report By EPA Leaves Out Data On Climate Change," Andrew Revkin and Katharine Seelye (2003, p. A1) describe how a soon-to-be-released U.S. Environmental Protection Agency (EPA) report on the state of the environment had been edited by the White House to "play down" a section describing the risks and causes of global warming. The editing eliminated references to many studies that concluded global warming is at least partly caused by concentrations of smokestack and tailpipe emissions, and that these elements could threaten human health and the health of the environment.

In the EPA report, White House officials deleted information from one study that concluded that global temperatures had risen sharply in the previous decade when compared to the previous 1,000 years. Officials substituted for this a reference to another study, partly financed by the American Petroleum Institute, questioning that conclusion. Also, in the report's summary statement on the climate, the White House removed the sentence "Climate change has global consequences for human health and the environment." This wording was replaced with the following: "The complexity of the Earth's system and the interconnections between its components make it a scientific challenge to document change, diagnose its causes and develop useful projections of how natural variability and human actions may affect the global environment in the future" (Revkin and Seelye 2003).

Drafts of the original EPA report, with the White House's changes, were leaked to the *New York Times*, along with internal agency memoranda in which some agency officials protested the changes. Two EPA officials, speaking under the condition of anonymity, confirmed that these documents were authentic. What overriding message is being communicated by the *New York Times* article?

In an interview responding to questions concerning the Revkin and Seelye article, the EPA's administrator, Christine Todd Whitman, stated, "As it [EPA] went

Chapter 29

through the review, there was less consensus on the science and conclusions on climate change. So, rather than go out with something half-baked, or not to put out the whole report, we felt it was important for us to get this out, because there is a lot of really good information that people can use to measure our successes." Also in response to questions about the article, James Connaughton, chairman on the White House advisory group Council on Environmental Quality, said, "It would be utterly inaccurate to suggest that this administration has not provided quite an extensive discussion about the state of the climate. Ultimately, EPA made the decision to not include the section on climate change because we had these ample discussions of the subject already" (Revkin and Seelye 2003).

Five days after the release of the *New York Times* article, another article appeared in the *Wall Street Journal* titled "EPA Is Upbeat, But Seeks Study of Rise in Asthma." This article, also about the EPA's report on the environment, was written by staff reporter John Fialka (2003). The article begins, "The U.S. Environmental Protection Agency issued its first overall assessment of the nation's environment, saying that the quality of the nation's air, water and land appears to be improving or 'holding its own' under the increasing stress of a growing economy and a growing population."

In the article, Fialka provides a selection of some of the report's major findings:

- "Overall air quality has improved, based on measures of six pollutants that have declined 25% over the last 30 years."
- "In 2002, 94% of the nation's population had access to community water systems that meet health-based drinking-water standards, as opposed to 79% in 1993."
- "Uncontrolled releases of toxic wastes have dropped 48% since 1988, and the amount of municipal solid waste recovered by recycling or composting has increased 1,100% in the last decade."

The Fialka article contains one sentence regarding White House editing of the EPA's report: "A two-page summary of the EPA's view was deleted after the White House and other federal agencies disagreed with the agency's assessment that man-made emissions of carbon dioxide and other gases are artificially warming the atmosphere." What overriding message is being communicated by the *Wall Street Journal* article?

Questions

1. Is this case study a primary or secondary source of information?
2. What overriding message is being communicated by the information in this case study?

3. What information is missing from this case study?

4. List two possible reasons (or motives) for the publishing of each of the two articles cited in this case.

5. Which of the following best describes the information presented in this study?
   a. Differing opinions regarding which parts of the EPA report are newsworthy
   b. Conflicts within the science community
   c. Differing political agendas
   d. Conflicts between scientists and politicians

6. List two reasons for your answer to question 5.

7. Based on this case study and your course readings, what criteria should one use to evaluate news stories?

Teaching Notes
Introduction and Background
This is an interrupted case where students work in small groups to analyze and critically evaluate the often political nature of news stories. The case is one of several used to teach the University at Buffalo's undergraduate information literacy course, ULC-257: Introduction to Library Research Methods. The subject matter of this case is also suitable for undergraduate-level courses in environmental studies, journalism, or political science.

One of the primary goals for the ULC-257 course is for students to hone their abilities to analyze and critically evaluate information. Using case studies that involve current or controversial news stories generates enhanced student interest and engagement as well as lively classroom discussions.

This particular case was developed from two newspaper articles that were published in different newspapers—the *New York Times* and the *Wall Street Journal*—about the release of an Environmental Protection Agency (EPA) report on the state of the environment. While the *New York Times* article discusses the White House editing of the report, eliminating several references to the causes and dangers of global warming, the *Wall Street Journal* article is more focused on the report's evidence of environmental improvements. Possible reasons for the differing viewpoints of these two news stories are investigated.

When this case is introduced, ULC-257 students have already learned about different types of information, the organization of information, and advanced methods for finding information. The case is used for a section of the course that is focused on analyzing and critically evaluating information. In the classes leading up to this case, students have learned how to distinguish between scholarly and more general books, periodicals, and websites; how to judge the authority of these sources; and

Chapter 29

what criteria should be used to evaluate them. Also, by this time of the semester, students are working in established small groups for classroom activities. They are aware of their responsibilities within their groups and accustomed to the case study method of learning course material.

Objectives
In completing this case, students will

- sharpen their skills to analyze and critically evaluate news stories,
- question the possible motivations for and influences on news stories, and
- identify specific criteria for evaluating news stories.

Common Student Misconceptions

- Scientific research is unaffected by political authority.
- News reporting on scientific research is unaffected by political authority.
- Prestigious news sources are not influenced by political biases.

Classroom Management

This case can be taught in a single 50-minute class period, as shown below in Table 29.1.

Table 29.1. Suggested Case Study Breakdown (50 min.)

Task	Time
Read case; review interrupted case questions	10 min.
Review questions 1–3; progressive disclosure	10 min.
Review question 4; progressive disclosure	10 min.
Review questions 5–6; progressive disclosure	10 min.
Review question 7	5 min.
Summarize for case learning objectives	2 min.
Show objectives that were covered	3 min.

Before work on the case begins, students are reminded to focus their attention on the case's informational aspects and avoid debating political ideologies. Students are, however, informed that they may supplement their answers, group discussions, and group reports with prior knowledge of the subject matter, but only if it is factual. The instructor helps ensure that group discussions do not go astray from the learning objectives or become overtly political. Students begin the class already seated in their established working groups and are given 5 minutes to read the case. The instructor uses think-pair-share and progressive disclosure methods (see below) for reviewing the case questions.

Interrupted Case Method

Students are instructed to stop reading after the third paragraph of the case and to consider the question "What overriding message is being communicated by the *New York Times* article?" At the end of the case and before a review of questions 1–7, students are asked to consider the question "What overriding message is being communicated by the *Wall Street Journal* article?" Think-pair-share (see below) is used to review these questions.

Think-Pair-Share

Students are asked to answer the case's questions individually, share those answers within their groups, decide as a group how to best answer each of the questions, and then report the group's answers to the class. Reporting responsibilities rotate within the group from answer to answer. To ensure proper focus on the case's informational aspects, the instructor visits each group discussion. Subtle intervention is sometimes required, though the instructor must remain unbiased and allow students to draw their own conclusions. The instructor then facilitates group reporting to the class.

Progressive Disclosure

Groups are asked to answer case questions 1–3 and report those answers to the class. Afterward, the instructor provides information that is not only related to the case but also is particularly relevant to questions 1–3, which segues into question 4. The same method of progressive disclosure is used after groups report on question 4, and also after groups report on questions 5 and 6. After reports on question 7, the instructor summarizes the case based on specific learning objectives.

The questions are designed to make students think critically about how information is presented in the two articles cited in the case. After group reports on questions 1–3, the instructor asks students to consider that the EPA report was released in June 2003, that George W. Bush was in his first term as president of the United States, and that the country and both major political parties were preparing for the 2004 presidential election. The instructor also reads from the transcript of a National Public Radio interview with Andrew Revkin (one of the authors of the *New York Times* article cited in the case):

> In the case of climate change, President Bush is really up against a hard place because, at least many analysts say, the administration is so tied to two things, the energy interests, coal in particular, and oil also, and they're so tied to their base, the political base, which is very conservative, and that base doesn't believe in climate change, or in global warming, and basically, he's kind of stuck. (Revkin 2003)

Chapter 29

After group reports on question 4, the instructor reads from an interview with Arthur O. Sulzberger, the publisher of the *New York Times*, and a letter written by Karen Elliott House, the publisher of the *Wall Street Journal*. Both of these passages are about the editorial philosophies of their respective newspapers.

Regarding the *New York Times*: "He [Sulzberger] prefers to call the paper's viewpoint 'urban.' He says the tumultuous, polyglot metropolitan environment The Times occupies means, 'We're less shocked,' and that the paper reflects 'a value system that recognizes the power of flexibility'" (Revkin 2003).

Regarding the *Wall Street Journal:* "Respected and successful newspapers, much like respected and successful people, must know who they are and where they stand. Thus, our editorial pages espouse a clear and consistent philosophy that we summarize as 'free people, free markets' and that encompasses a passionate belief in the virtue of individual liberties, free markets, free trade and even the free movement of people" (Okrent 2004).

The intent of this disclosure is for students to understand the intended audiences and editorial preferences for both publications. If students do not recognize the significance of these passages, the instructor informs them that the *New York Times* is generally regarded as a liberal news source and the *Wall Street Journal* as a conservative one.

Following group reports on questions 5 and 6, the instructor reads from an article titled "New Report Accuses Bush of Suppressing Research Data." The instructor stresses that the article was published in the internationally renowned, peer-reviewed journal *The Lancet*:

> Relations between the Bush administration and the scientific community continue to worsen. Last month, 62 of the nation's leading scientists endorsed a report that accused the administration of repeatedly misusing scientific data for political purposes and stacking advisory committees with political partisans. The signatories included 20 Nobel laureates and 19 recipients of the National Medal of Science, the federal government's highest award for scientific achievement. (House 2005)

For reports on question 7, a student volunteer is asked to write group answers on the whiteboard or blackboard. After the reports, the instructor summarizes group answers based on the learning objectives. The instructor may need to modify answers or add to them because students will be quizzed on the material at a later date. Quiz questions will not be related to the subject matter of the case or of the newspaper articles cited therein. Instead, students will be asked to list specific criteria for critically evaluating newspaper articles.

Web Version

This case and its complete teaching notes and references can be found on the website of the National Center for Case Study Teaching in Science at *http://sciencecases.lib.buffalo.edu/cs/collection/detail.asp?case_id=182&id=182*.

References

Fialka, J. *Wall Street Journal*. 2003. EPA Is Upbeat, But Seeks Study of Rise in Asthma. June 24.

Greenberg, D. 2004. New report accuses Bush of suppressing research data. *The Lancet* 363 (9412): 874.

House, K. *Wall Street Journal*. 2005. Letter From the Publisher: A Report to the *Wall Street Journal*'s World-Wide Readers. January 6.

Okrent, D. *New York Times*. 2004. Is the *New York Times* a Liberal Newspaper? July 25.

Revkin, A. 2003. "'N.Y. Times': EPA Report Omitted Global Warming Data." By M. Norris. NPR, June 19.

Revkin, A., and K. Seelye. *New York Times*. 2003. Report by the EPA Leaves Out Data on Climate Change. June 19.

U.S. Environmental Protection Agency (EPA). 2003. *EPA's draft report on the environment.* www.epa.gov/indicators/roe/html/roepdf.htm.

Chapter 30
Eating PCBs From Lake Ontario

Eric Ribbens

The Case

Read the following University at Buffalo press release.

Eating Lake Ontario Fish Linked to Shorter Menstrual Cycles; Consumption May Delay Pregnancy, UB Researchers Find

BUFFALO, N.Y. — Eating contaminated sport fish from Lake Ontario is associated with shortened menstrual cycles, epidemiologists from the University at Buffalo have found.

They also reported that the fish consumption was associated with a small, but statistically insignificant, delay in the time it took women to become pregnant.

The results are from two separate studies that are among the first to assess the dietary effect of low-level environmental exposure to organochlorines, heavy metals, and pesticides, all recognized reproductive toxicants, on the reproductive process in humans.

The studies are published in the *American Journal of Epidemiology*, which is dedicated to research by faculty members and graduates of the UB Department of Social and Preventive Medicine.

Women enrolled in New York State Angler Cohort provided the data for the studies. The cohort—composed of 10,518 male anglers, 918 female anglers, and 6,651 spouses or partners of male anglers—was formed in 1991 to provide a representative sample of fishing-license-holders between the ages of 18 and 40 from the 16 counties near Lake Ontario. The sample provides a population base for a variety of studies on the implications of Great Lakes contamination.

Chapter 30

Eating Great Lakes sport fish delivers a mixture of toxic chemicals, including PCBs, at a level estimated to be 4,300 times greater than through exposure in the air or via drinking water. Many of these chemicals accumulate in the body. Lake Ontario fish are reported to have more than twice the amount of dioxin, mirex, and PCBs than fish from the other Great Lakes, a finding that has resulted in the New York State Department of Health recommending that women of childbearing age eat no Lake Ontario fish. An earlier UB study showed that most anglers are aware of the advisory, but many don't know the specific recommendations for women and didn't change their habits because of the advisory.

For the two studies of fecundity and Lake Ontario fish consumption, researchers from UB's Department of Social and Preventive Medicine assessed both time-to-pregnancy, a measure that can reveal conception delays, and length and regularity of menstrual cycle, aspects that affect a woman's fecundity.

The time-to-pregnancy study, headed by Germaine Buck, PhD, UB associate professor of social and preventive medicine, involved 874 women who were trying to become pregnant between 1991 and 1993. Trained telephone interviewers collected information on time-to-pregnancy. Information on duration and frequency of sport-fish consumption was collected when participants enrolled in the cohort in 1991.

Time-to-pregnancy data were based on women's answers to questions asking if they stopped using birth control to become pregnant; if they were attempting to prevent pregnancy in any way; and during which cycle they became pregnant after deciding to try to conceive.

Consumption data showed that 42% of the women of child-bearing age ate Lake Ontario fish and 10% reported eating fish for at least 7 years, dating back to a time when lake contamination was higher than in recent years. Researchers found a small conception delay for women who ate fish, but the effect was not statistically significant.

The study on the association of fish consumption and length of menstrual cycle, headed by Pauline Mendola, PhD, involved data from 2,223 women from the same cohort who reported menstrual-cycle length when they were re-interviewed in 1993.

Results showed that eating sport fish from Lake Ontario more than once a month was associated with a menstrual cycle 1.1 days shorter than the cycles of women who did not eat sport fish. Among women who experience regular menstrual cycles, the reduction was half a day.

Women in the group with moderate to high PCB exposure due to consumption of Lake Ontario fish showed a 1.3-day average reduction in menstrual cycle compared to women who did not eat fish. The average reduction for women who experience regular cycles was half a day.

Mendola said that while these small decreases in menstrual-cycle length are not currently a major public-health concern, the findings may indicate that these environmental contaminants have an effect on hormone production, notably estrogen production, which could have larger implications.

Source: www.buffalo.edu/news/execute.cgi/article-page.html?article=34130009

Your Assignment

Write a critique of this report. In the next class period, you will hand in your written critique and should be prepared to discuss your opinions. Included in this critique should be answers to the following questions:

1. What was the research question for each study?
2. Describe the data collected by the 1991 New York angler study. What questions did the researchers ask the participants?
3. Describe the data collected by the time-to-pregnancy study. What questions did the researchers ask the participants?
4. Describe the data collected by the menstrual-cycle study. What questions did researchers ask the participants?
5. This was an observational study, not an experimental study. What are the advantages of the observational approach as compared to an experiment? What are the disadvantages?
6. What do we know about the statistical analysis of the data?
7. What did the researchers determine about the relationship between PCB consumption and the amount of time it takes to become pregnant?
8. What did the researchers determine about the relationship between PCB consumption and the length of menstrual cycles?
9. Were these conclusions valid in your viewpoint? If not, why not?
10. What information is missing that would help you evaluate this research project?
11. What would you change or add if you were in charge of answering the research questions?

Chapter 30

Teaching Notes

Introduction and Background

This case may be used in many courses, although I originally designed it for a biostatistics class. The news release is an actual news release describing the work of Mendola, Buck, and Vena (1997). I use it not because it is a good or bad example of scientific reporting, but because a discussion of this news release illuminates several important aspects of experimental design and interpretation and places statistics within the context of these components of the scientific method. The case uses one hour of class time and produces a written assignment from each student.

One of my goals for my students is to have them understand that statistical analysis is simply a tool to help them in the process of conducting research. I want them to realize that statistics are not magical devices yielding infallible answers, but rather tools that assist researchers by quantifying differences between sample treatments and producing probability values that describe the likelihood that differences exist in the population being sampled. I also want them to discover the natural interplay between the design and analysis of research projects.

This case also could be used as a component of a course examining how the media report science or in a course examining the effects of environmental contaminants. Another version of this case even more clearly emphasizes the role of the media in shaping our understanding of the science. This is an interactive case for first-year students using a PowerPoint presentation and personal response systems ("clickers") to answer questions that are embedded within the case (*http://sciencecases.lib.buffalo.edu/cs/collection/detail.asp?case_id=501&id=501*).

Objectives

- Reconstruct what researchers did on the basis of a news release
- Evaluate the statistical presentation of the results
- Identify potentially valuable missing pieces of information
- Delineate what additional steps they would take if they were conducting the research project

Common Student Misconceptions

- Scientists always find the answers they expect.
- Scientific inquiry is not ambiguous.
- Scientists never make mistakes.
- Science is an opaque process, and readers of research reports and articles cannot reconstruct what happened or critique the process or conclusions.

Classroom Management

I hand out the case one or two class periods before I intend to discuss it. I request that each student read the case and write a critique, which must include answers to each of the questions attached at the end of the news release ("Your Assignment"). I also tell them that we will discuss the case on the day their critique is due and suggest that they bring either a second copy of their critique or a set of notes about the questions so they can participate in the discussion in an informed way.

On the day of the discussion, I collect the critiques, remind students that this is an actual news release describing scientific research published in the *American Journal of Epidemiology*, and begin our discussion.

It is helpful to begin the discussion by asking questions about who wrote the news release. One of my students noted that "the manner in which the report presents the facts is somewhat ambiguous in places. This may well be due to the editing job, rather than the research method itself." This is an important point that may be missed by some students and can be optionally followed up later by leading a discussion about the interface between scientists and reporters.

I have found that discussions work well if we simply follow the questions at the end of the case. It is often helpful to write the students' answers to some of the questions on the blackboard, especially for some of the more open-ended questions, such as "What information is missing?" Generally, students are eager to contribute to the discussion once they have developed their critiques.

After identifying the overall research question, I spend about 15 to 20 minutes on the data collected by the researchers. Many of my students express confusion about the research design because the description in the news release is very terse. It is helpful to separate out the initial New York State Angler Cohort data from the two follow-up studies that are the focus of this news release. Students also may tend to be too superficial about the types of data collected, so it is important to bring out specific questions the interviewers asked in the discussion. For example, many students overlook the fact that the interviewers must have asked questions about both length of the menstrual cycle and regularity of the menstrual cycle.

Once we have identified the components of the data, I shift our discussion to the analysis. I do not expect my students at this point in the semester to indicate a specific statistical method that must have been employed, such as a multiple regression. However, I do expect that they explain that some sort of a test was conducted that produced a p value, or a probability of differences between the fish-eating women and those who did not eat fish.

Questions 7 and 8 are included primarily to make sure that students have grasped what the researchers claimed to have found. We address them briefly, but move on rapidly to spend more time on the remaining questions. Expect a wide range of answers to the question about the validity of the results. Some students will focus on missing information, such as whether or not the species of fish eaten makes a difference in PCB contamination levels. Others may express doubts about the meaningfulness of the results, pointing out that the observed differences are quite small. Some will point out that the determination of PCB contaminant levels is imprecise.

Finally, I shift the discussion to a combined analysis of missing information and what my students would do if given responsibility for this research. I find it helpful to combine these two questions in the discussion, so that when someone complains about missing information I can respond by asking what could be done to obtain the data. I also find that during this part of the discussion it is particularly helpful to list responses on the blackboard and perhaps to sort responses into different categories of concerns

One long-term benefit of using this case is that it encourages students to begin thinking independently. One of my students wrote, "This case study proved to have enlightened me. I have never taken the authority to dissect a study like this before. It is important to always question works and not just accept them, even if my questions do not always lead to the right answer."

Web Version

This case and its complete teaching notes, references, and answer key can be found on the website of the National Center for Case Study Teaching in Science at *http://sciencecases.lib.buffalo.edu/cs/collection/detail.asp?case_id=287&id=287*.

References

Mendola, P., G. M. Buck, and J. E. Vena. 1997. Consumption of PCB-contaminated freshwater fish and shortened menstrual cycle length. *American Journal of Epidemiology* 146 (11): 955–960. *http://aje.oxfordjournals.org/cgi/reprint/146/11/955.pdf*.

University at Buffalo. 1997. Eating Lake Ontario fish linked to shorter menstrual cycles, may delay pregnancy. News release. December 2. *www.buffalo.edu/news/execute.cgi/article-page.html?article=34130009*.

Section VII:
Ethics and the Scientific Process

The scientific enterprise depends on total honesty and a devotion to telling the truth. This is not to say that all scientists are pure of heart; periodically, a scientist is discovered to have falsified data for fame or fortune. These charlatans almost always are caught, because any claim that is sensational is sure to be checked. Naturally, there are individuals who delude themselves about their data and make grandiose claims. Others make honest and inadvertent errors. But one thing about the scientific enterprise invariably is true: If a truly important discovery is made, like the possibility of cold fusion, there will be overwhelming interest in duplicating the work. Extraordinary claims will be tested and retested because the implications can be revolutionary. The more important the claim, the more resolutely it will be challenged. Only the most robust ideas have any chance of long-term survival. It is a Darwinian process. Science is self-correcting.

On the other hand, if a scientist makes a minor contribution or a claim that fits into the prevailing framework of the field, errors can remain in the public domain for some time. It was many years before the claims of Cyril Burt's studies on twins were found to be flawed. Burt, a respected psychologist, claimed that the IQ of identical twins was similar even if they had been raised separately, in different families. He concluded that genetic factors were more important than environmental factors in determining intelligence. It was only after his death many years later that his studies were declared fraudulent by the British Psychological Society. Still, all subsequent work on this question confirms the correctness of his claims, although his own data remain suspect.

Until recently, ethics has not been an overt concern of scientists. The prevailing view was that scientists did not cheat. Before the late 1800s, scientists were few in number and often from wealthy families. This does not mean that they did not make extravagant claims, but scientists generally were seen as noble characters. They had little to gain and their reputation to lose. Indeed, that still is the case. A recent poll

Section VII

has shown that scientists are among the top of the American public's list of respected professions (Van Riper 2006).

As the temptation to make mischief has increased, we see more frequent revelations of scientists who have behaved unethically. But there are many more scientists, and they no longer are all well-to-do individuals with established aristocratic lineages. We now have thousands of scientists who are vying for jobs, tenure, and grant monies, and there is hot competition for the huge financial rewards that come to the inventors of novel techniques and the owners of valuable patents. Universities, once considered to be above the fray, now vigorously encourage entrepreneurial skills in their quest to reap the benefits of new scientific discoveries. Lawsuits and lawyers are part of the standard operating procedures of higher education. There are ethical land mines everywhere.

Many scientists act or speak as if they bear no responsibility for the use of their discoveries—as if their job is to uncover the secrets of the universe irrespective of the consequences. There are dramatic exceptions. Legend has it that Alfred Nobel, inventor of dynamite and a major manufacturer of the cannon and other armaments, created the Nobel Peace Prize to atone for the guilt he felt at having invented the explosives used in modern warfare (Jones 1983). Robert Oppenheimer, the physicist who headed the scientific team that created the atomic bomb used against Japan at the end of World War II, was profoundly conflicted over his role in this groundbreaking project (Oppenheimer 1947).

Discoveries are not made in a vacuum. They have consequences for society, and society therefore has something to say about them. Just because we *can* produce stem cells from embryos does not mean that we necessarily *should*—that is a question for thoughtful and reasoned debate. The same is true for cloning or manipulating our children's genes. Projects that start out with good intentions can turn out later to have been terribly wrong or misguided. As Oppenheimer and others have found, once a discovery is made, it is difficult to put the genie back in the bottle.

Ethics deals with questions of good and evil, right and wrong, virtue and vice, and justice and injustice. Traditionally, science educators have shied away from such topics, declaring them to be outside their expertise and best left to the philosophers. But we believe it is important to engage our students in ethical concerns. The cases in this section touch on some of the fundamental issues of ethics in science.

The case "Mother's Milk Cures Cancer?" features a young Swedish scientist, Dr. Svanborg, who discovered a piece of evidence that lent support to her hypothesis that human milk can protect against cancer. The story deals with the steps required to bring a drug to market and, more to the point of this section of our book, her deliberations over when to publish her findings and whether to join forces with a larger, more established research operation to get her work noticed, approved, and funded.

The next case, "Cancer Cure or Conservation," deals with the discovery that a chemical in the wood of the Pacific yew tree, Taxol, could retard the growth of some cancers. In this case, a fictionalized meeting of the Quinault Indian Nation Tribal Planning Committee becomes an occasion for a discussion and vote on whether to "lease" tribal land to Bristol-Myers for bio-prospecting of the tree or to leave the land as it is. The conflicts are between the needs of environmental protection and those of cancer patients.

The case "A Rush to Judgment?" takes us into the world of a psychology lab. A young undergraduate discovers that her older student colleague has breached certain ethical and procedural standards in conducting her research. The young student now is in a quandary as to how to handle the situation, which has implications for the future careers of her friend and the supervising professor.

The next case, "How a Cancer Trial Ended in Betrayal," delves into a true story and has students designing a clinical trial that mimics one done by the original scientists. In a role-playing exercise, the students uncover a secret violation of a blind test and manipulation of data to favor a positive result in treating skin cancer.

"Bringing Back Baby Jason: To Clone or Not to Clone" introduces students to the controversial issue of cloning. An accident leads to the loss of a child and the parents are presented with an opportunity to take their lost child's cells and clone him. The case leads students to learn about our present knowledge of cloning and to consider the ethical questions surrounding such a course of action.

The next case, "Selecting the Perfect Baby: The Ethics of 'Embryo Design,'" takes us deeper into the question of genetic manipulation. We are introduced to the Shannons, a couple who wish to have a child, but the prognosis is not promising because an earlier child has a genetic condition, Fanconi's anemia, with severe disabilities. Because the parents carry the trait and do not want to have another disabled child, their physician suggests harvesting multiple eggs from the wife for in vitro fertilization. Then, with careful testing of the embryos through development, he could select one without the genetic "defect." This procedure involves risk, expense, emotion, and the destruction of multiple embryos to get the perfect child. Should they do it?

The case "Studying Racial Bias: Too Hot to Handle?" is unusual in that it takes students behind the scenes to see how researchers design ways to test a hypothesis in an ethical manner. Today, any research that deals with human subjects must be approved by an Institutional Review Board (IRB), which passes judgment on whether the study meets federal and institutional ethical guidelines. The case consists of examining the IRB's ethical standards to determine whether to approve a student proposal examining racial bias in the hiring procedures of local businesses.

The final case, "Bad Blood: The Tuskegee Syphilis Project," deals with an experimental protocol that never would be approved today. It relates the story of investigators working in the early 20th century who, with the best of intentions, set out to study the progress of the sexually transmitted disease syphilis in African American men. Originally, there was no effective treatment for the affliction; years later, penicillin became available—yet the study continued and the men did not receive treatment. How this situation could evolve is a tragic example of how science can go wrong.

The scientific enterprise is limited by the mores and expectations of society, but for good or ill, it is continually pushing back the ethical frontiers. Dozens of examples come to mind where our moral limits are tested. Robert Oppenheimer was head of the scientific group that produced the atomic bomb and ushered in the era of modern warfare and nuclear energy. Jonas Salk is credited with eliminating polio from the world, and public health officials have brought sanitation, vaccines, and medicines to the world, only to see the world's population explode and, with it, increased famine. The developers and dispensers of birth control and in-vitro fertilization have shaped our modern attitudes toward sex and family life. Companies have commercialized the practice of genetically modifying crops, which has increased farm productivity but also has reduced plant diversity, producing monocultures of wheat, corn, banana, and the like that can potentially be wiped out by a single parasitic disease. These folks and hundreds of others are pushing us into uncharted ethical areas, confronting us with problems we have never seen before. We can't hide or ignore them. We created these moral dilemmas, and they won't go away.

References

Jones, G. 1983. The trial of Alfred Nobel. *The New Scientist* 20: 189-191.

Oppenheimer, R. 1947. "Physics in the Contemporary World." Arthur D. Little Memorial Lecture at M.I.T., Cambridge, MA, November.

Van Riper, T. 2006. America's most admired professions. Forbes.com. *www.forbes.com/2006/07/28/leadership-careers-jobs-cx_tvr_0728admired.html*.

Chapter 31

Mother's Milk Cures Cancer?

Linda L. Tichenor

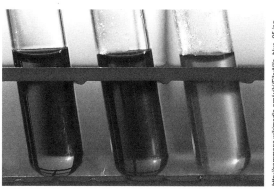

The Case

Part I: The Discovery

The winter days had become too short. It always seemed like twilight in this country so close to the Arctic Circle. The sky remained gray even in the midday light. It was three o'clock, and the coffee break Catherine Svanborg had just taken had refreshed her somewhat. As she prepared to run the last samples of the day, she took her glasses off and rubbed her eyes, then rested her head in the palm of her cool hands for a few minutes. Somehow she just couldn't seem to shake the mid-December depression she was feeling. She could hear the students arguing in the laboratory next door. They seemed to be quite distracted from their usual afternoon daily lab meeting where the day's findings were discussed and plans were made for the next day's work. A sharp knock at her door jarred her into her formal posture reserved for colleagues and students.

"Dr. Svanborg?"

Maybe it was the new medication she was taking, but Dr. Svanborg just couldn't seem to focus. Or maybe it was the deadline coming up in two weeks for the NIH grant she was preparing. She would be spending Christmas alone again, as she gave way to the last-minute details involved in writing her yearly report of findings. Ever since joining Lund University as a research associate in 1990, she had been trying to build a reputation as a serious authority in immunology. Recognition in the form of a grant from a prestigious organization would place her in that category.

"Can this wait until tomorrow?" she asked.

The student replied that it couldn't wait—that she might find what was going on in the next room very disturbing. The student was one of Dr. Svanborg's most promising researchers. He had been working on the effects of various protein extracts of human breast milk on virally infected tissue cultures and was making good progress toward isolating factors that activated the immune response toward the retrovirus.

"Dr. Svanborg! Really, you must come look at this tissue culture of cancer cells I've got under the fluorescent microscope. Something very weird is happening, and we can't agree on what it is!"

In the next few minutes, he showed her something that would send shivers of excitement through her body—like the time she received the letter announcing her election as a bona fide fellow to the Royal Academy of Science of Sweden.

Svanborg looked into the microscope. Did she actually see what she thought she was seeing? Another look confirmed that the cancer cells were indeed "committing suicide"! Without looking away, she drew a lab stool under her to steady her trembling knees. For the next half-hour, she stared into the scope, and yes, something was causing the cancer cells to "explode" from within.

The news her student had brought seemed completely unrelated to the topic he was actually researching. He had been looking for how mother's milk fights viruses in tissue culture. The virus is introduced into the cancer cells (or the culture) and treated with milk components to see if there is an increase of viral destruction. What he observed—the cancer cells "committing suicide"—wasn't typical because normally they reproduce forever without dying, a kind of cell-line immortality. That immortality makes cancer cells a good medium for tissue culture. Normal human cells commit suicide every day. The process of cellular suicide is called *apoptosis*. What the graduate student found was a protein factor in the milk that induced the cultured cells to die.

Questions About Science

1. Why are cancer cells or abnormal cells used in tissue cultures even when researchers are not studying cancer? Explain how a cancer cell differs from a normal cell.

2. What is known about the life expectancy of most normal cells? Review the cell cycle and the mechanisms for regulating normal cell division. What happens when the regulatory mechanisms go awry in normal cell division?

3. What is the scientific name for when a cell commits suicide? Why would cells commit suicide normally?

4. Predict what would happen if normal cells did not commit suicide during fetal development.

Part II: Response of the Scientific Community

As good as this news sounded, Dr. Svanborg knew it would be extremely difficult to convince her science colleagues that the work of her student was for real. First of all, she normally worked with micro-organisms that cause infectious diseases. Moreover, with only three undergraduate virology students and two doctoral students, her laboratory wasn't considered large by most standards. She knew her lab didn't qualify as a "high-profile, big science institute." Examples of institutes with that level of prestige were the NCI (National Cancer Institute) and NIH (National Institutes of Health). So how would she go about explaining and reporting these recent findings to the scientific community and maintain her reputation?

Dr. Svanborg decided to call her trusted friend and colleague David Solomon, a cancer researcher at NCI, and share the discovery with him.

His response was immediate. "Cathy, don't do it! Don't risk being labeled an unreliable researcher by the scientific community!"

David continued, "You know that novel ideas in science always challenge the current paradigm, and acceptance is slow. Your work will not be taken seriously by the scientific community initially. And you are not at a stage in your career to take that kind of a risk." His advice was to postpone reporting her findings until she had more evidence.

"You know," he added, "if this work had come from a well-known lab like mine here at the NCI, you'd have reporters calling six days to Sunday wanting your story. You'd have scientists eager to collaborate, and grant money to do whatever you wanted. But it's coming from your small lab in a foreign country. It's like General Motors versus a mom-and-pop garage operation."

Catherine was able to take his criticism in stride because he had been her ally for more than 25 years. As graduate students, they had worked together under an important researcher who used immunotherapy to treat cancer. Besides, she knew better than to risk her reputation by reporting a discovery without hard evidence to support her conclusions.

Questions About Science and Politics

1. What is meant by high-profile and big science research?

2. What and where is the National Cancer Institute? Where does the funding come from to support the work of the institute? Discuss how scientific research is funded in Sweden. How might funding be different there than in the United States?

3. Why did Dr. Solomon say that novel ideas in science always challenge the current paradigms? Isn't science supposed to be about new ideas? Try to justify what he meant by his statement.

4. What does the discovery of cancer-killing effects of breast milk while looking for effects of milk on viruses tell you about the process of science?

Part III: The Research

In her background research of the literature, Dr. Svanborg discovered a piece of evidence from 1995 that lent support to her hypothesis that human milk can protect

Chapter 31

against cancer. The study showed that the risk of childhood lymphoma is nine times higher in bottle-fed infants than in breast-fed infants. She and her students wondered if there were some connection between the two situations.

Svanborg and her group began to analyze breast milk more thoroughly and eventually discovered that the actual component of breast milk that was killing cancer cells is a protein called alpha-lactalbumin (sometimes called alpha-lac). In January 1999, Svanborg's team finally released results demonstrating that in the acid environment of an infant's stomach, the normal alpha-lac protein changed shape and transformed into a killer of cancer cells (or other potentially harmful cells, such as pneumonococcus bacteria). Her research group named the altered protein *HAMLET*, for Human Alpha-lactalbumin Made Lethal to Tumor cells. By genetically altering bacterial cells, they were able to mass-produce the factor, as is commonly done in the production of human insulin.

Now that they had produced *HAMLET* in sufficient quantities for research, it was ready to be tested on animals with tumors and then on human subjects with cancer. The team believed that the substance should not be toxic to animals because it was a naturally occurring protein in breast milk. If found to be useful in cancer treatment, this discovery would represent a great advance over the toxic cancer drugs currently in use, with their high risk of negative side effects.

Questions About Science

1. What are the contents of human breast milk, and how does it differ from commercially available milk from cows and soybeans?

2. Discuss the immune components of breast milk.

3. How does the fact that infants raised on breast milk were nine times less susceptible to lymphoma than bottle-fed infants demonstrate that there may be a link between something in human milk and cancer-cell destruction? Could there be something other than the actual breast milk that lowered the rate of lymphoma in breast-fed children? How can a scientist be sure that the link is actually related to breast milk? Discuss how these links are actually substantiated by the scientific community.

4. How does protein shape relate to function? How might shape be altered other than the changing pH of an environment?

5. How do researchers mass-produce a protein such as *HAMLET* through genetic engineering?

6. Describe a mechanism whereby a protein actually causes cell death in tumor cells and bacteria.

7. What are the disadvantages of modern chemotherapy in curing cancer? Why would a naturally occurring protein be a great advance in treating cancer?

Section IV: From the Lab to the Pharmacy

To produce the quantities of material needed for animal research, Dr. Svanborg's research group used genetically engineered *HAMLET* factor in bacterial cells. The methods they used were similar to the way insulin is produced for mass marketing. Once a protein has been isolated, a copy of the DNA can be made and inserted into a plasmid. Plasmids are vectors of the gene of interest that are placed into bacterial cells. Once a line of bacterial cells with the transformed plasmid is reproducing successfully, they are grown in large vats like the ones used when microbrewing beer. The protein product is extracted from the vats and tested first on animals. If the animal tests go well, human testing can then begin.

Human testing normally is done in three stages. The first stage is performed on animals to test for safety and merit. Then a limited number of people will participate in a study to see if tumor cells are killed in humans. The final test is conducted with a large group study involving many trials and controls. This type of testing may take several years before final approval of the substance is granted for pharmaceutical use.

If the alpha-lac (*HAMLET*) factor can kill cancer cells in humans, then it will not be long before the pharmaceutical companies will get involved, but they must be convinced that the work is worth their attention. Often, naturally occurring drug products are labeled "orphan drugs" and are not marketed to the general public. Would this alpha-lac component of human breast milk be treated as such?

All in all, Dr. Svanborg was happy with the progress they had made. "When we started doing research here in this little town in this little country on the edge of the known world," she mused to herself, "few people were aware of our work. Now that this enormous opportunity has come to us, we want the world to know it."

Questions About Science and Economics

1. Why would a scientist like Dr. Svanborg be reluctant to publish too soon an idea that breast milk fights cancer?

2. Why is testing on humans done in three stages? Why can't the process be hurried up if it is known that a substance can kill cancer cells and ultimately cure many people with a dreaded disease? What is the procedure for clinical testing of a potential new therapeutic agent in the United States and other countries if the treatment is categorized as an "orphan"?

3. Why might drug companies be reluctant to support research on *HAMLET*? How might the caution of drug companies actually delay scientific progress?

4. How would drug companies come to support a laboratory such as Dr. Svanborg's?

5. Is the alpha-lac discovery patentable?

6. If Dr. Svanborg and a drug company were to team up, who would claim the rights to this treatment methodology?

Chapter 31

7. If drug companies do not decide to pursue the development of the alpha-lac treatment, could or should cancer patients take matters into their own hands and attempt to purchase human breast milk directly from lactating women? What implications might that practice raise?

Teaching Notes

Introduction and Background

In the June 1999 issue of *Discover* magazine, Peter Radetsky raises important questions about the nature of scientific work in his article "Got Cancer Killers?" The article brings to the attention of the lay public the concept of serendipity in scientific research.

The naive student, especially the college freshman, does not realize that many scientific discoveries are made during the process of experimentation and that those discoveries are coincidental to the answer that the researcher actually seeks. Students view science as an orderly, controlled, and rational process. That scientists hope for surprises is usually not taught in laboratory science. This is the joy of creativity in science!

It is important for nonscience majors to understand how politics and culture play a vital role in how science proceeds. Another surprise to students is that scientific research is provisional and a product of social and cultural interaction. This case allows students to discover that scientific research depends, in part, on peer approval and social hierarchy—that is, scientists succumb to external pressures to guide their research.

In addition to the social issues raised above, this case study possesses rich and timely content and relates to more than one subdiscipline of biology and biochemistry. The content of the case includes cell biology (especially control of cell division), apoptosis, immunity development, cancer, cancer research, microbial biology, genetic engineering, and breast feeding in humans. The level of detail in the content is left for the instructor to determine. The case can be focused on at least five paths of discussion listed below in the learning objectives. Instructors should determine the target audience and tailor the case to suit their learning objectives.

Objectives

Upon completion of the case study, the student will understand the following:

- Not only the science content underpinning the case, but also how content areas in biology are interrelated—that is, how content knowledge of the immune system, cell cycle regulation, cancer biology, and microbiology must all be integrated to do the type of research in this case

- The process of scientific research and how it often includes serendipity, provisionality, and human bias

- How funding for scientific research and new discoveries are linked
- That the conduct of scientific research as well as the reporting of science findings are partly determined by economic and political pressures
- How pharmaceutical products are developed from laboratory discoveries to drugs for the consumer

The case is written to be used in all college levels. However, the discovery of the provisional nature of science has the most influence on the novice science student. The social issues involved in science include the nature of scientific work, social constraints of doing science, gender and science, politics and science, and pharmaceutical businesses. The case has been broken up into four parts to allow a distinction between the scientific content and the political, social, and economic aspects of scientific research.

Common Student Misconceptions

- Important research can be done only in large laboratories with a large budget.
- All successful research is done with careful planning and foregone conclusions.
- It is unlikely that mother's milk would have any effect on cancer cells.
- It is unlikely that students will make any discoveries; that is professors' work.
- Cancer cells go through a totally different life cycle than normal cells.
- Scientists do not understand much about what causes cancer.

Classroom Management

A novel approach to understanding how scientists think is to have a panel discussion. The panel would consist of science professionals drawn from faculty or industry representing different disciplines in the sciences. Through the examination of different perspectives, students can determine how accidental discovery and the politics and cultural aspects of science might have affected scientists personally during their careers in research.

Class Period 1 (50–75 minutes): Organizing the Panel

A brainstorming session allows students to prioritize what they wish to learn from the panel of experts in the first class period. Then the students prepare thoughtful questions to ask panel members in the second class period. The ideal panel size is six members. Having more members dilutes the contributions of each, but having fewer than four does not lend itself to a varied discussion.

The questions, which are written by the students, may include some of the discussion questions provided at the end of each section of the case. Students should

Chapter 31

take the opportunity to ask the panel members to address their personal experiences in scientific research.

The instructor should assist the students in selecting panel members based on the types of questions developed by the students. It is easier if the instructor coordinates the panel members with a specific time and day for the discussion. The topics to be discussed should also be given to each panel member.

Class Period 2 (50–75 minutes): Running the Panel Discussion

The instructor or student-moderator introduces the panel members by describing how their expertise fits the discussion. Each panel member presents a three-minute summary of his or her background in scientific research. Once the introductions have been made, students direct their questions to individual panel members. Students will have selected the questions to be asked in advance, and each student group or individual student is allowed to ask one question in turn. The moderator keeps the discussion focused and facilitates the order of the questioning. The moderator should also be sure that each group or individual student has participated. At the conclusion, the instructor makes an assignment based on the panel discussion. The assignment allows students the opportunity to reflect on the outcomes of the panel discussion and demonstrate their communication skills, either written or oral, depending on the instructor's course objectives.

Class Period 3 (50–75 minutes): Discussing the Science Content and Political Issues

In the third class period, content can be discussed more thoroughly. The content in the case may be used to launch a discussion on the relevance of the cell cycle, what controls the stages, and what determines whether cell division will proceed. Sometimes instructors discuss the stages of mitosis in great detail without putting the process in the larger context of the cell cycle. It is important to discuss the checkpoints that allow each of the stages of the cell cycle to continue through to mitosis and what kinds of signals may alter the normal patterns of cell division. The term *apoptosis* is introduced and a discussion of normal cell death can follow.

Normal tissue grown in a culture has a limited number of cell divisions, whereas transformed (tumor) tissue will divide without limitation. These concepts should be discussed in conjunction with why cancer cells, rather than normal cells, are used in tissue culture.

A discussion of cancer as relates to genetics "gone awry" should be anticipated. Students will have many side questions regarding cancer as a human disease because many will have family members who have died from or are still living with the disease. The instructor should be prepared to respond to those questions but also redirect the discussion to accomplish the learning objectives of the case. Treatments such as radiation and chemotherapy, with their side effects, should be discussed. Expect a discussion of the role of the immune system and immunotherapy. Depending on the class, there can be an extensive discussion about what happens at the molecular level and the importance of protein shape.

The politics of science are rarely discussed in science textbooks. Students may wish to interview the panel members on what they consider to be big science and high-profile research, as well as their personal experiences in research, grant-writing, and funding efforts and pressures.

Finally, the discussion will invariably lead to the reasons why alpha-lac is not already on the market. Do pharmaceutical companies feel that investing in this naturally occurring product is worth their effort monetarily? A discussion of what are known as orphan drugs and why they never reach the market is good at this point, along with conversations about the role of the Food and Drug Administration (FDA).

Class Period 4 (Optional): Project

The project should allow the student to analyze information and draw conclusions to be presented in a fourth class period. Ideally, students would present conclusions orally; however, if devoting a fourth period seems excessive, a formal paper may be used. Again, the course objectives should be taken into consideration when the instructor crafts the assignment. The topics chosen can be strictly scientific or focus on political and economic issues.

Web Version

This case and its complete teaching notes and references can be found on the website of the National Center for Case Study Teaching in Science at *http://sciencecases.lib.buffalo.edu/cs/collection/detail.asp?case_id=397&id=397*.

Chapter 32

Cancer Cure or Conservation

Pauline A. Lizotte and Gretchen E. Knapp

The Case

Jim Redwood had been summoned by the Quinault Indian Nation Tribal Planning Committee to return to the long house for an important vote. The tribal council elected in 1995 had been approached by the U.S. Forest Service to "lease" its remaining land to Bristol-Myers Squibb for the bio-prospecting of the Pacific yew (Taxus brevifolia Nuttall), which was needed for the development of the new drug Taxol. Taxol was one of a few drugs that showed incredible promise in the treatment of rapidly growing cancers.

"Great," he thought, "I've got a backlog of sick patients in the clinic already and now I've got to rush over there to vote on the land issue. Why can't they ever plan these things better?" But Jim knew his last thought was just a way to blame someone else for his lack of help at the clinic. The tribal elders had planned this meeting well in advance so the expert advisors would have time to converge on their small reservation of a little more than 200,000 acres in the Southwest corner of the Olympic Peninsula on the Pacific coastal area of Washington. "They're probably all there, waiting for me," he thought. It was going to be one of those days.

Jim hadn't realized when he was elected to the Tribal Planning Committee how important the position would become. Who could have anticipated that one of the smaller understory trees, the Pacific yew, would become the primary source of a "miracle drug" for cancer patients? He only hoped that the drug would be developed in time to help his mother, who had recently been diagnosed with metastatic breast cancer.

Jim took a deep breath. The other 10 members of the planning committee were already seated. The only seat remaining was his, front and center as the moderator. He swiftly took his place, gave a quick nod to the other committee members, raised his arms, closed his eyes, and began with the traditional opening prayer. "Oh, Great Spirit, we ask for Your guidance today as we consider the important issues before us.

Chapter 32

Grant us wisdom in deciding the fate of the land You have given Your children." He lowered his arms and sat down.

"We will begin by introducing our expert advisors. On my left is Dick Shaffer, assistant director of Timber Management for the U.S. Forest Service Regional Office in Portland, Oregon. Next to him is Dr. Gordon Cragg, chief of the Natural Products Branch of the National Cancer Institute. On my far right is Dr. Mark Plotkin, executive director, Ethnobiology and Conservation Team, Smithsonian Institution. Next to him is Hal Hartzell Jr., vice president of the Native Yew Conservation Society. Mr. Shaffer will begin with his statement."

"Thank you, Dr. Redwood. Let's just get to the bottom line. Women are dying every day of ovarian and breast cancer. The Forest Service places a high priority on helping Bristol-Myers Squibb in every way we can, legally and environmentally. This Taxol stuff that comes from the Pacific yew is the best thing to come along for treating these women. Let's get in there and harvest those trees. Who knows how many lives are being lost while we waste time debating this."

Dr. Gordon Cragg chimed in. "I agree, Mr. Shaffer. We need to get started since the process takes some time to extract the Taxol. The current yield of Taxol is 1 gram per 30 pounds of bark, assuming a 73% recovery rate. Since these trees are not that large, that means we need 1.5 trees for every gram of Taxol. A patient typically requires 500 milligrams per course of treatment, with four courses necessary, for a total of 2 grams per patient. That's three trees per patient and there are about 40,000 women needing this treatment now."

"Do you use the whole tree to get this medicine or only certain parts? Must the tree be killed to get it?" asked Billie Rainfeather, a tribal member.

Dr. Plotkin responded. "You know, that's a legitimate concern. It's primarily in the bark, and only 10% of the yew population in your forested areas are the size preferred by bark collectors, trees 10 inches or larger in diameter. I think that if Bristol-Myers Squibb plans on meeting that level of demand in the form of bark from the tree, the species would be in great jeopardy. As I'm sure you know, removing the bark automatically kills the tree. So, we risk eradicating trees of this yew species with any significant dimension, and endanger the future of the species by removing the seed source for future generations."

Hal Hartzell jumped in. "Yes, Mark, and that's not all. The U.S. Fish and Wildlife Service estimates that three to four million large yew trees remain on federal lands. To treat a year's worth of ovarian and breast cancer patients combined, that's 150,000 treatments and would require the bark from 1.5 million trees. How often can you justify harvesting that much?"

"Can't the Forest Service figure out some way to grow this tree on a tree farm?" asked Billie. "I really hate to see our land ruined. Do we know what this removal will do to the rest of the plants and animals? Our people use this tree to make traditional ceremonial crafts like bows, arrows, masks, and other items. We also depend on the forest for deer and elk and other game animals to feed our families. How will we be able to live off the land if it is ruined to harvest one kind of tree?"

Hal added, "Yew habitat is old growth forest. Remove that and yew species will decline. The yew is difficult to start from seed, and it produces few seeds, which only germinate after passing through the gut of some animal. It's a shade tolerant species, slow growing, and is a favored browse food for elk and deer. Harvesting these trees would definitely impact the populations of some animals in this area."

Billie turned to Dick Shaffer. She decided to get to the heart of the matter. "The cancer cure is good, but is that the real reason you want our land? You know, there is always some doubt in our minds when the government wants Indian lands. I'll bet the drug company stands to make a big profit from this drug. How much compensation are you offering the tribe? We have done a better job in the last 80 years managing our tribal lands than you folks did for the 150 years before that. Will our land be able to recover from the loss of this tree?"

The crowd in the long house was getting unruly. Jim decided it was time to call for a vote. "It's time for the planning committee to consider the advice of the experts. We will reconvene in two hours and let you know our decision."

Jim had heard so many contradictory statements, his head was spinning. How should he vote?

Questions

1. What tree does the drug company Bristol-Myers Squibb want to harvest?
2. What important substance is contained within the tree?
3. What part of the tree contains the substance?
4. Why is the drug Taxol valuable to humans?
5. Where does the tree grow?
6. What "function" does the tree have in its natural environment?
7. How many trees must be harvested to treat one person?
8. How do the Quinault Indians traditionally use the tree?
9. What percentage of the yew trees have a diameter smaller than the optimum size for collection?
10. Why can't the drug company just collect the bark?
11. Why can't the drug company grow the yew tree in a tree farm?

Chapter 32

Teaching Notes

Introduction and Background

This case is based on the controversies surrounding the harvesting of the Pacific yew from 1989 to 1997 to develop paclitaxel (Taxol), a revolutionary anticancer drug. The information contained in the dialogue ascribed to all characters except the Native Americans is taken from either direct quotations or published works. The Native American dialogue is entirely fictional.

The conflicts appeared on the surface to be between environmental protection and cancer chemotherapy. Ethnobotanists from the National Cancer Institute collected *Taxus brevifolia Nuttall* (Taxaceae) bark in 1962 as part of a nationwide program to discover natural-product drugs. The Quinault, a Pacific Native American tribe, used yew tea to treat various ailments. Wild populations of *T. brevifolia* grow in the national forests managed by the USDA Forest Service in Washington and Oregon. After cytotoxicity was demonstrated in 1964, a team of chemists affiliated with the Research Triangle Institute in Research Triangle Park, North Carolina, obtained pure drug in 1966. Preliminary antitumor studies at NIH demonstrated paclitaxel's chemotherapeutic efficacy in some forms of cancer. The chemical structure, a complex diterpene with an unusual oxetane ring and an ester side chain, was reported in 1971 after a worldwide competition to determine its structure.

By the late 1970s, scientists finally understood the drug's ability to promote the irreversible assembly of tubulin into microtubules. Taxol was the first anticancer drug known to act by promoting microtubule assembly rather than preventing it. The first results of the clinical trials were reported in 1989, and Taxol was fast-tracked through the FDA's regulatory procedures. Taxol and Taxotere, a derivative, have been approved for the treatment of metastatic breast cancer and in clinical trials have demonstrated activity against cancers of the ovary, lung, esophagus, bladder, endometrium, and cervix, as well as Kaposi's sarcoma and lymphoma.

Bristol-Myers Squibb was given exclusive rights to provide Taxol from *T. brevifolia* under a cooperative research and development agreement with the U.S. government in 1991. The public outcry against the proposed harvesting of *T. brevifolia* trees and shrubs in federal lands in California, Idaho, Montana, Oregon, and Washington led to the Pacific Yew Act of 1992 and increased awareness and regulation of bio-prospecting worldwide.

Taxol has been semi-synthesized (Pezzutto 1996), but biologists soon discovered that *T. brevifolia* was not the only source of Taxol. A novel endophytic fungus, *Taxomyces andreanae*, appears to produce Taxol (Stierle, Strobel, and Stierle 1993). Taxol also has been discovered in the African Fern Pine (*Podocarpus gracilior Pilger*), suggesting a phylogenetic affinity between Podocarpus and Taxus (Stahlhut et al. 1999). In 2003, Canadian tree physiologist Stewart Cameron began testing methods to grow Ground Hemlock (*Taxus canadensis*) sustainably. Cameron and Smith (2003) hope this project "can generate 200 kilograms of paclitaxel (Taxol) annually from nursery-grown biomass, which would treat about 100 thousand women a year."

This case was designed to expose students to basic conservation biology concepts by examining the competing needs among scientists and others in a real-life science-and-society scenario. We especially hoped to cultivate student appreciation for the complexity of scientific work and its interactions with public policy. Science isn't something that happens just in a lab. It is a process that can lead to life-changing effects both for humans and the organisms that share our habitat.

Objectives

- To learn about the risks and benefits of bio-prospecting
- To learn about ethnobotany and its importance to the drug industry
- To examine the direct and indirect benefits of *T. brevifolia* in the Pacific Northwest ecosystem
- To learn about the anticancer drug Taxol and the controversy surrounding its production
- To explore the complexity of scientific work and its interactions with public policy
- To gain a better appreciation of the complexities involved in land-use decisions

Common Student Misconceptions

- Discoveries by scientists that will benefit large groups of people need to be implemented regardless of the claims of indigenous peoples.
- Cancer is an incurable disease.
- American Indians have exclusive control over their own land.
- Real science is done in the lab and has nothing to do with the public.

Classroom Management

This case can be approached from several different angles but basically is composed of two major issues: the biological and ethical, which are intimately intertwined. Both issues are suitable for a nonmajors course but we would recommend focusing on one at a time.

The biological issues are highlighted in the objectives that look at risks and benefits of bio-prospecting, ethnobotany, drug development, and land use. As a precase assignment, different groups of students could be assigned to bring in brief definitions or explanations of each of these topics. As the students proceed through the case and answer the questions, the group of students responsible for obtaining the information would volunteer their answers at that time. This would facilitate the understanding of the specific objectives.

If the instructor prefers to investigate the ethical issues, other examples of land management could be used for comparison purposes. Our own historical concept of "manifest destiny" and the federal government's takeover of Native American lands in the early part of our country's history serve as a great comparison. The recent conflict in South Africa between the minority white landholders and the majority African people, while not exactly parallel, illustrates the conflicts that can arise over land use and ownership. Other ethical issues that could be considered might include whether we should save only species that are of benefit to humans now or in the future or should we look instead for sustainability of an ecosystem?

We developed this case for use in an undergraduate-level introductory biology course for nonmajors at our institutions. It also could be used in a science, technology, and society (STS) course, an environmental science course, or a general education course such as Foundations of Inquiry, which introduces students to the ways of knowing in various disciplines.

We found that one 50-minute class period was sufficient, although a second class period could be used to examine other topics the students find interesting, such as the chemistry of Taxol and its unique effect on cancer in humans.

We set up the case through a 10-minute kick-off exercise that begins with a think-pair-share activity. In this activity, students consider answers by themselves and write them down, then share and compare those answers with a partner. In this exercise, students were asked to jot down as many uses of plants as possible. Posted on the overhead was the question "How many different ways are plants involved with your life?" Students volunteered answers, which we wrote on the whiteboard. Predicted responses included textiles and clothing, food, sporting goods, containers, decorations, dyes, drugs, perfumes, building materials, musical instruments, furniture, and cosmetics. We drew the categories from student responses to organize their answers. Students were asked to determine which items were direct benefits of nature and which were indirect benefits. Can direct benefits be in conflict with indirect benefits?

We segued into the case through a mini-lecture. The two key points were that (1) more than 25% of today's medicines are plant-based, and (2) more than 70% of forests in the United States have been logged (often clear cut) since the 1920s.

We handed out the case and instructed students to jot down answers to the case study questions as they read, directing half of the class to answer questions 1–5 and the other half to answer 7–11. Both sets of students were directed to answer question 6. Students then read the case.

After allowing sufficient time for students to read the case and answer the questions assigned, our opening question to the class was "What are the issues that Jim needs to consider?" We again used the think-pair-share activity to create a list of issues. We randomly called on students to provide the answers and list them on the whiteboard.

As issues arose, we provided additional information via overheads. These included a map showing the location of wild populations of *T. brevifolia* in North America, a map of the location of Native American and federal lands in the Pacific Northwest, the structure of the Taxol molecule, and photos depicting *T. brevifolia*. All of this information allowed us to help students define concepts such as biodiversity, sustainable ecology, and bio-prospecting.

Student discussion among pairs and larger groups was lively. Several students reflected on how their emotional concerns about a family member with cancer shaped their views. Others asked if the drug could be synthesized, and we briefly explained the challenges of total Taxol synthesis. We took five minutes to wrap up the main points brought up in their discussion before moving on to a class vote. The main points were written on the whiteboard. The class then voted as to how many were for or against Jim voting to lease the land to the drug company. Students were asked to provide support (i.e., evidence from the case) for their positions. Because this is an open-ended case, there were multiple "right" answers.

Assignments

As stated earlier, groups of students could be assigned to look up specific vocabulary terms (*bio-prospecting, ethnobotany, drug development,* and *land use*) before reading the case. This would facilitate the learning of the concepts.

A suggestion for a follow-up assignment to explore biological issues would include students being required to bring a list of 20 plants (or animals) used to develop medicines. Other possible assignments are to have students (1) choose one of these "medicinal" organisms and research its background, (2) find the chemical structure and mechanism of action of a specific drug in humans, (3) explore how a particular organism fits into its environment and how its removal would affect other species in the ecosystem, or (4) examine why plants produce chemicals and how those chemicals interact with other organisms.

To further develop a concern for ethical issues, instructors could have students (1) write up a historical look at the development of quinine for use in treating malaria, (2) discuss the collection of species from environments that are sensitive to disturbance, or (3) discuss the ethical issues of bio-prospecting in general.

For an exciting, in-depth look at how ethobotany is carried out in the field, the instructor might wish to show all or part of the video *The Shaman's Apprentice*, which follows Mark Plotkin's exploits in the Amazon. This would be a good follow-up focusing on how botany and bio-prospecting affect the culture of indigenous people in the Americas.

Chapter 32

Web Version

The case and its complete teaching notes and references can be found on the website of the National Center for Case Study Teaching in Science at *http://sciencecases.lib.buffalo.edu/cs/collection/detail.asp?case_id=198&id=198*.

References

Cameron, S. I., and R. F. Smith. 2003. The development of *Taxus canadensis* as a domesticated crop for paclitaxel (Taxol) production. Paper presented at the Association of Applied Biologists Symposium on Medicinal Plants, London, April.

Pezzutto, J. 1996. Taxol production in plant cell culture comes of age. *Nature Biotechnology* 14: 1083.

Stahlhut, R., G. Park, R. Petersen, W. Ma, and P. Hylands. 1999. The occurrence of the anti-cancer diterpene taxol in *Podocarpus gracilior Pilger* (Podocarpaceae). *Biochemical Systematics and Ecology* 27: 613–622.

Stierle, A., G. Strobel, and D. Stierle. 1993. Taxol and taxane production by Taxomyces andrenae, an endophytic fungus of the Pacific yew. *Science* 260: 214–15.

Chapter 33
A Rush to Judgment?

Sheryl R. Ginn and Elizabeth J. Meinz

The Case

The Players

Stefanie Perry originally planned to major in English but changed her mind after enrolling in Dr. Lee's general psychology class. She thought Dr. Lee was a wonderful teacher and asked Dr. Lee to serve as her academic adviser. Noting Stefanie's enthusiasm, Dr. Lee asked Stefanie to work in the psych lab this semester. Stefanie knew it would really help her in her research methods class, plus she figured it wouldn't hurt to bond with a professor, especially one she liked as much as Dr. Lee. Stefanie was excited about the job in the psychology laboratory even though at this point her only real jobs had been setting up the equipment for Dr. Lee's experiments and distributing and collecting informed consent forms. Last week, Jolene, Dr. Lee's senior lab coordinator, had asked Stefanie to help her with her senior project. She also invited Stefanie to the rush party for her sorority. Stefanie was very excited about the invitation. It meant that, in all likelihood, she would receive a bid for Beta Alpha Delta.

In consultation with Dr. Lee, Jolene had designed her senior project to examine the effects of gender and test administration format on spatial abilities. Male and female undergraduate students signed up to participate in the study for extra credit in their introductory psychology classes. The experimental protocol called for students to be tested in one of two formats: traditional paper and pencil (TPP) or via computer administration (CA). In the paper-and-pencil format, students would be presented with two parts of Vandenburg and Kuse's Mental Rotations Test (MRT). Each part of the MRT contains 10 items, 5 items per page, and administration is timed. Participants are presented with one target and four choice answers for each item on the test. Their instructions are to select the two choices that are mentally rotated versions of the targets and mark their answers on the test itself. In the computerized version that Jolene developed for this project, participants would be presented with one problem at a time and indicate their choices on the numeric keypad.

Chapter 33

Something of an overachiever, Jolene is juggling the research project and her coursework with being recruiting chair of Beta Alpha Delta, president of the psychology club, and secretary of the outdoor recreation club. Jolene has a 3.82 GPA, participates in the honors program at her university, and has been accepted into a PhD program in neuropsychology at State, Dr. Lee's alma mater.

Dr. Lee also has a very hectic schedule and is eagerly anticipating the end of the semester. This busy assistant professor is currently teaching a course overload and is not looking forward to grading four sets of upcoming final exams; there are also those three senior research projects and Jolene's senior honors project to supervise. Dr. Lee has established a close relationship with Jolene and considers her to be one of the best students the department has ever had. This has made it easy to place increasing reliance on her to handle administrative duties related to lower-level classes. In anticipation of the upcoming tenure review in the fall, Dr. Lee has been trying to finish two research papers this spring but is no closer to this goal than he was in January. It doesn't look like that book will get finished this year either.

The Situation

Jolene had to meet with the chair of the honors program on the day she was to test group CA but forgot to tell Stefanie that she wouldn't be in the lab. Stefanie was in the lab when the group to be tested arrived, but not having heard from Jolene, she was forced to let the understandably upset students leave after 10 minutes. Jolene knew that her friend Matthew had a computer science lab on Wednesday afternoons, so she asked the professor of the class if she could talk to the students about her research project and recruit participants. Students who agreed to participate filled out the consent forms, so she tested those students in the computer lab on campus. Nineteen of the students who participated were men. The quiet room made Jolene yawn on this rainy afternoon, and she was happy when the last student completed the project so that she could go home and take a nap.

That Friday night Stefanie arrived at the rush party a few minutes early. When she walked into the formal room, she found about 20 young women who were rushing the sorority all filling out the MRT with pencils bearing the sorority insignia. Some of the women had written their names in large letters at the top of their sheets, most likely in an attempt to help Jolene remember their names. Quite a few of the women were laughing about the test and calling out the answers to one another. Jolene was not present when Stefanie arrived, but returned shortly thereafter and seemed surprised to find Stefanie in the room. Although she was carrying a folder full of consent forms in her backpack (she had run back to the lab to grab them because she realized she had forgotten them), Jolene decided not to hand them out because she didn't want Stefanie to notice her mistake.

The Dilemma

Two weeks later, Stefanie attended the psychology club meeting where the seniors were presenting their data to psychology faculty and students. Stefanie became increasingly confused during Jolene's presentation. Jolene's description of her research methodology did not accurately reflect the procedures Stefanie had observed. Jolene reported that her data confirmed her hypothesis that men would outperform women regardless of the type of test administration. Furthermore, Jolene reported that scores in the TPP group were better than scores in the CA group. Stefanie knows that Dr. Lee and Jolene were planning to present these data at the annual meeting of the American Psychological Association in August.

Questions

1. What kinds of problems are inherent in Jolene's research project?
2. How would these problems affect the research results?
3. How would you solve these problems?
4. What should Stefanie do?
5. What would the consequences of these actions be for Jolene? For Dr. Lee? For Stefanie?

Teaching Notes

Introduction and Background

This directed case was designed for use in a sophomore-level research methods course as an introduction to the sections on ethics in research as well as research design. It could also be used in an introductory psychology course or in other introductory-level science courses as a critical-thinking exercise to illustrate the research process in science.

In this case, ethical issues associated with the treatment of research participants, public misrepresentation of scientific research, and supervision of student research assistants by faculty are explored through the eyes of the main characters. The main characters include two college students and a professor, all of whom will draw sympathy from readers because of the all-too-familiar hectic pace of their lives. By examining the way in which a research project was conducted, students also begin to explore the basic principles of good experimental design (e.g., equivalent groups, controlled testing environments).

One feature of this case is the ability to use it solely to discuss research ethics or as a springboard for a discussion of experimental design, or for both purposes.

Chapter 33

Objectives
Upon completing this case, students should

- understand the fundamental ethical principles that apply to research with human participants: informed consent, freedom from harm, freedom from coercion, anonymity, and confidentiality;
- understand the obligations of faculty and student researchers with respect to the participants, the research, and each other;
- be able to identify threats to internal validity in a research project; and
- be able to design an ethically and methodologically sound research project.

We believe that no preclass preparation is necessary on the part of the students and do not assign readings for this topic. We assign the appropriate chapters in the students' textbooks after the discussion to reinforce issues examined in the case. Other sources can also be introduced, such as the APA Ethical Principles of Psychologists and Code of Conduct (*www.apa.org/ethics/code2002.html*).

Common Student Misconceptions

- It is a breach of ethics to inform a professor about a colleague's possible misdeeds; students don't "rat on" other students.
- If you are conducting research on human subjects, it is enough to get their verbal consent before you carry out the research; you do not need written documentation.
- It is permissible to let people involved in a psychological study compare their answers, even though that is not in the experimental protocol.
- It doesn't make any difference if you are dealing with women or men in research as long as the sample size is large.

Classroom Management
Case analysis can be completed in a 50-minute class period (as described below) if 10 minutes are spent on the initial reading and the instructor is careful to spend about 5–10 minutes on each of the questions.

- Students are given a copy of the case at the beginning of the class period and given about 10 minutes to read it.
- Ask students the first question on page 291: *What kinds of problems are inherent in Jolene's research project?* As students respond to each question in turn, their responses should be written down categorically. One strategy is to write

methodological problems on one side of the blackboard and ethical problems on the other.

- Ask students: *How did the problems you've identified affect the research results?* Write the results on the blackboard or on separate sheets of a flip chart.

- Ask students: *If Jolene could redo the study, what would you suggest she do to solve the problems you have identified?* Write the results on the blackboard or separate sheets of a flip chart.

- Ask: *What should Stefanie do?* Write responses on the board. Alternatively, you may wish to ask students to list all of the possible things that Stefanie could do. As a follow-up, ask: *What are the consequences for her actions?* Write the students' responses next to each of the actions they listed above.

- Wrap up by saying: *You have clearly identified a number of problems inherent in Jolene's research project. What do you think the consequences should be for Jolene? For Dr. Lee? For Stefanie?*

To summarize the case for the class, you can close with the following:

Today you have learned about the importance of ethical conduct in research and proper adherence to research protocols on the validity of scientific results. In general, researchers should strive to treat their participants ethically and groups should be as equivalent as possible. These two features are necessary to actually have some trust in your research results.

This summary could be referred to later as a springboard for discussion of threats to internal validity and Type I and Type II errors. A Type I error occurs when the null hypothesis is rejected when it is in fact true; a Type II error occurs when the null hypothesis is not rejected when it is in fact false.

If your university has an honor code, your students may be more sensitive to the fact that Jolene's actions may have constituted a violation of your honor code. In our experience, a limited number of students wanted to talk about Dr. Lee's role in the ethical violations of the case. The instructor may initiate a discussion of the tenure process and the influence of this incident on Dr. Lee's tenure status. It has also been our experience that students tend to refer to Dr. Lee as a "he" even in departments in which there is an even, or female-heavy, gender distribution among faculty. If this is the case, the roles of women in science and the reasons for the students' gender attributions can be discussed.

Web Version

This case and its complete teaching notes, references, and answer key can be found on the website of the National Center for Case Study Teaching in Science at *http://sciencecases.lib.buffalo.edu/cs/collection/detail.asp?case_id=250&id=250.*

Chapter 34
How a Cancer Trial Ended in Betrayal

Ye Chen-Izu

The Case
Part I: Background

> *Birmingham, Alabama*—After Bob Lange spent eight weeks rubbing an experimental cream [BCX-34, from the prominent biotech company BioCryst] on the fiery patches on his body, researchers at the University of Alabama at Birmingham told him the drug was defeating the killer inside him. He felt grateful. "I believed it," he recalls. "I actually thought I might be cured."
>
> But it was a lie. The drug had no effect on Lange's rare and potentially fatal skin cancer. And the two key people testing the drug knew it. Lange and 21 other patients were victims of fraud—a scheme made possible by the close tie between the university and the state's most prominent biotech company.
> —*The Baltimore Sun*, June 24, 2001

In this case study, we will conduct a small-scale "clinical trial" in class to simulate the real clinical trial that was conducted by the University of Alabama at Birmingham and the biotech company BioCryst to study the effects of an experimental drug, BCX-34, in treating (malignant) cutaneous T-cell lymphoma, a skin cancer.

Objectives
The objectives of this case study include the following:
- Learn the basics of scientific research in a clinical trial.
- Learn the principles of the scientific method.
- Consider the ethical issues involved in clinical trials.

Chapter 34

After reading the newspaper article, please write concise answers to the following questions:

1. What was the disease being treated?
2. What was the drug being tested?
3. What was the *hypothesis* underlying the clinical trial?
4. What kind of *experiments* could be used to test the hypothesis?

Part II: The Main Characters

The following are the main characters involved in the clinical trial that was conducted by the University of Alabama at Birmingham and the biotech company BioCryst. Please pay special attention to the job descriptions for the clinician and the scientist. You will be asked to play these roles in a simulated clinical trial.

Bob Lange and 21 other patients who had a rare and potentially fatal form of skin cancer were participating in the clinical trial of BCX-34 directed by Dr. W. Mitchell Sams Jr., chairman of the Department of Dermatology at the University of Birmingham at Alabama.

Harry W. Snyder Jr., MD—Dr. Snyder was the scientist who ran the BCX-34 studies for BioCryst. Job description for the scientist:

- Prepare the experimental drug for clinical use. Half of the tubes should contain the experimental drug and half should contain the placebo.

- Generate random codes for the tubes. Keep a record (the key) on the code of the tube and whether it contains the experimental drug or a placebo. This key is confidential and should only be seen by the scientist during the clinical trial.

- At the end of the clinic trial, obtain the clinical record from the clinician and compare it with the key to correlate the treatment effect and the presence or absence of the experimental drug in the treatment.

Renee Peugeot, RN—Ms. Peugeot, a nurse, was given "total responsibility" by Dr. Sams to conduct the company-funded study of BCX-34 in clinics. Job description for the clinician:

- Distribute coded tubes of cream to patients to rub on their skin lesions. Keep a record on the tube of the code and to which patient the tube is distributed. The codes on the tubes are randomized and do not contain information on whether a particular tube contains the experimental drug or a placebo.

- Monitor the size of skin lesions by tracing the circumference. Record changes in the size and colors of the lesion during the treatment.

- Keep a clinical record of the treatment effect on the patient. The clinical record is confidential and should be seen only by the clinician during the clinic trial.

Part III: Clinical Trial

The class will be separated into two groups. One group will play the role of the clinician (the clinician group), and the other group will play the role of scientist (the scientist group). One person from each group will be asked to play the role of Peugeot (a clinician) and Snyder (a scientist), respectively. This subgroup will be referred to as the Peugeot-Snyder group. The people in this subgroup will receive special instructions provided by the instructor.

Scientist Group Exercise

1. Assign codes (e.g., using three-digit numbers such as 867 or 705) to 10 index cards that represent tubes of cream in the clinical trial.

2. Randomly pick five cards to contain the experimental drug and 5 cards to contain the placebo. Keep a record (the "Key" for the clinical trial) of the code on each card and whether it contains the experimental drug or placebo. *Keep the key you generate confidential during the clinical trial.*

3. Give the coded cards to the clinician group.

Wait for the clinician group to hand over the clinical record. In the meantime, read about the clinician group exercise.

Clinician Group Exercise

Wait for the coded cards from the scientist group. In the meantime, read about the scientist group exercise.

1. After receiving the coded index cards from the scientist group, give one card to each "patient," who is represented by one sheet of paper that contains two photos of skin lesions that your instructor will give you.

2. Monitor the size of the skin lesions in the photo by tracing the circumference. Compare the lesion size *before* and *after* the treatment. Record the effect of treatment on each patient. *Keep the clinical record you generate confidential during the clinical trial.*

After the scientist group and the clinician group have completed the above tasks, the two groups convene to open the key and correlate the treatment effects with the presence or absence of the experimental drug in the treatment.

Compare the results obtained by the large group (scientist group and the clinician group) with that obtained by the Peugeot-Snyder group. Discuss any discrepancies that might exist between the two.

Teaching Notes
Introduction and Background
After reading a newspaper story that documents a fraudulently conducted clinical trial involving a treatment for cutaneous T-cell lymphoma, students simulate their own small-scale "clinical trial" in the classroom. The simulation involves a secret breaching of a blind test and manipulation of data to favor a positive BCX-34 effect in treating the cancer. In one step of the trial simulation, students are presented with photographs of skin lesions from patients (these photographs are available on the website of the National Center for Case Study Teaching in Science—see Web Version). Each patient has before and after photos showing the effects of topical application of BCX-34. To determine progress, students measure the relative sizes of the skin lesions by tracing the circumferences.

This case study provides students with an opportunity to learn about the complexities and issues associated with clinical trials, particularly studies that may involve the treatment of very serious medical conditions. A clinical trial is a scientific investigation of a drug's effects on humans. It is the final step in the development of a new drug, following the study of the drug's pharmacological and toxic effects on laboratory animals. This case study presents a comprehensive tool for teaching students about both the scientific methods and the ethical issues involved in clinical trials.

The case is based on a real clinical trial conducted by the University of Alabama at Birmingham and the biotech firm BioCryst to study the effects of an experimental drug, BCX-34, in treating (malignant) cutaneous T-cell lymphoma, a skin cancer. To give students a taste of the scientific research process, the case includes a simulation of the clinical trial so that students can personally experience a small-scale research project. This classroom exercise can be fine-tuned to emphasize various aspects of a clinical trial.

Developed for first- or second-year college students, the case focuses on the scientific method, with special attention to the issues of objectivity and ethics in scientific research. As such, it would be a good case for an introductory science course for science majors or for a nonmajors science course. The material in the case study can be adapted to emphasize other topics, such as the pathophysiology and treatment of cancer. The case can also be tailored to specific student populations. For example, health profession students could be provided with more information on the drug approval process (preclinical and clinical trials), T-cell lymphoma, and the effects of BCX-34.

Objectives

- To demonstrate the basics of drug testing in a clinical trial. The simulated clinical trial is designed so that students gain first-hand experience in conducting a scientific research project in a classroom setting.

- To teach the principles of the scientific method. By guiding students through the research project step by step, one can teach them about the principles of the scientific method used along the way.

- To discuss the ethical issues involved in clinical trials. The fraud committed in the real clinical trial for BCX-34 demonstrates a typical example of unethical conduct in scientific research. This case provides an excellent opportunity to discuss factors that may influence the quality of scientific research.

Common Student Misconceptions

- Scientific data are objective and unaffected by personal bias.
- Scientific data are always unambiguous and give rise to clear conclusions.
- Using blind tests is an optional step in a scientific study.

Classroom Management

This case study can be used in classes of various sizes. A basic unit for conducting the simulated clinical trial, described in detail below, includes 10 students in the scientist group, 10 in the clinician group, one student playing the role of Snyder, and one playing the role of Peugeot. If the class is smaller, students would need to see more "patients"; if the class is larger, duplicate sets of handouts could be used for two or more clinical trial units.

To help prepare the students for the case study, briefly provide them with some background information in the form of a mini-lecture or a separate handout on cutaneous T-cell lymphoma and BCX-34:

- *Cutaneous T-cell lymphoma* is a disease in which T-cells of the lymph system become cancerous and affect the skin. Cutaneous T-cell lymphoma usually develops slowly over many years. In the early stages of the disease, the skin may become itchy and dry, and dark patches may develop on the skin. As the disease progresses, tumors may form on the skin, a condition called mycosis fungoides. As more of the skin becomes involved, the skin may become infected. The disease can spread to lymph nodes or other organs in the body, such as the spleen, lungs, or liver.

- *BCX-34*, or 9-(3-Pyridinylmethyl)-7H-9-deazaguanine, acts by inhibiting purine nucleoside phosphorylase, thereby preventing the activation of T-cells (for more information, see Bantia et al. 1996). BCX-34 also inhibits the replication of infected T-cells.

Chapter 34

Give the students Part I of the case, which presents a brief newspaper story about a clinical trial that was conducted to test the effects of the experimental drug BCX-34 on cutaneous T-cell lymphoma. Ask the students to read the newspaper account of the clinical trial and answer the questions that follow. In class, discuss the hypothesis of the clinical trial based on the background knowledge they were provided on cutaneous T-cell lymphoma and BCX-34.

Next, give students Part II, which provides more details of the clinical trial. Discuss in class the experimental design phase of a human clinical trial.

Then give students Part III, which contains the materials for the simulated clinical trail. Separate the class into two groups. One group plays the role of the *clinician*, the other the role of the *scientist*. Give the scientist group 10 index cards that represent the tubes of cream in the clinical trial. Give the clinician group the 10 sheets of skin lesion photos. Each sheet represents one "patient" and contains two images of a skin lesion before and after treatment. Each image pair is based on a single photo whose size has, in some cases, been diminished or cropped to illustrate "progress."

Ask one volunteer from each group to play the roles of Peugeot (a clinician) and Snyder (a scientist). These students each receive a duplicate set of the index cards and skin lesion photos and are directed by the instructor in private that they are to breach the blind test and manipulate the data to favor a positive BCX-34 effect in treating the cancer.

Below are the instructions for the two groups.

Scientist Group Exercise

1. Assign codes (e.g., using three digit numbers such as 867 or 705) to 10 index cards. During the classroom simulation, these index cards represent tubes of cream that will be given to the patients. The number of cards may vary with the number of students. Please note that because the number of cards needs to be limited to a manageable size in this exercise, the sample size may not be large enough to obtain statistically significant results. However, this does not affect the goal of this exercise, which is to experience scientific research in action.

2. Randomly pick five cards to represent the experimental drug and five cards to represent the placebo. Keep a record of the codes and whether they represent the drug or not (the key, or randomization schedule). **Keep the key you generate confidential during the clinical trial.**

3. Give the coded index cards to the clinician group. Wait for the clinician group to hand over the clinical record. In the mean time, read about the Clinician Group Exercise.

Clinician Group Exercise

1. While waiting for the coded index cards from the scientist group, read about the Scientist Group Exercise.

2. After receiving the coded index cards (tubes of cream) from the scientist group, give one card to each "patient." Each patient is represented by one sheet of skin lesion photos.

3. Monitor the size of skin lesions in the photo by tracing the circumference. Each patient has two photos of skin lesion, showing the lesion *before* and *after* the treatment.

4. Record the effect of treatment on each patient. Keep the clinical record you generate confidential during the treatment.

5. At the end of the clinical trial, present the clinical record to the scientist group.

Have the two groups get together to analyze data and discuss the results from the simulated clinical trial. Then ask the two students playing the parts of Snyder and Peugeot to present their results. In class, discuss the discrepancies that might exist between the two sets of results.

To wrap up, summarize the stages of scientific investigation:

1. Proposal of hypotheses based on previous knowledge
2. Design of experiments to test the hypotheses
3. Experimentation, data analysis, and summarization of results
4. Communication of results to other people
5. Evaluation of evidence and determination of whether the previous hypotheses need to be modified, followed by development of further hypotheses

The following are suggested follow-up questions:

1. Does the method used to determine lesion size provide an objective measurement?

   Because there is no clear boundary between a lesion and the surrounding normal tissue, determining the circumference of lesions depends largely on a given clinician's judgment.

2. What method was used to minimize the subjectivity that might be involved in the experiments?

   Discuss in detail the double-blind test technique commonly used in clinical settings. Emphasize the need for objectivity in scientific research.

Chapter 34

Students will naturally be interested in what actually happened in the clinical trial conducted by BioCryst and the University of Alabama at Birmingham. Peugeot and Snyder were married to one another and owned stocks of BioCryst. A double-blind test was not carried out. Snyder told Peugeot which tube contained the drug. Peugeot forged the data by using the ambiguity in measuring the lesion size. Snyder also modified the data to favor a positive drug effect. The couple manipulated the clinical trial to gain approval of the experimental drug to increase the value of their stock in the company. Eventually, Peugeot and Snyder were convicted of defrauding the U.S. Food and Drug Administration (FDA). Dr. Sams, under whose direction the clinical trial was conducted, was banned from testing drugs for the FDA and eventually retired from the University of Alabama at Birmingham. In addition, the National Institutes of Health accused the university of poor oversight and suspended enrollment of patients in 550 studies. Investors in BioCryst lost an estimated $34 million after the bogus data were discovered.

Web Version

This case and its complete teaching notes, photos, and answer key can be found on the website of the National Center for Case Study Teaching in Science at *http://libweb.lib.buffalo.edu/cs/collection/detail.asp?id=233&case_id=233*.

References

Bantia, S., J. Montgomery, H. G. Johnson, and G. M. Walsh. 1996. *In vivo* and *in vitro* pharmacological activity of the purine nucleaside phosphorylase inhibitor BCX-34: The role of GTP and dGTP. *Immunopharmocology* 35 (1): 53–63.

Birch, D. M., and G. Cohn. *Baltimore Sun.* 2001. How a cancer trial ended in betrayal. June 24.

Chapter 35
Bringing Back Baby Jason: To Clone or Not to Clone

Jennifer Hayes-Klosteridis

The Case

Ted McMasters sat alone in the waiting area of the Green Spring Valley shock trauma unit. It had been an hour since the state police marched into his law office on Calvert Street to tell him that his wife, Julie, and two-year old son, Jason, had been in a serious car accident and were en route via helicopter to the shock trauma unit. "It's the best shock trauma unit on the coast," he remembered hearing one of the officers say ... like that mattered.

Ted knew that Julie was on her way home from the babysitter's when the accident happened. Ironically, she left work early because she wasn't feeling well. They both hoped she was pregnant again. The state police told him she was stopped at a red light when an 18-wheel flatbed carrying a load of bricks slammed into the passenger side of her Audi A6 station wagon. She and Jason had been rushed to shock trauma within an hour of the accident. Both were in critical condition during the helicopter transfer and were rushed into surgery immediately upon arrival.

Ted stood shaking in the waiting room of the fourth-floor surgical unit. He felt cold and wondered if it was the room temperature or the shock of the accident. Not knowing the status of Julie's and Jason's health was too painful for words. As he waited for news, he reflected on his life.

Ted and Julie wanted a child from the day they were married. They both came from large families, and having children just seemed like the natural thing to do. It was such a long road trying to get pregnant with Jason ... but ten years and countless miscarriages had finally produced their son. Ted knew this baby would be their last chance for the "typical American family." Of course he blamed himself. He should have frozen more sperm after his testicular cancer. Who knew that the radiation therapy would be so devastating to his ability to have children? It hadn't been easy for them, but Jason was well worth the wait! He was a healthy,

Chapter 35

bright child with big brown eyes and a sprinkling of freckles across his nose. He loved to climb and always had scuff marks on his knees. "I wonder if this new baby will be like Jason," Ted had wondered. ...

"Mr. McMasters?" the sound of a doctor's voice interrupted his musings. "I am Dr. Wilcox. I have been attending to your wife. She is in stable condition in the intensive care unit. We managed to repair her punctured right lung and broken femur. Right now her prognosis is good. Um, sir, did you know that your wife was pregnant?"

"She just had a pregnancy test yesterday. We were still waiting on results. Why?"

"Please sit down, Mr. McMasters," the doctor said gently. "I am so sorry. The stress from the accident caused your wife to hemorrhage. She lost the baby. Fortunately, the damage to her uterus is insignificant, so you will be able to have more children."

Stunned and devastated, Ted answered in a quavering voice, "I'm sterile."

"Oh, I am so sorry. Please forgive my assumption."

"Doctor, my baby boy was also in the accident with his mother. The pediatrician said they thought he had some internal bleeding, but they were taking him up to surgery and didn't have any further information."

"There is no easy way to tell you this. I'm sorry. Jason's injuries were quite substantial. His heart stopped beating while he was on the operating table. We were unable to revive him."

Ted slumped in his chair silently, tears streaming down his face. How could this be? How could Jason be gone? His big, brown-eyed baby boy, gone?

"Mr. McMasters, I know this may not be the best time to mention this, but ..."

"Yes, my wife and I are organ donors. We would ..."

"Oh, I'm not suggesting organ donation. I am suggesting you contact a friend of mine, Dr. Bossilier. She is the director of Clonaid, the first human cloning company."

"What are you saying, doctor?"

"I am saying that you and your wife could have Jason back, alive and healthy, by making a clone of him from his cells. He would be Jason's identical twin, a chance for you to continue your family tree."

"Whaaaat? I'm not sure I understand what you're saying."

"I'm saying that it may be possible for you to clone your son. I realize this all comes as quite a shock right now and you have to discuss your options with your wife. If you are interested, here's my card."

Ted watched the doctor walk from the room and contemplated what he should do.

Can you help Ted make his decision?

Questions

1. What is cloning? From a scientific standpoint, what procedures would be used to clone a human?

2. Is cloning humans different procedurally from cloning animals? Are there any physiological risks to cloning humans? Are these risks different for animals? Why or why not?

3. What is imprinting? What role does imprinting play in cloning humans? Animals?

4. How does cloning differ from sexual reproduction?

5. What roles do meiosis and mitosis play in maintaining the integrity of the genetic code during cloning? During sexual reproduction?

6. Will Jason II be "identical" to Jason I? Why or why not?

7. What is the difference between a somatic cell and a sex cell?

Teaching Notes

Introduction and Background

Until the mid-1990s, American society relegated cloning to the ranks of science fiction. With the completion of the Human Genome Project and the creation of Dolly the sheep and other mammalian clones, the potential for the production of human clones moves from fiction to the foreground of reality. Public opinion changed from quiet skepticism to outright concern. Today the public raises ethical, medical, political, economic, and religious questions about the potential uses and misuses of this new technology.

For our students to be able to make informed opinions about cloning technology, it is important for them to consider the societal issues in connection with the genetics and cell biology that make human cloning a real possibility. Often in our undergraduate science courses there is little opportunity to engage in didactic exchanges about current events while teaching the basic facts and concepts. Case-based instruction provides faculty the opportunity to do both.

This case study is a dilemma case that explores the concept of human cloning. While this case presents a fictitious scenario, it is grounded in fact. In September 1999, Charleston attorney and former state delegate Mark Hunt and his wife, Tracey, began contemplating cloning their infant son, Andrew, who died after a heart surgery procedure. The Hunts felt that the duplicate child would provide them a sense of peace over losing their only child. To pursue their goal of producing a clone of Andrew, the Hunts provided at least $500,000 to Clonaid president Brigitte Bossilier to develop a laboratory and purchase equipment. In 2001, the Hunts ended their relationship with Bossilier and Clonaid, citing a "loss of confidence" in her ability to produce a clone of their dead son.

Chapter 35

This case was designed for use in an undergraduate genetics course at the University of Maryland, College Park. The course is offered each semester to more than 400 students in a large lecture hall setting. Students are required to attend a three-hour discussion each week. The course is open to students of all majors, but enrollment is largely composed of life sciences majors. Students enrolled in the course must have completed one year of introductory biology and inorganic chemistry and have at least an average level of understanding of basic cell biology and genetic mechanisms. This case was assigned to the students in the discussion sections near the end of the fall semester.

Objectives

- To introduce students to the concept of human cloning
- To develop an understanding of the basic genetic concepts underlying the cloning process, including imprinting, mitosis, meiosis, asexual reproduction, and sexual reproduction
- To encourage students to consider the scientific and social aspects of human cloning
- To have students integrate key course terms with the concept of human cloning, including *dominant, recessive, co-dominance, partial dominance, recessive lethal, pleiotropic, intergenetic suppressor, intragenetic suppressor, recessive lethal, epistasis, penetrance, expressivity, imprinting, totipotency, pleuropotency, stem cell, embryonic stem cell, enucleation, biotechnology, in vitro fertilization, cloning,* and *transgenic organisms*
- To enhance students' critical-thinking skills

Common Student Misconceptions

- A cloned mammal will be a carbon copy of the original mammal.
- Inheritance patterns are based exclusively on single-gene inheritance modes.
- A cloned human will have the same personality as the original person.

Classroom Management

Students were assigned this case, either online through a password-protected course management system or as a printed copy, one week in advance of the case discussion to enhance lecture material pertaining to imprinting, cloning, and biotechnology. Three graduate teaching assistants were trained to conduct the classroom discussion of this case. Students were required to read the case and answer the associated questions as preparation for an in-class discussion. On the day of the case discussion, students were put into groups of four to discuss the question, "What should Ted consider in making a decision about whether to clone his son or not?" Students were allowed 30 minutes to explore the related issues. At the end of this time, the graduate teaching assistants asked a member from each group to present one key concern to the class. As the students presented their concerns, the graduate teaching assistants encouraged the other students to comment on the concern presented, elaborate on related issues, and discuss key terminology. At the end of the discussion period (70 minutes), the students were asked to return to their groups for 10 minutes to decide what Ted should do based on the in-class discussion.

This case should encourage undergraduate genetics students to explore the feasibility of human cloning from a scientific and ethical standpoint. To fully comprehend this possibility, students will need to understand state-of-the-art cloning technology, the problems encountered in cloning animals, and the potential scientific problems associated with human cloning. Underlying this process are the basic concepts they should have learned throughout the semester-long course.

Web Version

This case and its complete teaching notes, references, and answer key are available on the National Center for Case Study Teaching in Science website at *http://sciencecases.lib.buffalo.edu/cs/collection/detail.asp?case_id=404&id=404*.

Chapter 36
Selecting the Perfect Baby: The Ethics of "Embryo Design"

Julia Omarzu

The Case

The research team assembled quietly in the lab. There were some difficult decisions to be made today. Kelly, a new research assistant, looked forward to the discussion. Privately, she hoped Dr. Wagner and the rest of the team would agree to help the couple that had appealed to them.

"Good morning, everyone," Dr. Wagner said to begin the meeting. "We have a lot to talk about. I'll summarize this case for those of you who may not have had time to read the file. Larry and June Shannon have been married six years. They have a four-year-old daughter named Sally who has been diagnosed with Fanconi's anemia. Sally was born without thumbs and with a hole in her heart. Shortly after her birth, she began suffering symptoms related to impaired kidney function and digestion that have only increased in severity. Fanconi's anemia is a progressive disease that often results in physical abnormalities and a compromised immune system. Sally needs a lot of special care and has already had several surgeries. She can't digest food normally or fight off infections as easily as a normal child would. If she doesn't receive a bone marrow transplant, she will develop leukemia and die, most likely within the next three to four years. Neither Larry nor June had any clue they were both carriers of this disease."

"A frightening diagnosis," said Kevin, a research technician.

"Difficult to live with, as well. Not only will they probably lose this child, they must be crushed about the possibility of having another child with this illness," commented Liz Schultz, the team's postdoctoral researcher in gynecology and fertility.

"Exactly their problem," continued Dr. Wagner. "The Shannons are interested in having another child and have approached us regarding pre-implantation genetic diagnosis (PGD). They are aware of the risks and the odds of success. They are anxious to begin the process as soon as possible."

Chapter 36

"Kelly, you're new to the team, so let me summarize the PGD process for you. It's a three-step process, with chances of failure and complications at each step. First, in vitro fertilization (IVF) is performed. Some of June's ova would be removed and fertilized with Larry's sperm outside of June's womb. If this procedure works, we should have several viable, fertilized embryos. Our second step is to perform genetic analysis on the embryos, removing a cell from each and testing for the presence of the Fanconi's anemia genes. If we find embryos that are free of Fanconi's, we can then perform the third step, which would be implanting the healthy embryos back into June's uterus."

"Wait a minute," said Kelly. "How many embryos are we talking about? They just want one child, not a half dozen."

Dr. Wagner laughed. "Yes, I know. But during the in vitro fertilization and implantation processes, we almost always have embryos that do not survive. There is only about a 23% chance of any implanted embryo thriving. There is a better chance for a positive outcome when we remove and fertilize multiple ova. In this particular case, the odds of a multiple pregnancy are very small, given the limitations on the ova we will be able to implant."

"Okay, I know I don't understand all of this. But how can Mrs. Shannon produce that many eggs all at the same time?" asked Kelly. "She wouldn't normally do that, would she?"

"No," said Liz. "So before we even begin any of these procedures, June would have to take hormones to increase the number of ova she releases. As Dr. Wagner said, there are risks involved with every step of this procedure. Hormone therapy can have some side effects, including mood and cognitive effects. Some women suffer physical complications as well, although this is relatively rare. There are some studies that link hormone therapy to increased risks of ovarian cancer, although there is other research that contradicts that."

"Plus," Dr. Wagner added, "along with the risks to June, there is no guarantee that the procedure will be successful. Many couples must undergo the IVF procedure more than once before the implantation is successful in producing a healthy, full-term baby. In this case, it will be even more complicated because we cannot use all of the fertilized embryos but must limit ourselves only to those that are free of Fanconi's anemia."

"But we've done several of these types of procedures with a pretty high rate of success," said Kevin. "Why should this one be different? You've screened the couple, right, and you said they're aware of the risks?"

"Yes, but this case is very complicated." Dr. Wagner sighed. "The Shannons have requested not only a Fanconi's-free child, but one that will be a perfect bone marrow match for Sally. Sally's illness may be treated with a transplant of healthy cells into Sally's bone marrow. Because Fanconi's patients are so fragile, however, the donor's cells have to be a near perfect match, and that's hard to find. Siblings are the best

Selecting the Perfect Baby: The Ethics of "Embryo Design"

bet. In the meantime, Sally's condition is deteriorating. The Shannons naturally want to give Sally as many years of normal life as possible so they want to take aggressive action. They want to cure Sally's disease by planning and creating another child with specific genetic markers."

"How would that work?" asked Kelly.

"You've heard of stem cell research?" began Liz. "Stem cells are special cells that can produce all the different organs and tissues of the human body. They are found in embryos or fetuses, and are usually obtained for research from embryos that die or are rejected in fertility procedures. That is the kind of research that has been so politically controversial. But a less potent type of stem cell is also found in adult humans and can also be obtained from umbilical cord blood. If we were to help the Shannons and the procedure was successful, the blood from their new baby's umbilical cord could be used for Sally's bone marrow transplant, resulting in no injury at all to the baby and a possible cure for the worst symptoms of Sally's illness."

"The Shannons are suggesting that we perform the PGD procedure as we normally do, but select only those embryos that are both free of Fanconi's anemia *and* are also a perfect match for Sally," said Dr. Wagner. "This presents some real ethical dilemmas for us. We have never tried this before. People have had PGD done to detect and prevent a variety of illnesses in their children, just as we have done here before. But what we are proposing now would be selecting for a specific combination of genetic traits, a combination that will not benefit the planned child but will save an existing child. We will be selecting an embryo and then using it essentially as a blood donor for its sibling. It will be umbilical cord blood, which would be discarded anyway, but it's still a controversial procedure. If we agree, it also means we will be destroying embryos that are perfectly healthy, but are just not a match for Sally. I'm interested in pursuing this, but these are serious issues to consider. Not the least of which is that we may have trouble getting it approved. Before I run it past the review board, I want to know how you all feel about trying it."

"Well, I say go ahead with it. It will be a genetic breakthrough. In time, we'll be able to prevent all kinds of problems with this procedure. Why not start now?" urged Kevin.

Another doctor on the team who had remained silent nodded in agreement.

"I'm not sure yet how I feel about this," said Liz. "I feel a little uncomfortable with the precedent this might set. We'll be opening the door to who knows what type of genetic selection. Do we want the responsibility for that?"

A couple of others on the team seemed to side with her.

"Yes," said Kelly. "But think about the poor Shannons. And especially Sally. Does she deserve to suffer just because we're arguing about ethical problems of the future?"

"Well, it sounds like we all need to talk about this some more before we can reach a real consensus," Dr. Wagner concluded. "I don't want to start on a case this important without everyone's agreement."

Chapter 36

Questions

1. How could baby Sally inherit Fanconi's anemia even though neither parent suffers from it?
2. What other illnesses or developmental disabilities can be inherited in this way?
3. What are the odds that the Shannons' second child would also have this disease?
4. What are the basic processes of IVF and PGD?
5. What risks are involved in this whole procedure?
6. How could a sibling's blood help cure Sally?
7. How could PGD be used to create a sibling to help cure Sally?
8. What is so unusual about the PGD proposed by the Shannons?
9. What are some ethical issues related to the use of IVF? What are some ethical issues related to the use of PGD? What do you think about those issues?
10. What do you think the research team should do? What should the Shannons do?

Teaching Notes

Introduction and Background

This dilemma and discussion case was designed for an introductory course in developmental psychology but it can be used in many introductory science classes. Although the debate and doctors described are fictional, the case is based on actual events from the late 1990s that were extensively reported in the public press and in a documentary film.

In 1994, Jack and Lisa Nash had a daughter, Molly, who inherited a rare genetic disorder called Fanconi's anemia. By having another child with specific genetic markers, the Nashes could use stem cells from the new baby's umbilical cord blood to effectively cure Molly. Their search for doctors to provide this type of pre-implantation genetic diagnosis and treatment was controversial. Screening their embryos to eliminate the genetic disorder in a second child was not the problem. The controversial step was to eliminate some healthy embryos and implant only those that matched Molly's needs. Eventually they were successful in obtaining the treatment. Molly now has a little brother whose umbilical cord blood was used to treat her. Currently, she appears to be doing fine.

Prior to my use of cases in the classroom, I used the story of Jack and Lisa Nash to initiate student discussion. Students were eager to debate the ethical issues of genetic manipulation and fertility treatment. I observed that in previous semesters students easily identified with the parents in the story and with the suffering child. I wanted them to approach the issue from the scientist's point of view, so I wrote the fictional research team debate to frame the story.

Objectives

- To demonstrate a basic understanding of how developmental disorders can be transmitted genetically, including the differences between disorders triggered by recessive genes, X-linked genes, and genetic mutation
- To explain in vitro fertilization and pre-implantation genetic diagnosis, including basic risks involved with the procedures
- To consider and discuss ethical issues involved in these procedures

Common Student Misconceptions

- There is always a clear right or wrong answer in science.
- At least one parent must directly display a trait or characteristic for a child to inherit it.
- Fertility treatments are generally simple and successful procedures.

Classroom Management

In my course, students read the case and the questions that accompany it individually; we then discuss the material as a class during the next meeting. Students later complete individual follow-up papers. This case could also be assigned for small-group discussion. I have avoided this option because I find that students often have strong opinions on reproductive issues. Conflicts between students are not uncommon and I prefer to have them take place when I am mediating the whole-class discussion.

Students complete the case as homework and use their text and other sources to help them with any background information on genetics they might need. We spend the next hour (at least) of class time discussing the case. I usually begin the discussion by asking students what the Shannons want and what their possible options are. This can lead to a discussion of recessive-linked disorders and a calculation of the risk of having another child with Fanconi's anemia. Students can generate a list of options available to the couple, including the procedures outlined in the case. I focus on these procedures and discuss the ethical problems related to each. I round out the discussion by turning to the research team in the case. Students present their views on what the research team's dilemma is and the risks they run. We list their options and conclude by taking a vote on what the research team should do.

Various other assignments can be completed for this case. I have had students choose a genetic marker in their own family and draw a family tree tracing it through as many generations as they can. My students also complete an informed opinion paper after the discussion in which they address any or all of the following questions:

Chapter 36

If a problem were suspected during a pregnancy, would you want to know? Would you consider using IVF or PGD yourself? Why or why not? What do you think is the most important ethical issue associated with PGD? Describe both sides of the issue.

Web Version

The case and its complete teaching notes, references, and answer key are found on the National Center for Case Study Teaching in Science website at *http://science-cases.lib.buffalo.edu/cs/collection/detail.asp?case_id=347&id=347*.

Chapter 37
Studying Racial Bias: Too Hot to Handle?

Jane Marantz Connor

The Case
Part I: A Research Proposal
In a psychology of racism class at State University in River City, New Jersey, during fall 1998, students were watching a video called "True Colors" that had been produced and shown on the television show *Primetime Live* almost 10 years earlier. In the video, two young men, John and Glen, were shown pretending that they had just moved to St. Louis and were looking for an apartment and a job, ordinary activities of a person new to town. John and Glen appeared similar in every important way—dress, appearance, grooming, speech, and so on—except that John was white and Glen was black.

The video camera caught major differences in how they were treated. At an employment agency, for example, John was given several job leads while Glen was questioned with suspicion. At a housing complex, John was given a key and invited to check out an apartment while Glen was told there were no vacancies. At a car dealership, Glen was quoted a higher price and less favorable financing for the same car as John. On and on it went.

At the end of the video, the students discussed their reactions to what they had seen. The African American and Latino students in the class indicated that they were not surprised by what had been shown. They thought it was an accurate reflection of the experiences of people of color in the United States. The Asian American students agreed with them but were less strong in expressing their opinions. Most of the white students, however, reacted quite differently. Though they thought that the discrepancies in how John and Glen were treated in the video were striking and deplorable, they did not believe that those discrepancies were typical of everyday interactions. They thought that racism was pretty much a thing of the past. They had questions about how many agencies, apartments, or stores John and Glen and the camera crew had visited to obtain the segments that were included in the video. Knowing that the video had been made almost 10 years earlier and in a different region in the country,

Chapter 37

they also felt what they saw in the video would be quite different from what would occur in River City in 1998. In sum, watching the video did little to bridge the gap between the perceptions of white students and students of color regarding the existence and magnitude of racism in River City today, as Dr. Barbara Jones, the instructor, had originally hoped it might when she decided to show the video.

After class was over, four concerned students spoke with Dr. Jones, whose research interests included prejudice and racism.

"The video we saw today was provocative and attention-getting but was not controlled research. Would it be possible for us to apply the research skills we have been learning as psychology majors and investigate racism in River City?" asked Greg.

"I would really like to know if the discrimination we saw in the video today is a common occurrence or a rare event," said Lourdes.

With much enthusiasm, the students signed up for an independent study course with Dr. Jones the following semester, and by February 1999 they had designed an experiment to investigate the existence of racism locally. Following university guidelines, Dr. Jones and the students wrote a description of the research they wanted to do and how they would do it, and submitted the proposal to the Institutional Review Board (IRB) at State University.

This board has the responsibility to see that research involving human subjects is performed in an ethical manner.

Summary of the Proposal

Black students and white students will visit nearby stores and apply for jobs. The conversations will be recorded and the content of the tapes will be coded and scored by other students who will not know the race of the student in the tape or the store that was visited. After being scored, the tapes will be erased and the data from all of the stores will be analyzed and summarized as a group; no information about individual stores will be reported. From an analysis of the statements on the tapes and the number of phone calls, interviews, and job offers the students receive, it will be possible to evaluate whether the students' race affected their job opportunities.

Part II: Ethics and the Conduct of Research With Human Subjects

How Does Ethics Relate to the Conduct of Research With Human Subjects?

Although scientific research has produced much of value to humankind, the rights of human participants in research have not always been protected by researchers. One infamous example is the heinous research done on prisoners in concentration camps by Nazi doctors during World War II. Another is the Tuskegee syphilis study with poor, rural black men, which was started in the 1940s, before an effective treatment for syphilis was known (see Chapter 38 in this book). To study the untreated course of syphilis, the men were not made aware of penicillin, which could have effectively treated their disease, even long after it was generally available.

To prevent such abuses, Congress created the National Commission for the Protection of Human Subjects of Biomedical and Behavioral Research in 1974. This commission was charged with (a) identifying the ethical principles that should underlie the conduct of biomedical and behavioral research involving human subjects and (b) developing guidelines that should be followed to ensure that such research is conducted in accordance with those principles. The Belmont Report, issued by the commission in 1979, recommended three basic principles to serve as the foundation for the evaluation of the ethical conduct of research.

Ethical Principles

a. Respect for Persons: Individuals are to be treated as autonomous persons "capable of deliberation about personal goals and acting under the direction of such deliberation."

b. Beneficence: Researchers have a responsibility to avoid harming research participants and to maximize possible benefits while minimizing possible harm.

c. Justice: Those who enjoy the benefits of research should also bear its burdens. It is not just, for example, that poor persons or minority people should be the subjects of research while the advantage of the knowledge gained is used to benefit people with money or positions of power in society.

The Belmont Report also identified three major applications of these principles.

Applications

a. Informed Consent: "Respect for persons requires that subjects, to the degree that they are capable, be given the opportunity to choose what shall or shall not happen to them. This opportunity is provided when adequate standards for informed consent are satisfied." Such consent must be based on (a) subjects having suitable information about the research presented in a comprehensible fashion and (b) subjects' participation being voluntary and free of coercion or undue influence.

b. Systematic Assessment of Risks and Benefits: "Benefits and risks must be 'balanced' and shown to be 'in a favorable ratio.'"

c. Selection of Subjects: On both an individual level and a social level, the burdens and benefits of research should be justly distributed—that is, potentially beneficial research should not be conducted only with advantaged persons and "risky" research with disadvantaged ones.

How Are These Principles Applied to Specific Research Protocols?

The Belmont Report was adopted as a statement of policy by the U.S. Department of Health, Education and Welfare, and, in 1981, federal regulations were issued to

Chapter 37

implement this policy (Title 45, Part 46 of the Code of Federal Regulations, Protection of Human Subjects, known as 45 CFR 46). All institutions receiving federal money are required to have an Institutional Review Board (IRB) to ensure that all research conducted at those institutions that involves human subjects, whether supported by federal money or not, is consistent with these guidelines. The Office of Protection from Research Risks (OPRR) within the National Institutes of Health oversees these IRBs and can impose severe sanctions, including the loss of all federal money, on institutions that do not comply with these regulations. The parts of these regulations most relevant to the present case are printed in Figure 37.1.

Objectives of This Case Study
Although much of the work of IRBs is routine, there are times when difficult decisions must be made. It is the purpose of this case study to give you some experience in applying federal regulations in such a difficult case. This case is fictionalized, but it is adapted from a case that really did occur.

Question
Do you think the research proposal described in Part I is consistent with federal guidelines for the approval of research with human subjects and the ethical principles outlined in the Belmont Report?

Part III: How Did the IRB Evaluate the Proposal?
The IRB at State University consisted of a lawyer, a philosopher who specialized in medical ethics, a vice president for university relations, a psychology professor, a minister, and another administrator who chaired the committee. It included four white males and two white females. The committee voted unanimously to disapprove the study. A summary of their concerns follows.

Mary Cooper (the lawyer): To me the proposal smacks of entrapment. Since it is against the law to discriminate in hiring based on race, it seems to me that they are creating a situation where store owners or employees could be charged with illegal behavior. All of this is to be done without either the consent of the owners or employees or their knowledge that they are even participating in a research project. You can't get much sneakier than that!

Patricia Barton (vice president for university relations): Frankly, I am concerned about the potential harm of this research for town-gown relations and the local community in general. We have enough conflicts with the local community: out-of-control fraternity parties; differences in interactional style between our urban students and the more rural local residents; and economic, political, and cultural differences between the two groups. It's clear from the way the proposal is written that the students want to give the results of their supposed research to the local media. We really don't need local merchants and citizens seeing the university as trying to trick them

Figure 37.1. Guidelines From the *Code of Federal Regulations,* **Protection of Human Subjects (45 CFR 46)**

Definitions:

a. Research: A systematic investigation designed to develop or contribute to generalizable knowledge.

b. Human subject: A living individual about whom an investigator (whether professional or student conducting research) obtains (1) data through intervention or interaction, or (2) identifiable private information.

c. Minimal risk: The risks of harm anticipated in the proposed research are not greater than those ordinarily encountered in daily life or during the performance of routine physical or psychological examination or tests.

It is the responsibility of the IRB to approve research based on the committee's determination that the following requirements are satisfied:

a. Risks to subjects are minimized by using procedures which are consistent with sound research design and which do not unnecessarily expose subjects to risk.

b. Risks to subjects are reasonable in relation to anticipated benefits, if any, to subjects, and the importance of the knowledge that may reasonably be expected to result.

c. No research is conducted without the informed consent of the participants, unless the committee approves a waiver of the requirement of informed consent, as described below.

The IRB may waive the requirement for informed consent if it determines that

- the research involves no more than minimal risk to the subjects,
- the waiver will not adversely affect the rights and welfare of the subjects,
- the research could not practically be carried out without the waiver,
- whenever appropriate, the subjects will be provided with additional pertinent information after participation.

Chapter 37

and make them look bad. And if this research should get a wider audience, possibly being picked up by the Associated Press, how is that going to look? "River City, that racist city in New Jersey." It will hurt the university in so many ways—recruiting students, getting money from the legislature, keeping minority faculty. It could be a disaster.

Robert Blake (psychology professor): I just don't see how this research would add to scientific knowledge at all. Studies like this have already been done. We know racism exists. What new finding could they hope to obtain? It seems more like a stunt to me than an honest piece of scholarly inquiry. I also feel obliged to point out that Dr. Jones hasn't published in a scholarly journal for several years, and I do have serious concerns about her motivation for this research. Where is the science here? Is she just trying to stir up trouble on campus? I wonder if she isn't just using the students for her own purposes.

Tom Delaney (minister): How could the research yield meaningful information anyway? The black and white students might speak to different employees, and maybe a store really isn't doing any hiring. The black student would think he was discriminated against when he really wasn't.

Frank Smith (university administrator): I am very concerned about the tape recorder. Isn't it against the law to tape record a conversation without a person's knowledge? They say that the conversation will be anonymously scored and erased, but who is to say that the students might not make an extra copy or talk with their friends about what happened in the different stores. I really don't see any way to guarantee the confidentiality of the data collected, and exposure, as we have seen with the Denny's and Texaco cases, could really hurt businesses. Also, how do we know that the students won't make a scene if they are treated badly by a racist employee? It is our responsibility to protect *all* research participants, even racist ones. There could be name-calling, or even worse, if a student thought that he was the victim of an employee's or store owner's racism. Maybe that is even what the students want. All of these risks without the informed consent of the potential subjects! It could be a legal and political nightmare.

Sam Fisk (the philosopher-ethicist): I hate deception experiments of any kind. When subjects are being deceived as to the nature of an interpersonal interaction, you are violating a fundamental principle of how people should relate to each other honestly. And they are not even proposing to do any debriefing of the subjects, to try to remove the negative effects of the deception afterwards. Not that I think that is really possible anyway. A lie is a lie. How do you remove the negative effects of being lied to? But for me the lack of informed consent in this proposal is the most important thing. From the days of the Nuremberg trials, a fundamental principle that has emerged is that subjects must be aware of, and give their consent to, participating in research. This consent should be based on an understanding of the potential risks and benefits to themselves and science for such participation. Federal guidelines allow the conduct of research without informed consent only under the most restrictive conditions. There is clearly more than "minimal risk" to the prospective participants here and, frankly, I just don't see the benefits.

Part IV: The Response of the Supporters of the Proposed Research

Supporters of the research included the students, Dr. Jones, and the local chapter of the National Association for the Advancement of Colored People (NAACP). They were very upset with the decision of the IRB. A summary of their views follows.

Greg (undergrad): I can't believe they think the study is too risky. Some of the deception experiments I have seen conducted in the psychology department induce a lot more stress in the subjects and have greater possibility for harm than what we are proposing. I have seen some students reduced to tears in "approved" experiments. It's obvious to anyone that this research cannot be conducted if we have to obtain the informed consent of the subjects. So what they are really saying is that we just can't study racism in the community. What happened to freedom of speech and academic freedom?

Leslie (undergrad): I am angry and insulted that one of their reasons for disapproving the study is the concern that I and the other black students will not follow the research protocol and will deliberately violate the confidentiality of the responses we are collecting. Other students are entrusted with highly sensitive information in research projects, including information about suicidal tendencies, sexual orientation, and family problems. Has anyone ever heard of any other study being disapproved because the IRB thought the undergraduate assistants would not follow the research protocol? And what are they thinking we would do in the store? Punch someone? Talk about stereotypical thinking! As a black man, I encounter racism every day. How can the risks of harm to the employees in this experiment be greater than in everyday life, which is the definition of "minimal risk," when what we are proposing to do is an everyday behavior? If anything, the risks of harm to the employees should be lower than in everyday life since the other testers and I will be trained and committed to following the research protocol and to not responding violently, no matter how provocatively an employee might behave.

Lourdes (undergrad): I looked up the definition of *entrapment*. It is when an enforcement agency entices the commission of an illegal behavior for the purpose of prosecution. We are not an enforcement agency, and the data we are collecting will be summarized and reported anonymously. We will not be prosecuting anyone! Furthermore, we are simply applying for a job, an everyday behavior, and could hardly be said to be "enticing" anything. How could a lawyer say this is entrapment? I also checked the law on tape recording. In New Jersey, it is legal to record a conversation as long as one party to the conversation is aware it is being recorded. We could do the study without the tape recorder, if necessary, but it would compromise the scientific rigor of the study.

Chapter 37

Suzanne (undergrad): It's clear to me that all of this is just a bunch of excuses. What they're really trying to protect is white privilege. Black people only get the jobs that white people don't want. This has been going on for hundreds of years. Why should we expect it to stop now? And, to the extent that there is a risk of harm to the subjects, isn't it only just that the persons bearing the risk of harm are the ones who have profited all these years from racism and discrimination against black people? How can they say the study has no merit? Don't they think it's important for us to know what's going on in River City? Haven't they heard of replication in science? It's obvious they're not concerned with the welfare of the human subjects, which is their responsibility, but the image of the university.

Dr. Jones (psychology professor): They say that they can't approve the research because of the lack of informed consent. But the IRB approves observational studies in the community every year that don't include informed consent. I know, because many of them are studies done by my statistics students. Why is this different? If the identity of the stores sampled is not revealed, where is the potential for harm to the research participant? I'm also pretty upset that some of the lay members of the committee don't understand the role of randomization in experimental design, and the persons on the committee who do took no steps to enlighten them. The job of the IRB is to evaluate the relative risks and benefits of the research. Can you believe the head of the IRB told me he sees nothing wrong with having the risks and benefits of anti-racism research judged by an all-white committee?

Billie (president of the local NAACP): This is just one more instance of institutional racism—the system ensuring that black people are kept down. Either they don't know how badly my people are being hurt by unemployment and racism, or they don't care. Since the beginning of science there has been no shortage of experiments like the one done at Tuskegee, which obtained knowledge at the expense of black people. And then there are all the bogus studies that have been done in the name of science to prove that black people are inferior and we deserve the treatment we get. But when we have a proposal whose purpose is to study how the powerful maintain their power, all we hear is ethics this and ethics that. My tax dollars support this university, too. Why shouldn't the university do something for us once in a while?

Question

If your group were the IRB, would you approve the study as is, disapprove the study, or approve it contingent on certain specific changes in the research protocol? In the third case, what changes would you require?

Teaching Notes

Introduction and Background

This case was developed to be used in a seminar on prejudice and racism. It could also be used in a general diversity course, a research methods course, a social psychology course, or a laboratory course. The case involves having students evaluate a research proposal to determine if it is consistent with ethical principles and federal guidelines for the conduct of research with human subjects. The students are given a summary of the principles and guidelines as well as a summary of the proposal and a description of how the proposal arose.

The research proposal was written by four undergraduate students under the supervision of a psychology professor. The students wanted to study the existence of racial bias in employment in the local community by sending black and white students into retail stores inquiring about and applying for jobs. They planned to tape record the conversations so they could be scored by observers blind to the race of the student. After the tapes were scored, the results would be summarized anonymously and the tapes erased. The student investigators would also record how many students of the two races received follow-up phone calls and requests for an interview.

The university's IRB unanimously decided that the study should not be approved because it felt the research was not consistent with federal guidelines for the protection of the research participants. The concerns of both the IRB and the supporters of the proposed research are included.

Objectives

- To learn about the role of Institutional Review Boards (IRBs) in protecting the welfare of research participants
- To illustrate the major areas of concern that IRBs address in their reviews of proposed research: informed consent, risks, and benefits
- To discover the differences in perspectives between those with and without power in society and the subjectivity of the judgments that the IRB must make
- To clarify the difference between individual racism and institutional racism
- To develop listening and analytical skills, as well as an appreciation for different values and perspectives
- To illustrate how research can be used both to increase our understanding of human behavior and to make societal changes to address issues of social justice

Common Student Misconceptions

- Scientists can do their own experiments in a way they think is appropriate without oversight.

Chapter 37

- The federal government does not regulate the scientific process.
- Racial prejudice is largely gone in the United States.
- It is impossible to do experiments involving the behavior of people.
- There are no ethical guidelines on using humans in experiments as long as you do not do them any bodily harm.
- It is not necessary to get the consent of people who are involved in a study as long as they are not subjected to bodily harm.

Classroom Management

To complete this case, students need to have a reasonable understanding of the material on the role and functions of the IRB, the ethical principles that are supposed to guide the IRB, and the specifics of the proposed research. This material should be given out prior to the class in which the case will be discussed and some of the Preparation Questions (pp. 325–326) assigned as homework. If at all possible, the students should be given a mini-test of 5–10 questions at the beginning of class, with the students working in small groups to determine the correct answers. This would allow them to discuss their understanding of the material they have read.

The topic of the case, racial bias, is a controversial one that generates a great deal of student interest. Students may express strong feelings and exhibit emotional reactions to the case, and it is important that the facilitator be as knowledgeable as possible on the issues that the case raises. But it is also important that the facilitator remain neutral about which position he or she supports.

Students are randomly assigned to groups of five or six students at the beginning of the class period. In their groups, the students discuss their perspective on the research proposal and whether or not it should be approved by the IRB. Additional information is provided throughout the class period, and students may be asked at various points to vote secretly as to whether they think the study should be approved. The instructor should explain to the students that this is a complicated case and that it is appropriate for them to approach the discussions with their minds open to new facts, ideas, and perspectives.

They should feel perfectly free to change their minds, their votes, or the reasons for their votes as new information is presented and a variety of perspectives are shared.

The students should discuss the question of approving the research proposal based on whether or not it is consistent with the guidelines and ethical principles outlined in Part II of the case. They also have the option of approving the proposal contingent on certain modifications of the experimental procedure, but they need to specify what modifications they want. Be sure to designate one person as leader of the group and tell this person it is his or her responsibility to ensure that the discussion is based on the ethical issues and guidelines.

See steps and timing in Table 37.1.

Table 37.1. Procedure for "Studying Racial Bias: Too Hot to Handle?"

Part I: Mini-test, taken working in groups. Answers reviewed briefly. (Students have read Parts I and II of the case at home.)	10–20 min.
Part II: Groups discuss their opinions as to whether or not the study should be approved. Vote taken, ballots collected by instructor.	10–15 min.
Part III: Students read Part III of case and discuss their reactions to and evaluation of the IRB's comments. Vote taken, ballots collected by instructor.	10–15 min.
Part IV: Students read Part IV of case and discuss their reactions to and evaluation of the supporters' comments. Vote taken, ballots collected by instructor.	10–15 min.
Part V: Instructor puts summary of all three votes on board. Class discusses their reactions and conclusions.	10–15 min.

Preparation Questions

1. How is the Belmont Report related to the Code of Federal Regulations, Protection of Human Subjects? Why were they created?

2. What are the elements of informed consent?

3. Is informed consent always required by federal regulations? Is it considered extremely important? Why or why not?

4. What ethical principles are violated when deception is used? Why?

5. What is minimal risk? What do you think would be an example of a medical study that involves minimal risk? What is one that exceeds minimal risk?

6. What is an example of a behavioral (psychological) study that involves minimal risk? What is an example of one that exceeds minimal risk?

7. How are risks supposed to be related to benefits?

8. If a student surveys students in his dorm to find out what they think of campus food, is this research? Why or why not?

9. Suppose a student wishes to know if there is a relationship between gender and the type of car a person drives. He goes to a nearby traffic intersection and classifies each car as either a sedan, a coupe, a van, a sports car, a truck, a utility vehicle or "other," and notes whether the driver is a male or a female. According to 45 CFR 46, is this research? Who are the subjects? Is informed consent required?

10. Would the conduct of the study described in this case be consistent with the principle of justice? Why or why not?

Questions to Facilitate or Follow Discussion

1. What do you think are the major risks to the human subjects in this experiment? Is it within the purview of the IRB to be concerned about the risk of harm to town-gown (i.e., community-university) relations? What about the reputation of River City?

2. What do you think are the potential benefits of the proposed research? Are these potential benefits to the individual research participants or to society as a whole? How can you tell if the risks are "in balance" with the benefits?

3. Do you think a person's evaluation of risks and benefits might depend on his or her personal experience with racism? Does that mean that person is biased? Is everyone biased?

4. Do you think it is ethical to tape-record someone without his or her consent? In all cases, or only for research purposes?

5. Is it ethical to collect data from someone without the person's informed consent? How might people feel if they found out afterward or suspected that they had been a subject in a study without their knowledge or consent? Is this a violation of a fundamental human right? Should it ever be done?

6. What are the costs to society of racism and discrimination? Are people of color the only ones who should be concerned about this? Why or why not? Does society have a compelling need to get information about this problem?

7. Is it fair to have an evaluation of the merits of anti-racism research judged by an all-white committee? Why or why not? Why do you think the committee was composed in this manner? What did Billie mean when she called this an example of institutional racism?

8. When the university administrator talked about the student testers breaking confidentiality and making a scene, which students did he mean? What image might he have had in mind when he thought of this concern or risk?

9. If the black and white testers speak to different employees, does this make it impossible to interpret any differences in how they are treated? How does randomization control for variation in factors not deliberately under study in the investigation? Would this apply to other variables such as time of day, number of shoppers present, and so on?

10. One variation in how the study could be conducted is to identify a sample of, for instance, 100 stores and to randomly assign a black tester to visit 50 stores and a

white tester to visit the other 50. This is what is called a "between-store" design. When both the black tester and the white tester visit the same stores, this is called a "within-store" design. What are the relative advantages and disadvantages of each design?

Web Version

This case and its complete teaching notes and references can be found on the website of the National Center for Case Study Teaching in Science at *http://sciencecases.lib. buffalo.edu/cs/collection/detail.asp?case_id=458&id=458*.

References

Fix, M. E., and M. A. Turner. 1998. *National report card on discrimination in America: The role of testing.* Washington, DC: Urban Institute Press. *www.urban.org/publications/308024.html*.

U.S. Department of Health and Human Services. 1979. *Belmont Report: Ethical principles and guidelines for the protection of human subjects of research.* Washington, DC: U.S. Government Printing Office. *www.hhs.gov/ohrp/humansubjects/guidance/belmont.htm*.

U.S. House of Representatives. 2005. *Code of federal regulations, Protection of human subjects (45 CFR 46).* Washington, DC: U.S. Government Printing Office. *http://ohsr.od.nih.gov/ guidelines/45cfr46.html*.

Chapter 38
Bad Blood: The Tuskegee Syphilis Project

Ann W. Fourtner, Charles R. Fourtner, and Clyde Freeman Herreid

The Case

The Disease

Syphilis is a venereal disease spread during sexual intercourse. It can also be passed from mother to child during pregnancy. It is caused by a corkscrew-shaped bacterium called a spirochete, *Treponema pallidum*. This microscopic organism resides in many organs of the body but causes sores or ulcers (called chancres) to appear on the skin of the penis, vagina, and mouth, and occasionally in the rectum or on the tongue, lips, or breast. During sex, the bacteria leave the sores of one person and enter the moist membranes of their partner's penis, vagina, mouth, or rectum.

Once the spirochetes wiggle inside a victim, they begin to multiple at an amazing rate. (Some bacteria have a doubling rate of 30 minutes. You may want to consider how many bacteria you might have in 12 hours if one bacterium entered your body doubling at that rate.) The spirochetes then enter the lymph circulation, which carries them to nearby lymph glands that may swell in response to the infection.

This first stage of the disease (called primary syphilis) lasts only a few weeks and usually causes hard red sores or ulcers to develop on the genitals of the victim, who can then pass the disease on to someone else. During this primary stage, a blood test will not reveal the disease, but the bacteria can be scraped from the sores. The sores soon heal, and some people may recover entirely without treatment.

Secondary syphilis develops two to six weeks after the sores heal. Then flulike symptoms appear with fever, headache, eye inflammation, malaise, and joint pain, along with a skin rash and mouth and genital sores. These symptoms are a clear sign that the spirochetes have traveled throughout the body by way of the lymph and blood systems, where they now can be readily detected by a blood test (e.g., the Wassermann test). Scalp hair may drop out to give a "moth-eaten" look to the head. This secondary stage ends in a few weeks as the sores heal.

Chapter 38

Signs of the disease may never reappear, even though the bacteria continue to live in the person. However, in about 25% of those originally infected, symptoms will flare up again in late- or tertiary-stage syphilis. Almost any organ can be attacked, such as in the cardiovascular system, producing leaking heart valves and aneurysms (balloon-like bulges in the aorta that may burst, leading to instant death). Gummy or rubbery tumors filled with spirochetes and covered by a dried crust of pus may develop on the skin. The bones may deteriorate as in osteomyelitis or tuberculosis and may produce disfiguring facial mutilations as nasal and palate bones are eaten away. If the nervous system is infected, a stumbling foot-slapping gait may occur or, more severely, paralysis, senility, blindness, and insanity.

The Health Program

The cause of syphilis, the stages of the disease's development, and the complications that can result from untreated syphilis were all known to medical science in the early 1900s. In 1905, German scientists Erich Hoffman and Fritz Schaudinn isolated the bacterium that causes syphilis. In 1907, the Wassermann blood test was developed, enabling physicians to diagnose the disease. Three years later, German scientist Paul Ehrlich created an arsenic compound called salvarsan to treat syphilis. Together with mercury, it was either injected or rubbed onto the skin and often produced serious and occasionally fatal reactions in patients. Treatment was painful and usually required more than a year to complete.

In 1908, Congress established the Division of Venereal Diseases in the U.S. Public Health Service (PHS). Within a year, 44 states had organized separate bureaus for venereal disease control. Unfortunately, free treatment clinics operated only in urban areas for many years. Data collected in a survey begun in 1926 of 25 communities across the United States indicated that the incidence of syphilis among patients under observation was "4.05 cases per 1,000 population, the rate for whites being 4 per 1,000, and that for Negroes 7.2 per 1,000." (*Note:* All direct quotes in this case study are from Jones [1993].)

In 1929, Dr. Hugh S. Cumming, the surgeon general of the U.S. Public Health Service, asked the Julius Rosenwald Fund for financial support to study the control of venereal disease in the rural South. The Rosenwald Fund was a philanthropic organization that played a key role in promoting the welfare of African Americans. The fund agreed to help the PHS in developing health programs for southern African Americans.

One of the fund's major goals was to encourage their grantees to use black personnel whenever possible as a means to promote professional integration. Thus, the mission of the fund seemed to fit well with the plans of the PHS. Macon County, Alabama, was selected as one of five syphilis-control demonstration programs in February 1930. The local Tuskegee Institute endorsed the program. The institute and its John A. Andrew Memorial Hospital were staffed and administered entirely by African

American physicians and nurses: "The demonstrations would provide training for private physicians, white and colored, in the elements of venereal disease treatments and the more extensive distribution of antisyphilitic drugs and the promotion of wider use of state diagnostic laboratory facilities."

In 1930, Macon County had 27,000 residents, 82% of whom were African Americans, most living in rural poverty in shacks with dirt floors, no plumbing, and poor sanitation. This was the target population, people who "had never in their lives been treated by a doctor." Public health officials arriving on the scene announced they had come to test people for "bad blood." The term included a host of maladies and later surveys suggest that few people connected that term with syphilis.

The syphilis control study in Macon County turned up the alarming news that 36% of the African American population had syphilis. The medical director of the Rosenwald Fund was concerned about the racial connotations of the findings, saying "There is bound to be danger that the impression will be given that syphilis in the South is a Negro problem rather than one of both races." The PHS officer assured the fund and the Tuskegee Institute that demonstrations would not be used to attack the images of black Americans. He argued that the high syphilis rates were not due to "inherent racial susceptibility," but could be explained by "differences in their respective social and economic status." However, the PHS failed to persuade the fund that more work could break the cycle of poverty and disease in Macon County. So when the PHS officers suggested a larger scale extension of the work, the Rosenwald Fund trustees voted against the new project.

Building on what had been learned during the Rosenwald Fund demonstrations and the four other sites, the PHS covered the nation with Wassermann tests. Both blacks and whites were reached with extensive testing, and in some areas mobile treatment clinics were available.

The Experiment

As the PHS officers analyzed the data for the final Rosenwald Fund report in September 1932 and realized that funding for the project would be discontinued, the idea for a new study evolved into the Tuskegee Study of Untreated Syphilis in the Negro Male. The idea was to convert the original treatment program into a nontherapeutic human experiment aimed at compiling data on the progression of the disease on untreated African American males.

There was precedence for such a study. One had been conducted in Oslo, Norway, at the turn of the century on a population of white males and females. An impressive amount of information had been gathered from these patients concerning the progression of the disease. However, questions of manifestation and progression of syphilis in individuals of African descent had not been studied. In light of the discovery that African natives had some unique diseases (e.g., sickle-cell anemia—a disease of red blood cells), a study of a population of African males could reveal biological

Chapter 38

differences during the course of the disease. (Later, the argument that supported continuation of the study may even have been reinforced in the early 1950s when it was suggested that native Africans with the sickle-cell trait were less susceptible to the ravages of malaria.)

In fact, Dr. Joseph Earle Moore of the Venereal Disease Clinic of the Johns Hopkins University School of Medicine stated when consulted, "Syphilis in the Negro is in many respects almost a different disease from syphilis in the white." The PHS doctors felt that this study would emphasize and delineate these differences. Moreover, whereas the Oslo study was retrospective (looking back at old cases), the Macon Study would be a better prospective study, following the progress of the disease through time.

It was estimated that of the 1,400 patients in Macon County admitted to treatment under the Rosenwald Fund, not one had received the full course of medication prescribed as standard therapy for syphilis. The PHS officials decided that these men could be considered untreated because they had not received enough treatment to cure them. In the county there was a well-equipped teaching hospital (the John A. Andrew Memorial Hospital at the Tuskegee Institute) that could be used for scientific purposes.

Over the next months, in 1932, cooperation was ensured from the Alabama State Board of Health, the Macon County Health Department, and the Tuskegee Institute. However, Dr. J. N. Baker, the state health officer, received one important concession in exchange for his approval. Everyone found to have syphilis would have to be treated. Although this would not cure them—the nine-month study was too short—it would keep them noninfectious. Dr. Baker also insisted that local physicians be involved.

Dr. Raymond Vonderlehr was chosen for the field work that began in October 1932. Dr. Vonderlehr began his work in Alabama by spreading the word that a new syphilis control demonstration was beginning and that government doctors were giving free blood tests. Black people came to schoolhouses and churches for examination—most had never before seen a doctor. Several hundred men over 25 years old who had not been treated for "bad blood" and had been infected for more than five years were identified as Wassermann-positive. Cardiovascular problems seemed particularly evident in this population in the early days, reaffirming that African Americans might be different in their response to the disease, but nervous system involvement was not evident.

As Dr. Vonderlehr approached the end of his few months of study, he suggested to his superior, Dr. Clark, that the work continue for 5–10 years because "many interesting facts could be learned regarding the course and complications of untreated syphilis." Dr. Clark retired a few months later, and in June 1933, Dr. Vonderlehr was promoted to director of the Division of Venereal Diseases of the PHS.

This promotion began a bureaucratic pattern over the next four decades that saw the position of director go to a physician who had worked on the Tuskegee study.

Bad Blood: The Tuskegee Syphilis Project

Dr. Vonderlehr spent much of the summer of 1933 working out the study's logistics, which would enable the PHS to follow the men's health through their lifetimes. This included gaining permission from the men and their families to perform an autopsy at the time of death, thus providing the scientific community with a detailed microscopic description of the diseased organs.

Neither the syphilitics nor the controls (those men free of syphilis who were added to the project) were informed of the study's true objective. These men knew only that they were receiving treatment for "bad blood" and money for burial. Burial stipends began in 1935 and were funded by the Milbank Memorial Fund.

The skill of the African American nurse, Eunice Rivers, and the cooperation of the local health providers (most of them white males) were essential in this project. They understood the project details and the fact that the patients' available medical care (other than valid treatment for syphilis) was far better than that for most African Americans in Macon County. The local draft board agreed to exclude the men in the study from medical treatment when that became an issue during the early 1940s. State health officials also cooperated.

The study was not kept secret from the national medical community in general. Dr. Vonderlehr in 1933 contacted a large number of experts in the field of venereal disease and related medical complications. Most responded with support for the study. The American Heart Association asked for clarification of the scientific validity, then subsequently expressed great doubt and criticism concerning the tests and procedures. Dr. Vonderlehr remained convinced that the study was valid and would prove that syphilis affected African Americans differently than people of European descent. As director of the PHS Division of Venereal Diseases, he controlled the funds necessary to conduct the study, as did his successors.

Key to the cooperation of the men in the Tuskegee study was the African American PHS nurse assigned to monitor them. She quickly gained their trust. She dealt with their problems. The physicians came to respect her ability to deal with the men. She not only attempted to keep the men in the study; many times she prevented them from receiving medical care from the PHS treatment clinics offering neoarsphenamine and bismuth (the treatment for syphilis) during the late 1930s and early 1940s. She never advocated treating the men. She knew these treatment drugs had side effects. As a nurse, she had been trained to follow doctor's orders. By the time penicillin became available for the treatment of syphilis, not treating these men had become a routine procedure, which she did not question. She truly felt that these men were better off because of the routine medical examinations, distribution of aspirin "pink pills" that relieved aches and pains, and personal nursing care. She never thought of the men as victims; she was aware of the Oslo study: "This is the way I saw it: that they were studying the Negro, just like they were studying the white man, see, making a comparison." She retired from active nursing in 1965, but assisted during the annual checkups until the experiment ended.

Chapter 38

By 1943, when the Division of Venereal Diseases began treating syphilitic patients nationwide with penicillin, the men in the Tuskegee study were not considered patients. They were viewed as experimental subjects and were denied antibiotic treatment. The PHS officials insisted that the program offered even more of an opportunity to study these men as a "control against which to project not only the results obtained with the rapid schedules of therapy for syphilis but also the costs involved in finding and placing under treatment the infected individuals." There is no evidence that the study had ever been discussed in light of the Nuremberg Code, a set of ethical principles for human experimentation developed during the trials of Nazi physicians in the aftermath of World War II. Again, the study had become routine.

In 1951, Dr. Trygve Gjustland, then the current director of the Oslo study, joined the Tuskegee group to review the experiment. He offered suggestions on updating records and reviewing criteria. No one questioned the issue of contamination (men with partial treatment) or ethics. In 1952, the study began to focus on the study of aging, as well as heart disease, because of the long-term data that had been accumulated on the men. It became clear that syphilis generally shortened the lifespan of its victims and that the tissue damage began while the young men were in the second stage of the disease (Tables 38.1–38.3).

Table 38.1. 1963 Viability Data of Tuskegee Group

	Dead		Alive		Unknown	
	Number	Percentage	Number	Percentage	Number	Percentage
412 syphilitics	242	59	85	21	85	21
192 controls	87	45	66	34	39	20

Source: Rockwell, Roof Yobs, and Brittain Moore (1964)

Table 38.2. Abnormal Findings in 90 Syphilitics and 65 Controls

Abnormality	Syphilitics		Controls	
	Number	Percentage	Number	Percentage
Electrocardiographic	41	46	21	32
Cardiomegaly via x-ray	37	42	22	34
Peripheral neuropathy	12	13	5	8
Hypertension	38	43	29	45
Cardiac murmurs (aortic systolic)	24	27	20	31
Urine	28	36	21	33

Source: Rockwell, Roof Yobs, and Brittain Moore (1964)

Table 38.3. Aortic Arch and Myocardial Abnormalities at Autopsy

	Aortic Arch		Myocardial	
	Number	Percentage	Number	Percentage
140 syphilitics	62	44	48	34
54 controls	8	15	20	37
	$X^2\ P < 0.005$		$X^2\ P > 0.25$ not different	

Source: Caldwell et al. (1973)

In June 1965, Dr. Irwin J. Schatz became the first medical professional to object to the study. He suggested a need for PHS to re-evaluate their moral judgments. The PHS did not respond to his letter. In November 1966, Peter Buxtin, a PHS venereal disease interviewer and investigator, expressed his moral concerns about the study. He continued to question the study within the PHS network.

In February 1969, the PHS called together a blue ribbon panel to discuss the Tuskegee study. The participants were all physicians, none of whom had training in medical ethics. In addition, none of them were of African descent. At no point during the discussions did anyone remind the panel of PHS's own guidelines on human experimentation (established in February 1966). According to records, the original study had been composed of 412 men with syphilis and 204 controls. In 1969, 56 syphilitic subjects and 36 controls were known to be living. A total of 373 men in both groups were known to be dead. The rest were unaccounted for. The age of the

Chapter 38

survivors ranged from 59 to 85, one claiming to be 102. The outcome of this meeting was that the study would continue. The doctors convinced themselves that the syphilis in the Tuskegee men was too far along to be effectively treated by penicillin and that the men might actually suffer severe complications from such therapy. Even the Macon County Medical Society, now made up of mostly African American physicians, agreed to assist the PHS. Each was given a list of patients to follow.

In the late 1960s, PHS physician Dr. James Lucas stated in a memorandum that the Tuskegee study was "bad science" because it had been contaminated by treatment. PHS continued to put a positive spin on the experiment by noting that the study had been keeping laboratories supplied with blood samples for evaluating new blood tests for syphilis.

Peter Buxtin, who had left the PHS for law school, bothered by the study and the no-change attitude of the PHS, contacted the Associated Press. Jean Heller, the reporter assigned to the story, did extensive research into the Tuskegee experiment. When interviewed by her, the PHS officials provided her with much of her information. They were men who had nothing to hide. The story broke on July 25, 1972. The study immediately stopped.

Questions

1. Carefully analyze this case. When you examine the story and data tables, what information appears to have been gained from this study? That is, what kind of argument can be made for the benefits of the study?

2. What do you believe were the motives for the people to become involved in the study, specifically the subjects? PHS personnel? The Tuskegee staff? The Macon County physicians? Nurse Rivers?

3. What kind of criticisms can you offer of this study?

4. What were the factors underlying the cessation of the project?

5. Could this project (or one similar to it involving AIDS or radiation effects) be conducted today?

Teaching Notes

Introduction and Background

This case is a synopsis of events described by James H. Jones in his book *Bad Blood: The Tuskegee Syphilis Experiment* (all direct quotes in the case study are from this monograph). The Tuskegee study became an instant classic on the ethics of human experimentation once the Kennedy congressional hearings in 1973 occurred.

Even though this is a historical case, we can still learn important lessons about the evolution in our thinking on issues of science, human experimentation, and race—and how they are colored by our culture. In addition, we can emphasize certain long-standing principles of science that have not changed over the century; for example, proper controls are still seen as essential. Also, there are clear parallels in dealing with disease conditions within special segments of the population today (e.g., breast cancer, AIDS), which lead to special research projects with political and legal overtones.

Objectives

- To examine a classic medical case history involving scientific and moral issues
- To learn about syphilis
- To learn about the Nuremberg Code and regulations where human subjects are involved
- To understand how well-intentioned studies may go awry when medical advances and societal values shift
- To consider the meaning of informed consent in human studies
- To consider some of the difficulties inherent in working with ethnic and racial populations
- To evaluate the experimental design of a classic medical study

Common Student Misconceptions

- Syphilis is a disease that cannot be cured.
- Syphilis is a disease that differs in white and African American populations.
- It is ethical to study diseases in people without their consent or knowledge as long as you are providing good health care for them.
- Physicians can withhold treatment of patients to maintain the credibility of a scientific study.
- Over the course of a scientific study, it is reasonable to assume that the ethical standards will remain the same.

Classroom Management

This case is ideally suited for the classic case discussion format used for decades in business schools, although it can easily be adapted for small-group cooperative learning teams. It is divided into three parts: the disease, the public health program, and the experiment.

The section on the disease is a straightforward account of the symptoms of syphilis, and it normally figures only in a modest way in the discussion, as a backdrop to the case. The instructor can highlight the disease by an early focus on the disease symptoms, perhaps with graphic photos and a review of data and readings from original papers.

The section on the public health program should be viewed from the perspective of concern regarding the extent of disease in rural southern America, the need to establish a health vehicle to address the problems of disease, and the concern of an expanding civil rights movement regarding health care and health care professionals.

In writing this case, we have kept the account relatively straightforward, eschewing emotion-laden phrases, keeping in mind the science and ethics earlier in the century. It has a documentary feel to it; that is intentional. As a result, we hope to accomplish two things: (1) to keep the reader focused on the science first, and (2) to avoid the easy criticism that comes from second-guessing events that took place more than 50 years ago.

The documentary approach helps dampen the tendency of some individuals to use this case as a platform to deride racism without serious analysis. We choose to start with the least volatile of the issues, just as we would in the classroom.

Suggested Question Outline for In-Class Discussion
The Science

1. What kind of disease is syphilis?
2. What did we know about the disease in 1930?
3. What was the original purpose of the study? Was the goal accomplished?
4. How did the goals of the project change over time?
5. What was the logic behind the choice of subjects?
6. What kinds of data were collected in the project and what conclusions resulted from the work?
7. What kind of scientific criticisms of the research can we offer?

Human Experimentation

1. What benefits did the men gain from the experiment?
2. What evidence do we have that the men were harmed by their participation in the project?
3. Was it possible to inform the men about the true goals of the experiment, given their education status?
4. Given that men who participated in this study received health benefits, status, attention, and money, could they reasonably be expected to exercise good judgment about their participation in this project?

5. Are there circumstances you could imagine where informed consent would interfere with an experiment?

6. Are there any circumstances where the overall good of an experiment to society overrides the harm done to a small group?

The Racial Issue

1. What evidence do we have that race might have been a factor in the experiment?
2. What motivated the PHS investigators to choose Macon County as one of its study sites?
3. What differences were present in the experimental design of the Tuskegee and Oslo studies?
4. If the Tuskegee and Oslo studies had shown racial differences, how would that information have been used?
5. Is it reasonable to conclude that the administration of the Rosenwald Fund failed to fund the second PHS project because they identified racial bias in the work?
6. Is there any way to fund research on special groups in the U.S. population without running the risk of being accused of bias?
7. Given that certain segments of the population have special health problems, is there any way not to fund research on these groups without running the risk of being accused of bias?

Web Version

This case and its complete teaching notes and references can be found on the website of the National Center for Case Study Teaching in Science at *http://sciencecases.lib.buffalo.edu/cs/collection/detail.asp?case_id=371&id=371*.

References

Caldwell, J. G. et al. 1973. Aortic regurgitation in the Tuskegee study of untreated syphilis. *Journal of Chronic Diseases* 26: 187–199.

Jones, J. H. 1993. *Bad blood: The Tuskegee syphilis experiment.* New York: Free Press.

Rockwell, D. H., A. Roof Yobs, and M. Brittain Moore, Jr. 1964. The Tuskegee study of untreated syphilis: The 30th year of observation. *Archives of Internal Medicine* 114 (6): 792–798.

List of Contributors

Chen-Izu, Ye, assistant professor, Departments of Pharmacology, Internal Medicine, and Biomedical Engineering, University of California–Davis, Davis, CA

Colyer, Christa, professor, Department of Chemistry, Wake Forest University, Winston-Salem, NC

Connor, Jane Marantz, associate professor of human development (emeritus), Binghamton University, Binghamton, NY

Constible, Juanita, technical analyst, Coastal Louisiana Program, National Wildlife Federation, Baton Rouge, LA

DuRei, Kristie, former honors student, University at Buffalo, Buffalo, NY

Feenstra, Jennifer S., associate professor, Department of Psychology, Northwestern College, Orange City, IA

Ford, Rosemary H., associate professor, Biology Department, Washington College, Chestertown, MD

Fourtner, Ann W., science writer, Getzville, NY

Fourtner, Charles R., professor, Department of Biological Sciences, University at Buffalo, Buffalo, NY

Gallucci, Kathy, associate professor, Biology Department, Elon University, Elon, NC

Ginn, Sheryl R., instructor of psychology, Department of Social Sciences, Rowan-Cabarrus Community College, Concord, NC

Gow, Joan-Beth, assistant professor, Biology Department, Anna Maria College, Paxton, MA

Greuling, Ruth Ann, Office of the Provost, Northern New Mexico College, Espanola, NM

List of Contributors

Hager, Lisa D., associate professor, Psychology Department, Spring Hill College, Mobile, AL

Hayes-Klosteridis, Jennifer, director, Student Success Program, University of Maryland School of Nursing, Baltimore, MD

Herreid, Clyde Freeman, distinguished teaching professor, Department of Biological Sciences; Director, National Center for Case Study Teaching in Science, University at Buffalo, Buffalo, NY

Herreid, Ky F., web manager/editor, National Center for Case Study Teaching in Science, University at Buffalo, Buffalo, NY

Hollister, Christopher V., associate librarian, Arts & Sciences Libraries, University at Buffalo, Buffalo, NY

Holt, Susan, curriculum developer, Life Sciences Learning Center, University of Rochester, Rochester, NY

Hudecki, Michael S., research professor emeritus, Department of Biological Sciences, University at Buffalo, Buffalo, NY

Johnson, Dan, senior lecturer, Biology Department, Wake Forest University, Winston-Salem, NC

Kermick, Daniel, biology education major, Division of Biological and Health Sciences, University at Pittsburgh at Bradford, PA

Knapp, Gretchen E., adjunct professor, Biology Department, Illinois State University, Normal, IL

Lee, Richard E., Jr., distinguished professor, Department of Zoology, Miami University, Oxford, OH

Lizotte, Pauline A., instructor, Mathematics, Science and Health Careers, Manchester Community College, Manchester, CT

Maynard, Jacinth, assistant professor of mathematics, Division of Physical and Computational Sciences, University at Pittsburgh at Bradford, PA

List of Contributors

Meinz, Elizabeth J., associate professor, Department of Psychology, Southern Illinois University–Edwardsville, Edwardsville, IL

Merriam, Jennifer, associate professor, Biology Department, SUNY Orange County Community College, Middletown, NY

Meuler, Debra Ann, associate professor, Cardinal Stritch University, Milwaukee, WI

Muench, Susan Bandoni, professor, Biology Department, SUNY Geneseo, Geneseo, NY

Mulcahy, Mary Puterbaugh, associate professor of biology, Division of Biological and Health Sciences, University at Pittsburgh at Bradford, PA

Nava-Whitehead, Susan M., associate professor, Sciences and Education Department, Becker College, Worcester, MA

Omarzu, Julia, associate professor, Department of Psychology, Loras College, Dubuque, IA

Ribbens, Eric, associate professor, Department of Biological Sciences, Western Illinois University, Macomb, IL

Rowe, Matthew, professor, Department of Biological Sciences, Sam Houston State, Huntsville, TX

Sandro, Luke, high school teacher, Biology Department, Springboro High School, Springboro, OH

Schiller, Nancy A., librarian, Science and Engineering Library, Arts & Sciences Libraries; co-director, National Center for Case Study Teaching in Science, University at Buffalo, Buffalo, NY

Stanger-Hall, Kathrin, assistant professor, Department of Plant Biology, The University of Georgia, Athens, GA

Stonefoot, Sarah G., visiting assistant professor of photography & new media, Department of Art & Art History, Beloit College, Beloit, WI

List of Contributors

Taylor, Ann, associate professor, Department of Chemistry, Wabash College, Crawfordsville, IN

Terry, David R., assistant professor, Education Department, Alfred University, Alfred, NY

Tessmer, Michael, associate professor, Chemistry Department, Southwestern College, Winfield, KS

Tichenor, Linda L., associate professor, Biology Department, University of Arkansas at Fort Smith, Fort Smith, AK

Zavrel, Erik, MS/PhD candidate, Department of Biomedical Engineering, Cornell University, Ithaca, NY

Selected Bibliography on Case Study Teaching in Science

References on pages 345–359 are organized into the following sections:

I. Background

II. Writing Case Studies

III. Teaching With Cases

IV. Leading Case Discussions

V. Cases and Cooperative Learning/Small Groups

VI. The Research Basis for Teaching With Cases

VII. Teaching Critical Thinking in Science

I. Background

Herreid, C. F. 1994. Case studies in science: A novel method of science education. *Journal of College Science Teaching* 23 (4): 221–229.

Herreid, C. F. 1998. Sorting potatoes for Miss Bonner: Bringing order to case-study methodology through a classification scheme. *Journal of College Science Teaching* 27 (4): 236–239.

Herreid, C. F. 2007. *Start with a story: The case study method of teaching college science.* Arlington, VA: NSTA Press.

Lundeberg, M. A., B. B. Levin, and H. L. Harrington. 1999. *Who learns what from cases and how?* Mahwah, NJ: Lawrence Erlbaum Associates.

Wilkerson, L., and W. H. Gijselaers. 1996. *Bringing problem-based learning to higher education: Theory and practice.* San Francisco: Jossey-Bass.

Yadav, A., M. A. Lundeberg, M. DeSchryver, K. H. Dirkin, N. Schiller, K. Maier, and C. F. Herreid. 2007. Teaching science with case studies. *Journal of College Science Teaching* 37 (1): 34–38.

Bibliography

II. Writing Case Studies

Bennett, J., and B. Chakravarthy. 1978. What awakens student interest in a case? *Harvard Business School Bulletin* 54 (2): 12–15.

Cliff, W. H., and L. M. Nesbitt. 2005. An open or shut case? Contrasting approaches to case study design. *Journal of College Science Teaching* 34 (4): 14–17.

Herreid, C. F. 1994. Journal articles as case studies: *The New England Journal of Medicine* on breast cancer. *Journal of College Science Teaching* 23 (6): 349–355.

Herreid, C. F. 1997. What is a case? *Journal of College Science Teaching* 27 (2): 92–94.

Herreid, C. F. 1997/98. What makes a good case? *Journal of College Science Teaching* 27 (3): 163–165.

Herreid, C. F. 1999. Cooking with Betty Crocker: A recipe for case writing. *Journal of College Science Teaching* 29 (3): 156–158.

Herreid, C. F. 2000. And all that jazz: An essay extolling the virtues of writing case teaching notes. *Journal of College Science Teaching* 29 (4): 225–228.

Herreid, C. F. 2002. The way of Flesch: The art of writing readable cases. *Journal of College Science Teaching* 31 (5): 288–291.

Herreid, C. F. 2002. Twixt fact and fiction. *Journal of College Science Teaching* 31 (7): 428–430.

Herreid, C. F. 2005. Too much, too little, or just right? How much information should we put into a case study? *Journal of College Science Teaching* 35 (1): 12–14.

Leenders, M. R., and J. A. Erskine. 1989. *Case research: The case writing process*. London, ON: Research and Publications Division, School of Business Administration, University of Western Ontario.

Naumes, W., and M. J. Naumes. 1999. *The art & craft of case writing*. Thousand Oaks, CA: Sage Publications.

Schuwirth, L. W. T., D. E. Blackmore, E. Mom , F. Van Den Wildenberg, H. E. J. H. Stoffers, and C. P. M. Van Der Vleuten. 1999. How to write short cases for assessing problem-solving skills. *Medical Teacher* 21 (2): 44–50.

White, H. B. 1993. Research literature as a source of problems. *Biochemical Education* 21 (4): 205–207.

III. Teaching With Cases

Allchin, Douglas. 2000. How *not* to teach historical cases in science. *Journal of College Science Teaching* 30 (1): 33–37.

Bergland, M., K. Klyczek, and M. Lundeberg. 2006. Collaborative case-based study of genetic and infectious disease via molecular biology computer simulations and Internet conferencing. *International Journal of Learning* 13 (8): 149–154.

Bergland, M., M. Lundeberg, and K. Klyczek. 2006. Exploring biotechnology using case-based multimedia. *American Biology Teacher* 68 (2): 81–86.

Boehrer, J., and M. Linsky. 1990. Teaching with cases: Learning to question. *New Directions for Teaching and Learning* 42: 41–57.

Bradley, A. Z., S. M. Ulrich, M. Jones Jr., and S . M. Jones. 2002. Teaching the sophomore organic course without a lecture. Are you crazy? *Journal of Chemical Education* 79 (4): 514–519.

Brickman, P., S. Glynn, and G. Graybeal. 2008. Introducing students to case studies. *Journal of College Science Teaching* 37 (3): 12–16.

Brock, J., and J. Boehrer. 2001. Are cases taught, or do they teach themselves? *Journal of Policy Analysis & Management* 20 (2): 343–346.

Cabe, P. A., M. H. Walker, and M. Williams. 1999. Newspaper advice column letters as teaching cases for developmental psychology. *Teaching of Psychology* 26 (2): 128+ [3 p].

Camill, P. 2000. Using journal articles in an environmental biology course. *Journal of College Science Teaching* 30 (1): 38–43.

Chamany, K. 2006. Science and social justice: Making the case for case studies. *Journal of College Science Teaching* 36 (2): 54–59.

Cheng, V. K. W. 1995. An environmental chemistry curriculum using case studies. *Journal of Chemical Education* 72 (6): 525 [3p].

Christensen, C. R., A. J. Hansen, and L. B. Barnes. 1994. *Teaching and the case method*. 3rd ed. Boston: Harvard Business School Publishing Division.

Cliff, W. H. 2006. Case study analysis and the remediation of misconceptions about respiratory physiology. *Advances in Physiology Education* 30 (4): 215–223.

Cliff, W. H., and L. Nesbitt Curtin. 2000. The directed case method. *Journal of College Science Teaching* 30 (1): 64–66.

Cliff, W. H., and A. W. Wright. 1996. Directed case study method for teaching human anatomy and physiology. *Advances in Physiology Education* 15: Sl9–S28.

Conant, J. B. 1947. *On understanding science: An historical approach*. New Haven, CT: Yale University Press.

Cornely, K. 1998. Use of case studies in an undergraduate biochemistry course. *Journal of Chemical Education* 75 (4): 475–478.

Cornely, K. 2003. Content and conflict—The use of current events to teach content in a biochemistry course. *Biochemistry and Molecular Biology Education* 31 (3): 173–176.

Dayal, A. K., P. Van Eerden, L. Gillespie, N. T. Katz, L. Rucker, and J. Wylie Rosett. 2008. Case-based nutrition teaching for medical students. *Journal of Nutrition Education and Behavior* 40 (3): 191–192.

Delpier, T. 2006. CASES 101: Learning to teach with cases. *Nursing Education Perspectives* 27 (4): 204–209.

Dewprashad, B., C. Kosky, G. S. Vaz, and C. L. Martin. 2004. Using clinical cases to teach general chemistry. *Journal of Chemical Education* 81 (10): 1471–1472.

Dinan, F. J. 2002. Chemistry by the case: Integrating case teaching and team learning. *Journal of College Science Teaching* 32 (1): 36–41.

Bibliography

Dinan, F. J. 2005. Laboratory based case studies: Closer to the real world. *Journal of College Science Teaching* 35 (2): 27–29.

Duch, B. J. 1996. Problem-based learning in physics: The power of students teaching students. *Journal of College Science Teaching* 15 (5): 326–329.

Erskine, J. A., M. R. Leenders, and L. A. Mauffeffe-Leenders. 1981. *Teaching with cases.* London, ON: Research and Publications Division, School of Business Administration, University of Western Ontario.

Field, P. 2003. Senior seminar: Using case studies to teach the components of a successful seminar. *Journal of College Science Teaching* 32 (5): 298–301.

Fisher, E. R., and N. E. Levinger. 2008. A directed framework for integrating ethics into chemistry curricula and programs using real and fictional case studies. *Journal of Chemical Education* 85 (6): 796–801.

Folino, D. A. 2001. Stories and anecdotes in the chemistry classroom. *Journal of Chemical Education* 78 (12): 1615–1618.

Frisch, J. K., and G. Saunders. 2008. Using stories in an introductory college biology course. *Journal of Biological Education* 42 (4): 164–169.

Gallucci, K. 2006. Learning concepts with cases. *Journal of College Science Teaching* 36 (2): 16–20.

Guyer, R. L., M. L. Dillon, and L. Anderson. 2000. Bioethics cases and issues: Enrichment for social science, humanities, and science courses. *Social Education* 64 (7): 410–414.

Heister, L. E., and C. E. Lesher. 2007. Case-based learning in an upper level petrology laboratory class. *Journal of Geoscience Education* 55 (1): 80–84.

Henderson, L. 1998. A first period exercise in biochemistry: A pre-assessment case study. *Biochemical Education* 26 (2): 141–142.

Herreid, C. F. 1996. Structured controversy: A case study strategy. DNA fingerprinting in the courts. *Journal of College Science Teaching* 26 (2): 95–101.

Herreid, C. F. 1998. Return to Mars: How not to teach a case study. *Journal of College Science Teaching* 27 (6): 379–382.

Herreid, C. F. 1999. Dialogues as case studies: A discussion on human cloning. *Journal of College Science Teaching* 28 (4): 245–249.

Herreid, C. F. 2001. When justice peeks: Evaluating students in case method teaching. *Journal of College Science Teaching* 30 (7): 430–433.

Herreid, C. F. 2002. Using case studies in science—and still covering content. In *Team based learning: A transformative use of small groups*, ed. L. Michaelson, A. Knight, and L. Fink, pp. 109–118. Westport, CT: Praeger

Herreid, C. F. 2004. Why a "case-based" course failed. *Journal of College Science Teaching* 33 (3): 10–11.

Herreid, C. F. 2005. The interrupted case method. *Journal of College Science Teaching* 35 (2): 4–5.

Herreid, C. F. 2006. "Clicker" cases: Introducing case study teaching into large classrooms. *Journal of College Science Teaching* 36 (2): 43–47.

Herreid, C. F. 2007. The Boy Scouts said it best: Some advice on case-study teaching and student preparation. *Journal of College Science Teaching* 37 (1): 6–7.

Herreid, C. F. 2007. Trigger cases vs. capstone cases. *Journal of College Science Teaching* 38 (2): 68, 70–71.

Herreid, C. F., and K. DeRei. 2007. Intimate debate technique. *Journal of College Science Teaching* 36 (4): 10–13.

Hewlett, J. A. 2004. In search of synergy: Combining peer-led team learning with the case study method. *Journal of College Science Teaching* 33 (4): 28–31.

Hudecki, M. S. 2001. Alzheimer's disease under scrutiny: Short newspaper articles as a case study tool. *Journal of College Science Teaching* 31 (1): 57–60.

Hunt, P. 1951. The case method of instruction. *Harvard Educational Review* 21 (3): 2–19.

Hutchings, P. 1993. *Using cases to improve college teaching: A guide to a more reflective practice*. Washington, DC: American Association for Higher Education.

Irwin, A. R. 2000. Historical case studies: Teaching the nature of science in context. *Science Education* 84 (1): 5–26.

Kemp, H. V. 1980. Teaching psychology through the case study method. *Teaching of Psychology* 7 (1): 38+ [4p].

Konaklieva, M. 2000. Case studies combined with instrumental analysis in sophomore organic laboratory. *The Chemist* (July/Aug.): 13–16.

Knox, J. A. 1997. Reform of the college science lecture through storytelling. *Journal of College Science Teaching* 26 (6): 388–392.

Krauss, D. A., I. I. Salame, and L. N. Goodwyn. 2010. Using photographs as case studies to promote active learning in biology. *Journal of College Science Teaching* 40 (1): 72–76.

Leonard, J. A., K. L. Mitchell, S. A. Meyers, and J. D. Love. 2002. Using case studies in introductory psychology. *Teaching of Psychology* 29 (2): 142–144.

Lundeberg, M. A., K. Mogen, M. Bergland, K. Klyczek, D. Johnson, and E. MacDonald. 2002. Case it or else! Fostering ethical awareness about human genetics. *Journal of College Science Teaching* 32 (1): 64–69.

Millard, J. T. 2009. Television medical dramas as case studies in biochemistry. *Journal of Chemical Education* 86 (10): 1216–1218.

Mohrig, J. R., C. N. Hammond, and D. A. Colby. 2007. On the successful use of inquiry-driven experiments in the organic chemistry laboratory. *Journal of Chemical Education* 84 (6): 992–998.

Montes, I., A. Padilla, A. Maldonado, and S. Negretti. 2009. Student-centered use of case studies incorporating oral and writing skills to explore scientific ethical misconduct. *Journal of Chemical Education* 86 (8): 936–939.

Myers, R. S. 1991. Using news media to teach chemical principles. *Journal of Chemical Education* 68 (9): 769–772.

Bibliography

Ponder, M., and S. Sumner. 2009. Use of case studies to introduce undergraduate students to principles of food microbiology, molecular biology, and epidemiology of food-borne disease. *Biochemistry & Molecular Biology Education* 37 (3): 156–163.

Reicks, M., T. Stoebner, C. Hassel, and T. Carr. 1996. Evaluation of a decision case approach to food biotechnology education at the secondary level. *Journal of Nutrition Education* 28 (1): 33–38.

Ribbens, E. 2006. Teaching with jazz: Using multiple cases to teach introductory biology. *Journal of College Science Teaching* 36 (2): 10–15.

Richmond, G., and B. Neureither. 1998. Making a case for cases: The cholera epidemic, present and past. *American Biology Teacher* 60 (5): 335–342.

Rybarczyk, B. J. 2007. Tools of engagement: Using case studies in synchronous distance-learning environments. *Journal of College Science Teaching* 37 (1): 31–33.

Rybarczyk, B. J., A. T. Baines, M. McVey, J. T. Thompson, and H. Wilkins. 2007. A case-based approach increases student learning outcomes and comprehension of cellular respiration concepts. *Biochemistry and Molecular Biology Education* 35 (3): 181–186.

Sandstrom, S. 2006. Use of case studies to teach diabetes and other chronic illnesses to nursing students. *Journal of Nursing Education* 45 (6): 229–232.

Smith, A. C., R. Stewart, P. Shields, J. Hayes-Klosteridis, P. Robinson, and R. Yuan. 2005. Introductory biology courses: A framework to support active learning in large enrollment introductory science courses [using case studies]. *Cell Biology Education* 4 (2): 143–156.

Smith, R. A., and S. K. Murphy. 1998. Using case studies to increase learning and interest in biology. *American Biology Teacher* 60 (4): 265–268.

Styer, S. C. 2009. Constructing & using case studies in genetics to engage students in active learning. *American Biology Teacher* 71 (3): 142–143.

Simmons, S., M. Reicks, J. Weinsheimer, and J. Joannides. 2000. *Learning through uncertainty and controversy*. Minneapolis, MN: University of Minnesota.

Tarnvik, A. 2002. The multiple-case method. *Journal of College Science Teaching* 32 (2): 94–97.

Turgeon, A. J. 2007. Addressing problems encountered in case-based teaching. *Journal of Natural Resources and Life Sciences Education* 36: 134–138.

Van Hoewyk, D. 2007. Using a case-study article to effectively introduce mitosis. *Journal of College Science Teaching* 36 (6): 12–14.

Vosburg, D. A. 2008. Teaching organic synthesis: A comparative case study approach. *Journal of Chemical Education* 85 (11): 1519–1523

Wheatley, J. 1986. The use of case studies in the science classroom. *Journal of College Science Teaching* 15 (5): 428–431.

White, H. B. 1996. Addressing content in problem-based courses: The learning issue matrix. *Biochemical Education* 24 (1): 41–45.

White, H. B. 2003. Wrap-up assignments for problem-based learning problems. *Biochemistry and Molecular Biology Education* 31 (4): 260–261.

Wilcox, K. J. 1999. The case method in introductory anatomy and physiology: Using the news. *American Biology Teacher* 61 (9): 668–671.

Woody, M., S. Albrecht, and T. Hines. 1999. Directed case studies in baccalaureate nursing anatomy and physiology. *Journal of Nursing Education* 38 (8): 383–386.

Zhang, M., M. Lundeberg, T. J. McConnell, M. J. Koehler, and J. Eberhardt. 2010. Using questioning to facilitate discussion of science teaching problems in teacher professional development. *Interdisciplinary Journal of Problem-based Learning* 4 (1): 57–82.

IV. Leading Case Discussions

Christensen, C. R., D. A. Garvin, and A. Sweet, eds. 1991. *Education for judgment: The artistry of discussion leadership*. Boston: Harvard Business School Press.

Flynn, A. E., and J. D. Klein. 2001. The influence of discussion groups in a case-based learning environment. *ETR&D—Educational Technology Research and Development* 49 (3): 71–86.

Frederick, P. 1981. The dreaded discussion: Ten ways to start. *Improving College and University Teaching* 29 (3): 109–114.

Herreid, C. F. 2002. Naming names: The greatest secret in leading a discussion is using students' names. *Journal of College Science Teaching* 32 (3): 162–163.

Herreid, C. F. 2005. The business end of cases. *Journal of College Science Teaching* 35 (3): 12–14.

Lantz, J. M., and M. M. Walczak. 1997. The elements of a chemistry case: Teaching chemistry using the case discussion method. *The Chemical Educator* 1 (6): 1–22.

Miner, F. C. 1979. An approach for increasing participation in case discussions. *Journal of Management Education* 3 (3): 41–42.

Welty, W. M. 1989. Discussion method teaching: How to make it work. *Change* 21 (4): 41–49.

V. Cases and Cooperative Learning/Small Groups

Dinan, F. J. 2006. Opening day: Getting started in a cooperative classroom. *Journal of College Science Teaching* 35 (4): 12–14.

Dinan, F. J., and V. A. Frydrychowski. 1995. A team learning method for organic chemistry. *Journal of Chemical Education* 72 (5): 429–431.

Goran, D., and S. Braude. 2007. Social and cooperative learning in the solving of case histories. *American Biology Teacher* 69 (2): 80–84.

Herreid, C. F. 1998. Why isn't cooperative learning used to teach science? *BioScience* 48: 553–559.

Herreid, C. F. 1999. The bee and the groundhog: Lessons in cooperative learning—Troubles with groups. *Journal of College Science Teaching* 28 (4): 226–228.

Bibliography

Herreid, C. F. 2000. I never knew Joe Paterno: An essay on teamwork and love. *Journal of College Science Teaching* 30 (3): 158–161.

Meyers, S. A. 1997. Increasing student participation and productivity in small-group activities for psychology classes. *Teaching of Psychology* 24 (2): 105–115.

Michaelsen, L. K. 1992. Team learning: A comprehensive approach for harnessing the power of small groups in higher education. *To Improve the Academy* 11: 107–122.

Schwartz, P. L. 1989. Active, small group learning with a large group in a lecture theatre: A practical example. *Medical Teacher* 11 (1): 81–86.

Shulman, J. H., R. A. Lotan, and J. A. Whitcomb. 1998. *Groupwork in diverse classrooms*. New York: Teachers College Press.

Shulman, J. H., R. A. Lotan, and J. A. Whitcomb. 1998. *Facilitator's guide to groupwork in diverse classrooms*. New York: Teachers College Press.

Tien, L. T., V. Roth, and J. A. Kampmeier. 2002. Implementation of a peer-led team learning instructional approach in an undergraduate organic chemistry course. *Journal of Research in Science Teaching* 39 (7): 606–632.

VI. The Research Basis for Teaching With Cases

Aaron, S., J. Crocket, D. Morrish, C. Basualdo, T. Kovithavongs, B. Mielke, and D. Cook. 1998. Assessment of exam performances after change to problem-based learning: Differential effects by question type. *Teaching and Learning in Medicine* 10 (2): 86–91.

Albanese, M., and S. Mitchell. 1993. Problem-based learning: A review of literature on its outcomes and implementation issues. *Academic Medicine* 68 (1): 52–81.

Anderson, K., R. Peterson, A. Tonkin, and E. Cleary. 2008. The assessment of student reasoning in the context of a clinically oriented PBL program. *Medical Teacher* 30 (8): 787–794.

Antepohl, W., and S. Herzig. 1999. Problem-based learning vs. lecture-based learning in a course of basic pharmacology. *Medical Education* 33 (2): 106–113.

Arts, J., W. Gijselaers, and M. Segers. 2002. Cognitive effects of an authentic computer-supported, problem-based learning environment. *Instructional Science* 30 (6): 465–495.

Bowe, C. M., J. Voss, and H. Thomas Aretz. 2009. Case method teaching: An effective approach to integrate the basic and clinical sciences in the preclinical medical curriculum. *Medical Teacher* 31 (9): 834–841.

Caplow, J., J. Donaldson, C. Kardash, and M. Hosokawa. 1997. Learning in a problem-based medical curriculum: Students' conceptions. *Medical Education* 31 (6): 440–447.

Capon, N., and D. Kuhn. 2004. What's so good about problem-based learning? *Cognition & Instruction* 22 (1): 61–79.

Bibliography

Carrero, E., C. Gomar, W. Penzo, N. Fabregas, R. Valero, and G. Sanchez-Etayo. 2009. Teaching basic life support algorithms by either multimedia presentations or case based discussion. *Medical Teacher* 31 (5): 189–195.

Chernobilsky, E., M. DaCosta, and C. Hmelo-Silver. 2004. Learning to talk the educational psychology talk through a problem-based course. *Instructional Science* 32 (4): 319–356.

Demetriadis, S., and A. Pombortsis. 1999. Novice student learning in case based hypermedia environment: A quantitative study. *Journal of Educational Multimedia and Hypermedia* 8 (2): 241–269.

Dochy, F., M. Segers, P. VandenBossche, and D. Gijbels. 2003. Effects of problem-based learning: A meta-analysis. *Learning and Instruction* 13 (5): 533–568.

Dupuis, R. E., and A. M. Persky. 2008. Use of case-based learning in a clinical pharmacokinetics course. *American Journal of Pharmaceutical Education* 72 (2): 1–7.

Ertmer, P., T. Newby, and M. MacDougall. 1996. Students' responses and approaches to case-based instruction: The role of reflective self-regulation. *American Educational Research Journal* 33 (3): 719–752.

Gijbels, D., F. Dochy, P. VandenBossche, and M. Segers. 2005. Effects of problem-based learning: A meta-analysis from the angle of assessment. *Review of Educational Research* 75 (1): 27–61.

Grunwald, S., and A. Hartman. 2010. A case-based approach improves science students' experimental variable identification skills. *Journal of College Science Teaching* 39 (3): 28–33.

Hake, R. R. 1998. Interactive-engagement vs. traditional methods: A six-thousand-student survey of mechanics test data for introductory physics courses. *American Journal of Physics* 66 (1): 64–74.

Harland, T. 2003. Vygotsky's zone of proximal development and problem-based learning: Linking a theoretical concept with practice through action research. *Teaching in Higher Education* 8 (2): 263–272.

Hays, J., and J. Vincent. 2004. Students' evaluation of problem-based learning in graduate psychology courses. *Teaching of Psychology* 31 (2): 124–127.

Hayward, L., and M. Cairns. 1998. Physical therapy students' perceptions of and strategic approaches to case-based instruction. *Journal of Physical Therapy Education* 12 (2): 33–42.

Hmelo-Silver, C. 2004. Problem-based learning: What and how do students learn? *Educational Psychology Review* 16 (3): 235–266.

Hoag, K., J. Lillie, and R. Hoppe. 2005. Piloting case-based instruction in a didactic clinical immunology course. *Clinical Laboratory Science* 18 (4): 213–220.

Johnson, E., S. Herd, and K. Andrewartha. 2002. Introducing problem-based learning into a traditional lecture course. *Biochemistry & Molecular Biology Education* 30 (2): 121–124.

Bibliography

Keefer, M., and K. Ashley. 2001. Case-based approaches to professional ethics: A systematic comparison of students' and ethicists' moral reasoning. *Journal of Moral Education* 30 (4): 377–398.

Kesner, M., A. Hofstein, and R. Ben-Zvi. 1997. Student and teacher perceptions of industrial chemistry case studies. *International Journal of Science Education* 19 (6): 725–738.

Kopp, V., R. Stark, N. Heitzmann, and M. R. Fischer. 2009. Self-regulated *learning* with *case-based* worked examples: Effects of errors. *Evaluation & Research in Education* 22 (2–4): 107–119.

Kunselman, J., and P. Johnson. 2004. Using the case method to facilitate learning. *College Teaching* 52 (3): 87–92.

Lundeberg, M., K. Mogen, M. Bergland, K. Klyczek, D. Johnson, and E. MacDonald. 2002. Fostering ethical awareness about human genetics through multimedia-based cases. *Journal of College Science Teaching* 32 (1): 64–69.

Lundeberg, M. A., and G. Scheurman. 1997. Looking twice means seeing more: Developing pedagogical knowledge through case analysis. *Teaching and Teacher Education* 13 (8): 783–797.

Lundeberg, M. A., H. Kang, B. Wolter, R. delMas et al. 2011. Context matters: Increasing understanding with interactive clicker case studies. *Educational Technology Research and Development* 59 (5): 645–671.

Mayo, J. 2002. Case-based instruction: A technique for increasing conceptual application in introductory psychology. *Journal of Constructivist Psychology* 15 (1): 65–74.

Mayo, J. 2004. Using case-based instruction to bridge the gap between theory and practice in psychology of adjustment. *Journal of Constructivist Psychology* 17 (2): 137–146.

Prince, K., H. van Mameren, N. Hylkema, J. Drukker, A. Scherpbier, and C. van der Vleuten. 2003. Does problem-based learning lead to deficiencies in basic science knowledge? An empirical case on anatomy. *Medical Education* 37 (1): 15–21.

Reicks, M., T. Stoebner, and C. Hassel. 2003. Evaluation of a decision case approach to food biotechnology education at the secondary level. *Journal of Nutrition Education* 28 (1): 33–38.

Scaffa, M., and D. Wooster. 2004. Effects of problem-based learning on clinical reasoning in occupational therapy. *American Journal of Occupational Therapy* 58 (3): 333–336.

Schmidt, H. G., H. T. Van der Molen, W. W. R. Te Winkel, and W. H. F. W. Wijnen. 2009. Constructivist, problem-based learning does work: A meta-analysis of curricular comparisons involving a single medical school. *Educational Psychology* 44 (4): 227–249.

Segers, M., F. Dochy, and E. DeCorte. 1999. Assessment practices and students' knowledge profiles in a problem-based curriculum. *Learning Environments Research* 2: 191–213.

Stepich, D., P. Ertmer, and M. Lane. 2001. Problem-solving in a case-based course: Strategies for facilitating coached expertise. *Educational Technology Research & Development* 49 (3): 53–69.

Theroux, J., and C. Kilbane. 2004. The real-time case method: A new approach to an old tradition. *Journal of Education for Business* 79 (3): 163–167.

Tranvik, A. 2007. Revival of the case method: A way to retain student-centered learning in a post-PBL era. *Medical Teacher* 29 (1): 32–36.

Vaughan, D., C. DeBiase, and J. Gibson-Howell. 1998. Use of case-based learning in dental hygiene curricula. *Journal of Dental Education* 62 (3): 257–259.

Yadav, A., G. M. Shaver, and P. Meckl. 2010. Lessons learned: Implementing the case teaching method in a mechanical engineering course. *Journal of Engineering Education* 99 (1): 55–69.

VII. Teaching Critical Thinking in Science

Ahern-Rindell, A. J. 1998/99. Applying inquiry-based and cooperative group learning strategies to promote critical thinking. *Journal of College Science Teaching* 28 (3): 203–207.

Anderson, L. W., and D. Krathwohl, eds. 2001. *A taxonomy for learning, teaching, and assessing: A revision of Bloom's taxonomy of educational objectives.* New York: Longman.

Anderson, L. W., and L. A. Sosniak. 1994. *Bloom's taxonomy: A forty-year retrospective.* Chicago: The University of Chicago Press.

Bailin, S. 2002. Critical thinking and science education. *Science & Education* 11 (4): 361–375.

Bissell, A. N., and P. P. Lemons. 2006. A new method for assessing critical thinking in the classroom. *BioScience* 56 (1): 66–72.

Bloom, B. 1956. *Taxonomy of educational objectives: Handbook 1: Cognitive domain.* New York: Addison Wesley.

Brahler, C. J., I. J. Quitadamo, and E. C. Johnson. 2002. Student critical thinking is enhanced by developing exercise prescriptions using online learning modules. *Advances in Physiology Education* 26 (3): 210–221.

DaRosa, D. A., P. S. O'Sullivan, M. Younger, and R. Deterding. 2001. Measuring critical thinking in problem-based learning discourse. *Teaching & Learning in Medicine* 13 (1): 27–35.

Davson-Galle, P. 2004. Philosophy of science, critical thinking and science education. *Science & Education* 13 (6): 503–551.

Bibliography

Dori, Y., R. Tal, and M. Tsaushu. 2003. Teaching biotechnology through case studies: Can we improve higher order thinking skills of nonscience majors. *Science Education* 87 (6): 767–793.

Elder, L., and R. Paul. 2007. Critical thinking: The art of Socratic questioning. Part II. *Journal of Developmental Education* 31 (2): 32–33.

Ennis, R. H. 1962. A concept of critical thinking. *Harvard Educational Review* 32 (1): 81–111.

Ennis, R. H. 1989. Critical thinking and subject specialty: Clarification and needed research. *Educational Researcher* 18 (3): 4–10.

Ennis, R. H. 1991. Critical thinking: A streamlined conception. *Teaching Philosophy* 14 (1): 5–25.

Facione, P. A. 1998. *Critical thinking: What it is and why it counts*. Millbrae, CA: California Academic Press.

Feely, T. 1975. Predicting students' use of evidence: An aspect of critical thinking. *Theory & Research in Social Education* 3 (1): 63–72.

Friedel, C. R., T. A. Irani, E. B. Rhoades, N. E. Fuhrman, and M. Gallo. 2008. It's in the genes: Exploring relationships between critical thinking and problem solving in undergraduate agriscience students' solutions to problems in Mendelian genetics. *Journal of Agricultural Education* 49 (4): 25–37.

Friedel, C., T. Irani, R. Rudd, M. Gallo, E. Eckhardt, and J. Ricketts. 2008. Overtly teaching critical thinking and inquiry-based learning: A comparison of two undergraduate biotechnology classes. *Journal of Agricultural Education* 49 (1): 72–84.

Garratt, J., T. Overton, J. Tomlinson, and D. Clow. 2000. Critical thinking exercises for chemists: Are they subject-specific? *Active Learning in Higher Education* 1 (2): 152–167.

Gunn, T. M., L. M. Grigg, and G. A. Pomahac. 2008. Critical thinking in science education: Can bioethical issues and questioning strategies increase scientific understandings? *Journal of Educational Thought* 42 (2): 165–183.

Hager, P., R. Sleet, P. Logan, and M. Hooper. 2003. Teaching critical thinking in undergraduate science courses. *Science & Education* 12 (3): 303–313.

Herreid, C. F. 2004. Can case studies be used to teach critical thinking? *Journal of College Science Teaching* 33 (6): 12–14.

Hunkins, F. P. 1970. Analysis and evaluation questions: Their effects upon critical thinking. *Educational Leadership* 27 (7): 697–705.

Iwaoka, W. T., Y. Li, and W. Y. Rhee. 2010. Measuring gains in critical thinking in food science and human nutrition courses: The Cornell critical thinking test, problem-based learning activities, and student journal entries. *Journal of Food Science Education* 9 (3): 68–75.

Kamin, C., P. O'Sullivan, R. Deterding, and M. Younger. 2003. A comparison of critical thinking in groups of third-year medical students in text, video, and virtual PBL case modalities. *Academic Medicine* 78 (2): 204–211.

Krathwohl, D. R. 2002. A revision of Bloom's taxonomy: An overview. *Theory Into Practice* 41 (4): 212–218.

Kronberg, J. R., and M. Griffin. 2000. Analysis problems—A means to developing students' critical-thinking skills. *Journal of College Science Teaching* 29 (5): 34 [5p].

Krupat, E., J. M. Sprague, D. Wolpaw, P. Haidet, D. Hatem, and B. O'Brien. 2011. Thinking critically about critical thinking: Ability, disposition or both? *Medical Education* 45 (6): 625–635.

Lauer, T. 2005. Teaching critical thinking skills using course content material. *Journal of College Science Teaching* 34 (6): 34–37.

Marzano, R. J. 1998. What are the general skills of thinking and reasoning and how do we teach them? *The Clearing House* 71 (5): 268–273.

McLean, C. P. 2010. Changes in critical thinking skills following a course on science and pseudoscience: A quasi-experimental study. *Teaching of Psychology* 37 (2): 85–90.

McMillan, J. H. 1987. Enhancing college students' critical thinking: A review of studies. *Research in Higher Education* 26 (1): 3–29.

McNeal, A. P., and S. Mierson. 1999. Teaching critical thinking skills in physiology. *Advances in Physiology Education* 22 (1): S268–S270.

McPeck, J. E. 1990. Critical thinking and subject specificity: A reply to Ennis. *Educational Researcher* 19 (4): 10–12.

Mead, J. M., and L. Scharmann. 1994. Enhancing critical thinking through structured academic controversy. *American Biology Teacher* 56 (7): 416 [5p].

Morgan Jr., W. R. 1995. "Critical thinking"—what does that mean? Searching for a definition of a crucial intellectual process. *Journal of College Science Teaching* 24 (5): 336–340.

Narode, R., M. Heiman, J. Lochhead, and J. Slomianko. 1987. *Teaching thinking skills: Science*. Washington, DC: National Education Association.

Nisbett, R. E., G. T. Fong, D. R. Lehman, and P. W. Cheng. 1987. Teaching reasoning. *Science* 238: 625–631.

Nobitt, L., D. E. Vance, and M. L. DePoy Smith. 2010. A comparison of case study and traditional teaching methods for improvement of oral communication and critical-thinking skills. *Journal of College Science Teaching* 39 (5): 26–32.

Pascarella, E. T. 1989. The development of critical thinking: Does college make a difference? *Journal of College Student Development* 40 (5): 562–569.

Paul, R. W., and L. Elder. 1999. *Critical thinking handbook: Basic theory and instructional structures*. Dillon Beach, CA: Foundation for Critical Thinking.

Paul, R., and L. Elder. 2007. Critical thinking: The art of Socratic questioning. *Journal of Developmental Education* 31 (1): 36–37.

Bibliography

Paul, R., and L. Elder. 2008. Critical thinking: The art of Socratic questioning. Part III. *Journal of Developmental Education* 31 (3): 34–35.

Perkins, D. N., and G. Salomon. 1989. Are cognitive skills context bound? *Educational Researcher* 18 (1): 16–25.

Petress, K. 2004. Critical thinking: An extended definition. *Education* 124 (3): 461–466.

Pithers, R. T., and R. Soden. 2000. Critical thinking in education: A review. *Educational Research* 42 (3): 237–249.

Pizzini, E., S. K. Abell, and D. S. Shepardson. 1988. Rethinking thinking in the science classroom. *Science Teacher* 55 (9): 22–25.

Proulx, G. 2004. Integrating scientific method & critical thinking in classroom debates on environmental issues. *American Biology Teacher* 66 (1): 26–33.

Pushkin, D. B. 2000. Critical thinking in science—How do we recognize it? Do we foster it? *Perspectives in Critical Thinking* 110: 211–20.

Quinn, C., M. E. Burbach, G. S. Matkin, and K. Flores. 2009. Critical thinking for natural resource, agricultural, and environmental ethics education. *Journal of Natural Resources & Life Sciences Education* 38 (1): 221–227.

Quitadamo, I. J., C. J. Brahler, and G. Crouch 2009. Peer-led team learning: A prospective method for increasing critical thinking in undergraduate science courses. *Science Educator* 18 (1): 29–39.

Quitadamo, I. J., M. Kurtz, C. N. Cornell, L. Griffith, J. Hancock, and B. Egbert. 2011. Critical-thinking grudge match: Biology vs. chemistry—Examining factors that affect thinking skill in nonmajors science. *Journal of College Science Teaching* 40 (3): 19–25.

Rutledge, M. L. 2005. Making the nature of science relevant: Effectiveness of an activity that stresses critical thinking skills. *American Biology Teacher* 67 (6): 329–333.

Sandor, M. K., M. Clark, D. Campbell, A. P. Rains, and R. Cascio. 1998. Evaluating critical thinking skills in a scenario-based community health course. *Journal of Community Health Nursing* 15 (1): 21–29.

Smith, G. 2002. Are there domain-specific thinking skills? *Journal of Philosophy of Education* 36 (2): 207–227.

Swartz, R. J. 1997. Teaching science literacy and critical thinking skills through problem-based literacy. In *Supporting the spirit of learning: When process is content*, ed. A. L. Costa and R. M. Liebmann, pp. 117–141. Thousand Oaks, CA: Corwin Press.

Terry, D. R. 2007. Using the case study teaching method to promote college students' critical thinking skills. PhD diss., University at Buffalo, State University of New York.

Tiwari, A., P. Lai, M. So, and Y. Kwan. 2006. A comparison of the effects of problem-based learning and lecturing on the development of students' critical thinking. *Medical Education* 40 (6): 547–554.

Tyser, R. W., and W. J. Cerbin. 1991. Critical thinking exercises for introductory biology courses. *BioScience* 41 (1): 41–46.

Van Gelder, T. 2005. Teaching critical thinking: Some lessons from cognitive science. *College Teaching* 53 (1): 41–46.

Wolf, J., M. Stanton, and L. Gellott. 2010. Critical thinking in physical geography: Linking concepts of content and applicability. *Journal of Geography* 109 (2): 43–53.

Wyckoff, S. 2001. Changing the culture of undergraduate science teaching: Shifting from lecture to interactive engagement and scientific reasoning. *Journal of College Science Teaching* 30 (5): 306–312.

Yuretich, R. F. 2004. Encouraging critical thinking: Measuring skills in large introductory science classes. *Journal of College Science Teaching* 33 (3): 40–46.

Zipp, G. P., and C. Mailer. 2010. Use of video-based cases as a medium to develop critical thinking skills in health science students. *Journal of College Teaching & Learning* 7 (1): 1–4.

Appendix A: Case Summary and Overview

The cases in this book were chosen specifically to enhance the critical-thinking skills of students as the skills relate to science. They emphasize the scientific method and experimental design within a societal context and, as such, can be used in any science class. Below we list each case, provide a brief summary of the case, indicate topics and subjects touched upon (in addition to scientific method and experimental design), and identify the case type or method.

Chapter and Title	Summary	Topics	Subjects	Type or Method*
Chapter 5, Childbed Fever	Discusses Ignaz Semmelweis's efforts to remedy the problem of childbed fever in 19th-century Europe	Infectious disease Puerperal sepsis	Biology Epidemiology Medicine Microbiology Public health	Interrupted
Chapter 6, Mystery of the Blue Death	Examines John Snow's discovery of water-borne transmission of cholera in 19th-century London	Cholera Infectious disease Models of disease Populational thinking	Biology Epidemiology Medicine Microbiology Public health	Interrupted
Chapter 7, Salem's Secrets	Investigates possible causes for the mass hysteria that gave rise to the Salem witch trials in the 1600s	Ergot toxicity Fungal poisoning Mass hysteria Psychogenic illness	Biology Epidemiology Microbiology Sociology Psychology	Interrupted
Chapter 8, Bacterial Theory of Ulcers	Recounts the events that led Dr. Robin Warren and Dr. Barry Marshall to the bacterial theory of ulcers	Bacterial disease *Helicobacter pylori* Peptic ulcers	Biology Medicine Microbiology	Discussion

* See Key to Type or Method of Case on pp. 367–368.

Science Stories: Using Case Studies to Teach Critical Thinking

Appendix A

Chapter and Title	Summary	Topics	Subjects	Type or Method*
Chapter 9, Lady Tasting Coffee	Answers the question, Can a woman really distinguish between cups of coffee with milk added first or second, as she claims?	Probability Randomization	Statistics	Interrupted
Chapter 10, Memory Loss in Mice	Outlines the scientific process for implementing an animal model to study Alzheimer's	Alzheimer's disease Animal models Clinical trials Scientific funding	Biology Medicine Neuroscience	Discussion
Chapter 11, Mom Always Liked You Best	Asks, Do parents of baby coots show preferential feeding behavior?	Animal behavior Parental favoritism Preferential feeding	Biology Ecology Zoology	Interrupted
Chapter 12, PCBs in the Last Frontier	Covers research findings on the global transport of polychlorinated biphenyls	Bioaccumulation Persistent pollutants Sockeye salmon	Biology Chemistry Earth science Environmental science	Interrupted
Chapter 13, Great Parking Debate	Addresses the question, Do people leave their parking spaces more quickly if others are waiting?	Human behavior Observational research	Psychology Sociology	Interrupted
Chapter 14, Poison Ivy	Poses the question, Does jewelweed relieve the skin's allergic reaction to the toxin in poison ivy?	Allergic reaction Ethnobotany Natural products	Biology Ecology Plant science Toxicology	Interrupted

* See Key to Type or Method of Case on pp. 367–368.

Case Summary and Overview

Chapter and Title	Summary	Topics	Subjects	Type or Method*
Chapter 15, Extrasensory Perception?	Discusses whether there is scientific evidence for ESP	Paranormal Pseudoscience Telepathy	Psychology	Directed
Chapter 16, A Need for Needles	Provides a critical evaluation of the use of acupuncture	Acupuncture Complementary medicine Nervous system Pain management Placebo effect	Biology Medicine	Dilemma Discussion
Chapter 17, Love Potion #10	Poses the question, Is there evidence to support the scientific claims made in an ad for pheromones?	Pheromones Sexual attractants Science and the media Scientific literacy	Biology	Discussion
Chapter 18, The "Mozart Effect"	Asks whether evidence cited in an ad for a music CD supports the claim that listening to it enhances cognition and creativity	Principle of falsifiability Repeated measures design Within-subjects design	Psychology	Interrupted Directed
Chapter 19, Prayer Study	Evaluates the scientific validity of a study on the effects of intercessory prayer on cardiac patients	Pseudoscience Placebo effect	Biology Psychology Sociology	Discussion
Chapter 20, The Case of the Ivory-Billed Woodpecker	Asks, Has the ivory-billed woodpecker, thought to be extinct, really been "rediscovered"?	Scientific evidence Science and the media Scientific funding Extinction Habitat protection	Biology Ecology Environmental science	Interrupted

* See Key to Type or Method of Case on pp. 367–368.

Appendix A

Chapter and Title	Summary	Topics	Subjects	Type or Method*
Chapter 21, Moon to Mars	Addresses the arguments about whether the U.S. government should fund manned space exploration	Public policy Science and politics Scientific funding Space program	Space science	Public hearing Dilemma Role-play
Chapter 22, And Now What, Ms. Ranger?	Examines the controversy over teaching intelligent design in schools as a valid competing theory to evolution	Creationism Evolution Intelligent design	Biology	Dilemma Discussion Public hearing
Chapter 23, The Case of the Tainted Taco Shells	Explores issues related to genetically modified plants	Bioethics Biotechnology Evolution GMOs	Biochemistry Biology Food science Plant science	Interrupted Role-play
Chapter 24, Medicinal Use of Marijuana	Asks the question, Should marijuana be legalized?	Bioethics Medical marijuana Science and politics	Medicine Pharmacology Public health Sociology	Intimate debate
Chapter 25, Amanda's Absence	Examines how the FDA balances drug safety against medical needs	Bioethics Clinical trial Drug approval Risk-benefit analysis Vioxx	Biochemistry Biology Pharmacology	Interrupted Dilemma
Chapter 26, Sex and Vaccination	Addresses the controversy regarding compulsory vaccination of girls in Texas public schools against HPV	Bioethics Cancer Human papillomavirus STDs Vaccination	Biology Medicine Public health	Interrupted

* See Key to Type or Method of Case on pp. 367–368.

Case Summary and Overview

Chapter and Title	Summary	Topics	Subjects	Type or Method*
Chapter 27, Tragic Choices	Explores whether there is a connection between the MMR vaccine and autism	Bioethics Clinical study Measles Vaccination	Biology Epidemiology Medicine Public health	Interrupted Dilemma
Chapter 28, Ah-choo!	Examines if there is a connection between climate change and a rise in pollen allergies	Allergic reaction Asthma Hay fever Climate change	Biology Earth science Environmental science Public health	Analysis Role-play
Chapter 29, Rising Temperatures	Analyzes two articles from two different newspapers on an EPA report on the state of the environment	Climate change Science and the media Science and politics Scientific literacy	Earth science Environmental science Journalism	Analysis Discussion
Chapter 30, Eating PCBs From Lake Ontario	Provides analysis of a press release about research into the effects of eating contaminated fish	Statistical analysis Science and the media	Environmental science Epidemiology Journalism Public health Statistics	Discussion
Chapter 31, Mother's Milk Cures Cancer?	Explores the role of serendipity in scientific discovery and how research is funded	Cancer Clinical trial Drug approval Scientific funding Sociology of science	Biochemistry Biology Pharmacology	Dilemma Discussion
Chapter 32, Cancer Cure or Conservation	Discusses the controversy over the harvesting of Pacific yew to develop the anti-cancer drug Taxol	Bioprospecting Cancer Drug development Ethnobotany Risk-benefit analysis	Biology Ecology Environmental science Plant science	Dilemma Discussion

* See Key to Type or Method of Case on pp. 367–368.

Appendix A

Chapter and Title	Summary	Topics	Subjects	Type or Method*
Chapter 33, A Rush to Judgment?	Provides analysis of student responsibility in conducting research and reporting fraud	Human subjects Informed consent Research ethics	Psychology	Dilemma Discussion
Chapter 34, How a Cancer Trial Ended in Betrayal	Analyzes the issues associated with clinical trials	Bioethics Cancer Clinical trial Drug approval Scientific fraud	Biology Medicine Pharmacology	Role-play
Chapter 35, Bringing Back Baby Jason	Explores the genetic concepts and ethical issues involved in human cloning	Bioethics Biotechnology Cloning Epigenetics	Biology Genetics	Dilemma Discussion
Chapter 36, Selecting the Perfect Baby	Explores the science and issues of genetic manipulation and fertility treatments	Bioethics Biotechnology In vitro fertilization Pre-implantation genetic diagnosis	Biology Genetics	Dilemma Discussion
Chapter 37, Studying Racial Bias	Evaluates a research proposal for adherence to ethical principles and guidelines for using human subjects	Human subjects Informed consent Racism Research ethics Institutional review	Psychology Sociology	Analysis Discussion Role-play
Chapter 38, Bad Blood	Explores the ethics of human experimentation using the syphilis studies performed at the Tuskegee Institute from the 1930s–1970s	Bioethics Human experimentation Informed consent Research ethics	Biology Medicine Public health	Discussion

* See Key to Type or Method of Case on pp. 367–368.

Key to Type or Method of Case

Analysis case: This method is used to teach students the skills of analysis. The material is focused around answering questions such as, "What is going on here?" An analysis case frequently lacks a central character (though not always) and generally stops short of demanding that students make a decision (though, again, not always). Examples include a description of the *Valdez* oil spill, a collection of papers and data showing the possible effects of vitamin C on the common cold, or a selection of articles arguing whether the HIV virus is the causative agent of AIDS.

Dilemma case: A dilemma case presents an individual, institution, or community faced with a problem that must be solved. It often consists of a section that introduces the problem (and decision maker) at the moment of crisis. A background section fills in the information necessary to understand the situation, followed by a description of recent developments leading up to the crisis. Charts, tables, or graphs are often used to help lay the foundation for a solution. The teacher's role is to help students analyze the problem and consider possible solutions and their likely consequences.

Directed case: A directed case is designed primarily to enhance students' understanding of fundamental concepts, principles, and facts. The case usually consists of a short, dramatic scenario, accompanied by a set of directed questions that can be answered from the textbook or lecture. The questions are "closed-ended" (i.e., typically they have only one correct answer). Students prepare answers to the questions as homework in advance, which they provide in class when called on during the case discussion.

Discussion: The discussion method for teaching a case has long been used by business and law schools. On the surface, the method is simple: The instructor asks probing questions and students analyze the problem. Some instructors may question students using a strong directive approach, often called the Socratic method. Others may use a nondirective approach to class discussion, staying on the sidelines and acting more as facilitators while the students take over the analysis.

Interrupted case: In the interrupted case method, students are given a problem (case) to work on in small groups in stages. After the groups are given a short time to discuss the initial information, the instructor gives them additional information to analyze, apply, and discuss. This sequence is repeated several times as the problem gets closer to resolution. One of the great virtues of the method is the way it mimics how real scientists go about their work. Scientists do not have all of the facts at once; they get them piecemeal. This method of "progressive disclosure" is also characteristic of problem-based learning (PBL), but in the interrupted case method the case typically is accomplished in a single class period rather than over several days.

Appendix A

Intimate debate: Intimate debate is a powerful technique for exploring controversy. Two pairs of students face off across a small table, arguing first one side and then switching to argue the other side. At the end of the exercise, the students must abandon their artificial positions and try to come to a consensus on a reasonable solution to the problem. The advantages to this approach versus formal debate include time efficiency (multiple debates can take place simultaneously); dispassionate scrutiny (switching positions tends to reduce initial "buy-in" or commitment to a given side of the issue); greater participation by the reticent student (because there is no audience); and increased realism (the call for consensus opinion mimics real-world policymaking, where decisions must be made within given time frames and often with insufficient information).

Public hearing: A public hearing is another case format for exploring differing viewpoints on a topic. Public hearings are often structured by having students role-play a hearing board, which listens to presentations by different student groups. Alternatives to having students prepare the testimony include using the transcript of an actual public hearing or a script developed by the instructor that the students role-play.

Role-play: With role-playing, students assume (and may even act out) a role (possibly one that is contrary to their own position) to understand that role or the situations in which those assuming that role find themselves. This is often done in a case study in terms of adopting the role of a stakeholder in an issue to understand that particular person's or constituency's point of view.

Appendix B: Alignment With National Science Education Standards for Science Content, Grades 9–12

Content Standard	Chapter
Science as Inquiry	
Abilities necessary to do scientific inquiry	5, 6, 7, 8, 9, 10, 11, 12, 13, 14, 15, 16, 17, 18, 19, 20, 23, 27, 28, 30, 33, 34
Understanding about scientific inquiry	5, 6, 7, 8, 9, 10, 11, 12, 13, 14, 15, 16, 17, 18, 19, 20, 23, 24, 27, 28, 29, 30, 33, 34, 37, 38
Physical Science	
Structure of atoms	
Structure and properties of matter	
Chemical reactions	
Motions and forces	
Conservation of energy and increase in disorder	
Interactions of energy and matter	
Life Science	
The cell	8, 10, 22, 31, 34, 35, 36
The molecular basis of heredity	35, 36
Biological evolution	11, 14, 22, 34, 35, 36
Interdependence of organisms	11, 12, 22, 23, 32
Matter, energy, and organization in living systems	10, 16, 22, 34, 35
Behavior of organisms	7, 10, 11, 13, 15, 16, 23, 33, 37, 38
Earth and Space Science	
Energy in the earth system	
Geochemical cycles	
Origin and evolution of the earth system	22
Origin and evolution of the universe	22
Science and Technology	
Abilities of technological design	
Understanding about science and technology	8, 21, 23, 27, 31, 32, 35, 36

* See Key to Chapter Numbers on p. 370.

Appendix B

Content Standard	Chapter
Science in Personal and Social Perspectives	
Personal and community health	5, 6, 7, 8, 10, 14, 16, 23, 24, 25, 26, 27, 28, 30, 31, 32, 34, 35, 36, 38
Population growth	35, 36
Natural resources	7, 14, 21, 22, 23, 32
Environmental quality	12, 28, 29, 30, 32
Natural and human-induced hazards	6, 7, 8, 10, 12, 14, 23, 24, 25, 26, 27, 28, 34
Science and technology in local, national, and global challenges	5, 8, 12, 20, 21, 23, 24, 25, 26, 27, 28, 29, 32, 35, 36
History and Nature of Science	
Science as a human endeavor	5, 6, 8, 9, 10, 11, 12, 13, 14, 15, 16, 17, 18, 19, 20, 21, 22, 23, 24, 25, 26, 27, 28, 29, 30, 31, 32, 33, 34, 35, 36, 37, 38
Nature of scientific knowledge	5, 6, 7, 8, 9, 10, 11, 12, 13, 14, 15, 16, 17, 18, 19, 20, 22, 23, 24, 25, 27, 29, 30, 31, 33, 34, 35, 36, 37, 38
Historical perspectives	5, 6, 7, 8, 9, 14, 15, 16, 20, 21, 22, 25, 26, 27, 29, 31, 32, 38

Key to Chapter Numbers

Chapter	Title
5	Childbed Fever: A 19th-Century Mystery
6	Mystery of the Blue Death: John Snow and Cholera
7	Salem's Secrets: On the Track of the Salem Witch Trials
8	The Bacterial Theory of Ulcers: A Nobel Prize Winning Discovery
9	Lady Tasting Coffee
10	Memory Loss in Mice
11	Mom Always Liked You Best
12	PCBs in the Last Frontier
13	The Great Parking Debate
14	Poison Ivy: A Rash Decision
15	Extrasensory Perception?
16	A Need for Needles: Does Acupuncture Really Work?
17	Love Potion #10: Human Pheromones at Work?
18	The "Mozart Effect"
19	Prayer Study: Science or Not?
20	The Case of the Ivory-Billed Woodpecker
21	Moon to Mars: To Boldly Go … or Not
22	And Now What, Ms. Ranger? The Search for the Intelligent Designer
23	The Case of the Tainted Taco Shells
24	The Medicinal Use of Marijuana
25	Amanda's Absence: Should Vioxx Be Kept Off the Market?
26	Sex and Vaccination
27	Tragic Choices: Autism, Measles, and the MMR Vaccine
28	Ah-choo! Climate Change and Allergies
29	Rising Temperatures: The Politics of Information
30	Eating PCBs From Lake Ontario
31	Mother's Milk Cures Cancer?
32	Cancer Cure or Conservation
33	A Rush to Judgment?
34	How a Cancer Trial Ended in Betrayal
35	Bringing Back Baby Jason: To Clone or Not to Clone
36	Selecting the Perfect Baby: The Ethics of "Embryo Design"
37	Studying Racial Bias: Too Hot to Handle?
38	Bad Blood: The Tuskegee Syphilis Project

Appendix C: Evaluating Student Casework

(Based on Herreid 2001)

How do you grade students in classes with case teaching? There are a host of possibilities. Here we deal with only a couple. Let's start with the toughest.

Evaluating Class Discussion

Business school case teachers use this method all the time. It's not uncommon for them to base a student's final course grade on 50% class participation. And this with 50–70 students in a class! The idea sends shudders up the spines of most science teachers. Yet what's so tough about the concept? We constantly make judgments about the verbal statements of our colleagues, politicians, and even administrators. Why can't we do it for classroom contributions?

Most of our discomfort comes from the subjective nature of the act, something that we scientists work hard to avoid in our workaday world. It may be that we are even predisposed to becoming scientists because we are looking for a structured and quantifiable world. Flowing from this subjective quandary is the fact that we feel we must be able to justify our grades to the students. We are decidedly uncomfortable if we can't show them the numbers. This is one of the reasons that multiple-choice questions have such appeal for some faculty.

But let's take a look at how the business school people evaluate case discussion. Some of them try to do it in the classroom, making written notes even as the discussion unfolds, using a seating chart, and calling on perhaps 25 students in a period. As you might expect, this usually interferes with running an effective discussion. Other instructors tape-record the discussion and listen to it later in thoughtful contemplation. Most folks, however, sit down shortly after their classes with a seating chart in hand and reflect on the discussion. They rank student contributions into categories of excellent, good, or bad, or they may use numbers to evaluate the students from 1 to 4, with 4 being excellent. They may give negative evaluations to people who were unprepared or absent. These numbers are tallied up at the end of the semester to calculate the grade. And that's as quantified as it gets.

I especially like mathematician and philosopher Blaise Pascal's view of evaluation: "We first distinguish grapes from among fruits, then Muscat grapes, then those from Condrieu, then from Desargues, then the particular graft. Is that all? Has a vine ever produced two bunches alike, and has any bunch produced two grapes alike?" he asks. "I have never judged anything in exactly the same way," Pascal continues. "I cannot judge a work while doing it. I must do as painters do and stand back, but not too far. How far then? Guess ..." (Pascal 1958, p. 36).

Assignments

The simplest solution to casework evaluation is to forget classroom participation and grade everything on the basis of familiar criteria, such as papers or presentations. This puts professors back in familiar territory. Even business and law school professors use this strategy as part of their grades. I'm all for it. In fact, I always ask for some written analysis in the form of journals, papers, and reports. Along with an exam, these are my sole basis for grades. I don't lose sleep over evaluating class participation.

Exams

You can give any sort of exam in a case-based course, including multiple choice, but doesn't it make more sense to use a case for at least part of the exam? If you have used cases all semester and trained students in case analysis, surely you should consider a case-based test. Too often we test on different things than we have taught.

Peer Evaluation

Some of the best case studies involve small-group work and group projects. In fact, I strongly believe teaching cases this way is the most user-friendly for science faculty and the most rewarding for students. Nonetheless, even some aficionados of group work don't like group projects. They say, "How do you know who's doing the work?" Even if they ask for a group project, they argue against grading it. They rely strictly on individual marks for a final grade determination. I'm on the other side of the fence. I believe that great projects can come from teams, and if you don't grade the work, what is the incentive for participating? Moreover, employers report that most people are fired because they can't get along with other people. Not all of us are naturally team players. Practice helps. So, I'm all for group work, including teamwork during quizzes, where groups almost invariably perform better than the best individuals. But we have to build in safeguards such as peer evaluation.

"Social loafers" and "compulsive workhorses" exist in every class. When you form groups such as those in problem-based learning (PBL) and team learning (the best ways to teach cases, in my judgment), you must set up a system to monitor the situation. In PBL, it is common to have tutors who can make evaluations. Still, I believe it is essential to use peer evaluations. I use a method that I picked up from Larry Michaelsen in the School of Management at the University of Oklahoma.

At the beginning of every course, I explain the use of these anonymous peer evaluations. I show students the form that they will fill out at the end of the semester (see p. 374). Then they will be asked to name their teammates and give each one the number of points that reflects their contributions to group projects throughout the course. Suppose the group has 5 team members, and each person has 40 points to give to the other 4 members of his team. If a student feels that everyone has contributed equally to the group projects, then that person should give each teammate 10 points. Obviously, if everyone in the team feels the same way about everyone else, they all will get an average score of 10 points. Persons with an average of 10 points will receive 100% of the group score for any group project.

But suppose things aren't going well. Maybe John has not pulled his weight in the group projects and ends up with an average score of 8, and Sarah has done more than her share and receives a 12. What then? Well, John gets only 80% of any group grade and Sarah receives 120%.

There are some additional rules that I use. One is that a student cannot give anyone more than 15 points. This is to stop a student from saving his friend John by giving him 40 points. Another is that any student receiving an average of 7 points or less will fail my course. This is designed to stop a student from doing nothing in the group because he is simply trying to slip by with a barely passing grade and is willing to undermine the group effort.

Here are some observations after many years of using peer evaluations:

- Most students are reasonable. Although they are inclined to be generous, most give scores between 8 and 12.

- Occasionally, I receive a set of scores where one score isn't consistent with the others. For example, a student may get a 10, 10, 11, and a 5. Obviously, something is amiss here. When this happens, I set the odd number aside and use the other scores for the average.

- About one group in five initially will have problems because one or two people are not participating adequately or are habitually late or absent. These problems can be corrected.

- It is essential that you give a practice peer evaluation about one-third or halfway through the semester. The students fill these out, and you tally them and give the students their average scores. You must carefully remind everyone what these numbers mean, and if they don't like the results, they must do something to improve their scores. I tell them it is no use blaming their group

Appendix C

Peer Evaluation Form

Name_____ Group # _____

This is an opportunity to evaluate the contributions of your teammates to group projects during the semester. Please write the names of your group members in the spaces below and give them the scores that you believe they earned. If you are in a group of 5 people, you each will have 40 points to distribute. You don't give yourself points. (If you are in a group of 4, you'll have 30 to give away. In a group of 6, you'll have 50 points, etc.) If you believe everyone contributed equally to the group work, then you should give everyone 10 points. If everyone in the group feels the same way, you all will receive an average of 10 points. Be fair in your assessments, but if someone in your group didn't contribute adequately, give that person fewer points. If someone worked harder than the rest, give that person more than 10 points.

There are some rules you must observe in assigning points:

- You cannot give anyone in your group more than 15 points.
- You do not have to assign all of your points.
- Anyone receiving an average of fewer than 7 points will fail the course.
- Don't give anyone a grade that he or she doesn't deserve.

Group members   Points

1. _____   _____

2. _____   _____

3. _____   _____

4. _____   _____

5. _____   _____

6. _____   _____

Please indicate why you gave someone fewer than 10 points.

Please indicate why you gave someone more than 10 points.

If you were to assign points to yourself, what do you feel you deserve? Why?

members for their perceptions. They must fix things, perhaps by talking to the group and asking how to compensate for their previous weaknesses. Also, I will always speak privately to any student who is in danger. These practice evaluations almost always significantly improve the group performance. Tardiness virtually stops and attendance is at least 95%.

Assessing Critical Thinking

Because there are so many facets to critical thinking and so many different disciplines, it is not surprising that investigators interested in the topic have created their own techniques to attempt to measure changes. Here we will only focus on two methods that have general applicability for teachers using case studies.

Bissell and Lemons (2006) advocate creating a rubric system for a given course. They suggest that students answer short essay questions, which are then graded with a rubric based on Bloom's taxonomy. Here is a summary of their method, which was created for a course in biology. It will be evident that this approach can be used in any course, including those using case studies.

> Our methodology consists of several steps. First, we write questions that require both biological knowledge and critical thinking skills. Second, we document the particular content and critical-thinking skills required (e.g., application, analysis, synthesis) and then devise a scoring rubric for the question. Our scheme is a synthesis of the work of others who have devised rubrics that independently assess either content (Porter 2002; Ebert-May et al. 2003; Middendorf and Pace 2004) or critical-thinking skills (Facione et al. 2000). Third, we subject these questions to a test of validity by submitting them for review to colleagues who are experts in biology and biological education. Fourth, we administer the assessments to students and score them on the basis of the rubric that we established in advance. (Bissell and Lemons 2006, p. 68)

We find a more comprehensive approach to the measure of critical thinking in the work of Terry (2007). In an unpublished dissertation, he set out to study if the use of case studies would improve the critical-thinking skills of students in a small college. He had two classes taking a general science class. Over the course of a semester, he used several case studies in each class drawn from the website of the National Center for Case Study Teaching in Science. Before he taught each case, he tested the ability of each of the students to analyze some brief writing matter drawn from popular articles that dealt with the same subject of the case. He developed a rubric for scoring their ability to identify claims and supporting evidence made by the article. Then, after the case study was taught, he used the article again to measure improvement. This was repeated throughout the course for each of the cases.

Appendix C

In both classes, there was a definite improvement in critical-thinking skills between the pre- and post-case presentations. But here is an important point: The students who were not taught these subjects via the case method also improved. These students had been taught the material via multiple methods of presentation: discussion, video, and lecture. When Terry used the general assessment instrument, the Watson-Glaser Critical Thinking Appraisal, on both groups of students, he found that they had improved similarly. So what should we conclude? The case study approach certainly can be used to develop skills in critical thinking, but it is not the only method that can do the job.

Terry's basic method of appraisal is spelled out in his article in an upcoming 2012 issue of the *Journal of College Science Teaching*. His approach is based upon the suggestions of Tyser and Cerbin (1991). Here are the basic questions that Terry asked students to answer for each popular press article:

Identify one specific claim that is made in the assigned reading:
(Write the statement in your own words in the space below.)

Evaluate the following statements, check the one that applies, and provide justification in the space given:

❑ **The claim made in the article is valid.**
Summarize the relevant evidence from the article and explain why it is convincing.

❑ **The claim made in the article is not valid.**
Summarize the relevant evidence from the article and explain how it contradicts the claim.

❑ **The article does not contain sufficient support for the validity of the claim.**
Provide a specific example of additional evidence you would need to evaluate the claim.

This is a valuable tool that can be used in a variety of situations, especially when a grading rubric deals with only two issues, such as the one below: (1) Can the student identify a claim when she or he sees one? and (2) Can the student assess the validity of the evidence? Points are awarded based on a scale of 1–4.

Rubric for Assessing Claims and Evidence
Claim Identification
- No claim is identified; something other than a claim is identified. (0 points)
- Identified claim is not explicitly made in the article. (1 point)
- Appropriate claim is identified. (2 points)

Validity of Evidence
- No evidence is provided for justification; evidence is completely inappropriate. (0 points)
- Evidence is somewhat inappropriate; evidence does not support all aspects of claim; evidence is not from article. (1 point)
- Appropriate evidence from article is provided. (2 points)

Terry (in press) concludes:

The power of this technique lies in its ability to elucidate the kinds of claims and evidence that students rely on when they read science articles. A second major strength of the procedure is the fact that it measures general critical thinking skills in a content-specific manner. Third, it is a relatively straightforward instrument that can be created easily by teachers to measure at least one aspect of student critical-thinking ability. Relevant articles are readily available for virtually every conceivable scientific topic, and most can be modified to fit a particular group of students. With a sufficiently detailed rubric, the students' responses can point out misconceptions, fallacious logic, or general difficulties with the concept of appropriate evidence. With repeated use, this technique can provide specific information about how student thinking changes throughout a course, making it an effective tool for formative assessment.

References

Bissell, A. N., and P. P. Lemons. 2006. A new method for assessing critical thinking in the classroom. *BioScience* 56 (1): 66–72.

Ebert-May, D., J. Batzli, and H. Lim. 2003. Disciplinary research strategies for assessment of learning. *BioScience* 53 (12): 1221–1228.

Facione, P. A., N. C. Facione, and C. A. Giancarlo. 2000. The disposition toward critical thinking: Its character, measurement, and relationship to critical thinking skill. *Informal Logic* 20 (1): 61–84.

Herreid, C. F. 2001. When justice peeks: Evaluating students in case study teaching. *Journal of College Science Teaching* 30 (7): 430–433.

Appendix C

Middendorf, J., and D. Pace. 2004. Decoding the disciplines: A model for helping students learn disciplinary ways of thinking. *New Directions for Teaching and Learning* 98: 1–12.

Pascal, B. 1958. *Pascal's Pensees.* New York: E. P. Dutton.

Porter, A. C. 2002. Measuring the content of instruction: Uses in research and practice. *Educational Researcher* 31 (7): 3–14.

Terry, D. R. 2007. *Using the case study teaching method to promote college students' critical thinking skills.* PhD diss. University at Buffalo, State University of New York.

Terry, D. R. In press. Assessing critical thinking skills using articles from the popular press. *Journal of College Science Teaching.*

Tyser, R. W., and W. J. Cerbin. 1991. Critical thinking exercises for introductory biology courses. *BioScience* 41 (1): 41–46.

Index

*Page numbers printed in **boldface** type refer to figures or tables.*

A

A Need for Needles: Does Acupuncture Really Work?, 130, 141–146
  case and questions for, 141–144
  summary and overview of, 363
  teaching notes for, 144–146
    classroom management, 146
    introduction and background, 144–145
    objectives, 145
    student misconceptions, 146
  web version of, 146
A Rush to Judgment?, 269, 289–293
  case and questions for, 289–291
    dilemma, 291
    players, 289–290
    situation, 290
  summary and overview of, 366
  teaching notes for, 291–293
    classroom management, 292–293
    introduction and background, 291
    objectives, 292
    student misconceptions, 292
  web version of, 293
AAAS (American Association for the Advancement of Science), 26, 28, 29
ACS (American Cancer Society), 225, 227
Active learning techniques, 27–28
Acupuncture, 130, 141–146
Ah-choo! Climate Change and Allergies, 234–235, 247–252
  case and questions for, 247–250
    procedure, 247–250
    scenario, 247
  summary and overview of, 365
  teaching notes for, 250–252
    assessment, 251–252
    classroom management, 251
    introduction and background, 250
    objectives, 250–251
    student misconceptions, 251
  web version of, 252
Alaskan lakes, polychlorinated biphenyls in, 82–83, 109–112
Allchin, Douglas, xii
Allergies
  Ah-choo! Climate Change and Allergies, 234–235, 247–252
  Poison Ivy: A Rash Decision, 83, 121–128
Alpha-lactalbumin, 274–276, 279
Alvarez, Walter and Luis, 5–7
Alzheimer's disease, 82, 95–98
Amanda's Absence: Should Vioxx Be Kept Off the Market?, 185, 219–224
  case and questions for, 219–222
    drug withdrawals, 219–220
    prepared testimony, 221–222
    press release, 220–221
    review panel, 222
  summary and overview of, 364
  teaching notes for, 222–224
    classroom management, 223–224
    introduction and background, 222
    objectives, 222–223
    student misconceptions, 223
  web version of, 224
American Association for the Advancement of Science (AAAS), 26, 28, 29
American Cancer Society (ACS), 225, 227
American Medical Association, 167

Index

American Petroleum Institute, 253
American Psychological Association, 291, 292
Amyotrophic lateral sclerosis-Parkinson dementia complex in Guam, **12, 13,** 15
An Inconvenient Truth, 251
Analysis cases, 365, 366, 367
And Now What, Ms. Ranger? The Search for the Intelligent Designer, 184, 197–206
   case and questions for, 197–202
   summary and overview of, 364
   teaching notes for, 203–206
      classroom management, 204–206
      introduction and background, 203
      objectives, 203
      student misconceptions, 203
   web version of, 206
Anderson, L. W., viii, **ix**
Anesthesiology, 47
Angell, Marcia, 17
Animal studies
   Case of the Ivory-Billed Woodpecker, 131, 173–181
   Memory Loss in Mice, 82, 95–98
   Mom Always Liked You Best, 23, 82, 99–107
Announcement of scientific discoveries, 233–235
Antiscience, 169–171
Apoptosis, 272, 276
Aristotle, 26
Armstrong, Neil, 189
Arthur, Brian, 183
Asaro, Frank, 5
Asperger's syndrome, 243
Assignments, evaluation of, 372
Asteroid hypothesis of dinosaur extinction, 5–7
Asthma, 250, 254
Autism and MMR vaccine, 234, 237–245

B
Bacterial Theory of Ulcers: A Nobel-Prize-Winning Discovery, xi–xii, 36–37, 69–79
   case and questions for, 69–77
      conclusion, 76–77
      data analysis, **73,** 73–74
      Dr. Marshall, 70
      Dr. Warren, 69–70
      isolating bacteria, 72
      pilot study, 71–76
      presentation of findings, 74–75
      publication of results, 75–76
      Warren-Marshall partnership, 71
   summary and overview of, 361
   teaching notes for, 78–79
      classroom management, 79
      introduction and background, 78
      objectives, 78
      student misconceptions, 78
   web version of, 79
Bad Blood: The Tuskegee Syphilis Experiment, 336
Bad Blood: The Tuskegee Syphilis Project, 269–270, 329–339
   case and questions for, 329–336
      disease, 329–330
      experiment, 331–336, **334, 335**
      health program, 330–331
   summary and overview of, 366
   teaching notes for, 336–339
      classroom management, 337–339
      introduction and background, 336–337
      objectives, 337
      student misconceptions, 337
   web version of, 339
Baker, J. N., 332
Ballentine, N. H., 124
Bass, K. E., 162, 165
BCX-34 trial in cutaneous T-cell lymphoma, 295–302

Behe, Michael, 200
Belmont Report, 317–318
Beneficence, 317
Benson, Herbert, 168
Beta-amyloid and memory loss, 95–98
Bird parenting and chick plumage, 82, 99–107
Bissell, A. N., 375
Bizzozero, Giulio, 71
Bloom's taxonomy, viii, **ix,** 375
Bossilier, Brigitte, 304, 305
Breast Implants, **12, 13,** 17
Breast milk, protective effect against cancer, 268, 271–279
Bringing Back Baby Jason: To Clone or Not to Clone, 269, 303–307
    case and questions for, 303–305
    summary and overview of, 366
    teaching notes for, 305–307
        classroom management, 306–307
        introduction and background, 305–306
        objectives, 306
        student misconceptions, 306
    web version of, 307
Buck, Germaine, 262, 264
Burt, Cyril, 267
Bush, George W., 187–188, 193, 257–258
Buxtin, Peter, 336
Byrd, Randolph, 169

C

Caldwell, Roy, 1
Cameron, Stewart, 284
Campylobacter jejuni, 72
Cancer
    cervical, HPV vaccine and, 225, 227, 228
    data manipulation in clinical trial, 269, 295–302
    medicinal use of marijuana in, 213–214
    protective effect of breast milk against, 268, 271–279
    Taxol for, 268–269, 281–288
Cancer Cure or Conservation, 268–269, 281–288
    case and questions for, 281–283
    summary and overview of, 365
    teaching notes for, 284–287
        assignments, 287
        classroom management, 285–287
        introduction and background, 284–285
        objectives, 285
        student misconceptions, 285
    web version of, 288
Carbon dioxide, atmospheric, 247, 250, 251, 254
Carry, Jim, 238
Case of the Ivory-Billed Woodpecker, 131, 173–181
    case and questions for, 173–177
        background, 173–174
        e-mail exchange, 174–177, **175**
        main evidence, 174
    summary and overview of, 363
    teaching notes for, 177–180
        classroom management, 178–180
        introduction and background, 177, **178**
        objectives, 178
        student misconceptions, 178
    web version of, 181
Case of the Tainted Taco Shells, 184–185, 207–212
    case and questions for, 207–211
        project design: interest groups, 209–211
    summary and overview of, 364
    teaching notes for, 211–212
        classroom management, 212

Index

    introduction and background, 211
    objectives, 211–212
    student misconceptions, 212
  web version of, 212
Case studies. *See also specific cases*
  alignment with National Science Education Standards, xi, 369–370
  definition of, 28
  on ethics and the scientific process, xii, 267–339
  on experimental design, ix, 81–128
  historical, xi–xii, 35–79
  history as teaching method, 28
  interrupted case method, 23–24
  on science and media, xii, 233–266
  on science and society, xii, 183–230
  summaries and overviews of, xi, 361–366
  to teach concepts about the nature of science, 11–19
  to teach critical thinking, vii, x–xiii, 21–24, 28–32
  teaching methods using, 28–29
    constructivist learning model and, 31–32
    in groups, 28, 29
    problem-based learning, 28, 30–31
    reasons for effectiveness of, 29
  teaching notes for, xi
  type or method of, 361–368
  on unusual claims, xii, 129–181
  web versions of, xi
CDC (Centers for Disease Control and Prevention), 225–226
Cell biology, 272, 276, 278
Centers for Disease Control and Prevention (CDC), 225–226
Cerbin, W. J., 376
Cernan, Eugene, 189
Cervical cancer, 225, 227, 228

Chadwick, Edwin, 46–47
Childbed Fever: A 19th Century Mystery, xi, 35, 39–43
  case and questions for, 39–40
  summary and overview of, 361
  teaching notes for, 41–43
    classroom management, 42–43
    introduction and background, 41
    objectives, 41
    student misconceptions, 41–42
  web version of, 43
Cholera, xi, 35–36, 45–55
Claim identification, rubric for assessment of, 377
Class discussion, evaluation of, 371–372
Classical music, concentration, and creativity, 130, 159–165
Claviceps purpurea, 61
Climate change
  Ah-choo! Climate Change and Allergies, 234–235, 247–252
  Rising Temperatures: The Politics of Information, 235, 253–259
Cloning, 268, 269, 303–307
Code of Federal Regulations Protection of Human Subjects, 318, **319**
Cold fusion, 234, 267
Collins, Francis, 233
Columbus, Christopher, 190
Committee for the Scientific Investigation of Claims of the Paranormal, 2
Connaughton, James, 254
Conservation biology and drug development, 268–269, 281–288
Constructivist learning model, 31–32
"Context of discovery," 11, 19
"Context of justification," 11, 19
Coot parenting and chick plumage, 82, 99–107
Cosmos, 234

Council on Environmental Quality, 254
Cragg, Gordon, 282
Crawford, Lester M., 220
Creativity, 22
Credit for discoveries, 233
Crick, Francis, 200, 233
Critical thinking, vii, 25–32
   active learning and, 27–28
   assessment of, 375–376
   case study approach to teaching of, vii, x–xiii, 21–24, 28–32
   definition of, vii, 25
   discipline-specific, viii, 22, 26
   informal logic and, vii
   learning domains and, viii, **ix**
   modeling of, 23
   scientific method and, viii–xi
   as teaching objective, vii, 25–27
Critical Thinking: What Every Person Needs to Survive in a Rapidly Changing World, 27
Crook, M. D., 162, 165
Cumming, Hugh S., 330
Curiosity, 22
Cutaneous T-cell lymphoma, trial of BCX-34 in, 295–302

D
Darwin's Black Box, 200
Darwin's theory, 198–200
Data manipulation in clinical trial, 269, 295–302
Dawkins, Richard, 2
DEA (Drug Enforcement Administration), 213
Deer, Brian, 242
Democracy and Education, 26
Demon-Haunted World, The: Science as a Candle in the Dark, 129
Dewey, John, 26

Dickey, J., 16–17
Dilemma cases, 363–366, 367
Dinosaur extinction, asteroid hypothesis of, 5–7
Dioxin contamination, 18
Directed cases, 363, 367
Discussion cases, 362–366, 367
DNA structure, 233
Double Helix, The, 233
Dow Corning, 17
Dragon in My Garage, **12–14**, 14
Drug Enforcement Administration (DEA), 213
Drug testing in humans, 275
Drug withdrawal from market, 185, 219–224
Dualist thinking, 24, 54

E
Eadie, John, 100–103
Eating PCBs from Lake Ontario, 235, 261–266
   case and questions for, 261–263
   summary and overview of, 365
   teaching notes for, 264–266
      classroom management, 265–266
      introduction and background, 264
      objectives, 264
      student misconceptions, 264
   web version of, 266
Education policy, 29
"Embryo design," 269, 309–314
Embryonic stem cells, 311
Environmental Protection Agency (EPA), 18, 253–255, 257
Epidemic disease, 46
Epidemiology, 36, 46–47, 50–55
Ergot poisoning, 61, 64, 66–67
Erhlich, Paul, 330
Ethical principles, 317

Index

Ethics and the scientific process cases, xii, 267–270
   Bad Blood: The Tuskegee Syphilis Project, 269–270, 329–339
   Bringing Back Baby Jason: To Clone or Not to Clone, 269, 303–307
   Cancer Cure or Conservation, 268–269, 281–288
   How a Cancer Trial Ended in Betrayal, 269, 295–302
   Mother's Milk Cures Cancer, 268, 271–279
   A Rush to Judgment?, 269, 289–293
   Selecting the Perfect Baby: The Ethics of "Embryo Design," 269, 309–314
   Studying Racial Bias: Too Hot to Handle?, 269, 315–327
   summary and overview of, 365–366
Evaluating student casework, 371–377
   assessing critical thinking, 375–376
   assignments, 372
   class discussion, 371–372
   exams, 372
   peer evaluation, 372–375, **374**
   rubric for assessing claims and evidence, 377
Evolution vs. intelligent design, 184, 197–206
Exams, 372
Experimental design cases, ix, 81–83
   The Great Parking Debate, 83, 113–120
   Lady Tasting Coffee, 82, 85–93
   Memory Loss in Mice, 82, 95–98
   Mom Always Liked You Best, 23, 82, 99–107
   PCBs in the Last Frontier, 82–83, 109–112
   Poison Ivy: A Rash Decision, 83, 121–128
   summary and overview of, 362
Extrasensory Perception?, 130, 133–140
   case and questions for, 133–136
   summary and overview of, 363
   teaching notes for, 136–140
      classroom management, 137–140
      introduction and background, 136
      objectives, 136–137
      student misconceptions, 137
   web version of, 140

F

Family Photos, **12, 13,** 16–17
Fanconi's anemia, 269, 309–314
Farr, William, 36, 50
FDA (U.S. Food and Drug Administration), 97, 152, 185, 218, 219–224, 279, 302
Fialka, John, 254
Fish in Lake Ontario fish, reproductive effects of PCBs in, 235, 261–266
Fish Kill Mystery, **12, 13,** 15–16
Fisher, Ronald, xii, 82, 86–93
Fossil records, 16–17
Frontier in American History, The, 190

G

Galileo, 189
Gardasil and mandatory HPV vaccination, 185, 225–230
Generation Rescue, 242
Genetics
   Bringing Back Baby Jason: To Clone or Not to Clone, 269, 303–307
   Case of the Tainted Taco Shells, 184–185, 207–212
   Selecting the Perfect Baby: The Ethics of "Embryo Design," 269, 309–314
Germ theory of disease, 35
Gjustland, Trygve, 334

Global warming
   Ah-choo! Climate Change and Allergies, 234–235, 247–252
   Rising Temperatures: The Politics of Information, 235, 253–259
Gore, Al, 251
Gottlieb, Jack, 61
Great Parking Debate, 83, 113–120
   case and questions for, 113–116
      ratings of own and others' behavior, 116, **116**
      research question, 113–114
      study method, 114
      study results, 114–116
   summary and overview of, 362
   teaching notes for, 117–120
      classroom management, 117–120
      introduction and background, 117
      objectives, 117
      student misconceptions, 117
   web version of, 120
Greenhouse gases, 247, 250, 251, 254
Group projects, 28, 29, 30
   peer evaluation and, 372–375, **374**

H
Hake, Richard, 27
Hamilton, Linda, 100–103
HAMLET factor, 274–275
Hardcastle, William, 45
Harris, William, 167–168
Hartzell, Hal, Jr., 282–283
Hawking, Stephen, 234
Hay fever, 250
Healing and intercessory prayer, 18, 130–131, 167–172
Helicobacter pylori and gastric ulcers, 36, 72–78, **73**
Helium-3, 192
Heller, Jean, 336

Hillary, Edmund, 190
Historical cases, xi–xii, 35–37
   Bacterial Theory of Ulcers: A Nobel-Prize-Winning Discovery, xi–xii, 36–37, 69–79
   Childbed Fever: A 19th Century Mystery, xi, 35, 39–43
   Mystery of the Blue Death: John Snow and Cholera, xi, 35–36, 45–55
   Salem's Secrets: On the Track of the Salem Witch Trials, xii, 36, 57–68
   summary and overview of, 361
Hoffman, Erich, 330
House, Karen Elliott, 258
How a Cancer Trial Ended in Betrayal, 269, 295–302
   case and questions for, 295–298
      background, 295
      clinical trial, 297–298
      main characters, 296–297
      objectives, 295–296
   summary and overview of, 269, 295–302
   teaching notes for, 298–302
      classroom management, 299–302
      introduction and background, 298
      objectives, 299
      student misconceptions, 299
   web version of, 298, 302
HPV (human papillomavirus) vaccine, 185, 225–230
Human behavior, 83, 113–120
Human Genome Project, 233, 305
Human papillomavirus (HPV) vaccine, 185, 225–230
Human pheromones, 130, 147–157
Humoral model of disease, 45
Hypothetico-deductive method. *See* Scientific method

Index

I

Immune response to poison ivy, 121–128
In vitro fertilization (IVF), 269, 270, 310, 313, 314
Informal logic, vii
Informed consent, 317
Institutional Review Board (IRB), 269, 316, 318–324, **319**, 326
Instructional techniques, requiring active learning, 27–28
Intelligent design, 1–2, 184, 197–206
Intercessory prayer for healing, 18, 130–131, 167–172
International Space Station, 187
Interrupted case method, 23–24, 361–365, 367
Intimate debate cases, 215–218, 364, 368
IRB (Institutional Review Board), 269, 316, 318–324, **319**, 326
Iridium spike and dinosaur extinction, 5–7
IVF (in vitro fertilization), 269, 270, 310, 313, 314
Ivory-billed woodpecker sighting, 131, 173–181

J

Jewelweed for poison ivy, 123–127, **125, 126**
Johnson, D. W. and R. T., 28
Jones, James H., 336
Jones, John E., III, 204
Junk science, 169–171
Junk Science: What You Know That May Not Be So, 17
Jupiter Icy Moons Orbiter, 192
Justice, 317

K

Kennedy, John F., 188

Koch, Robert, 77
Krummel, E. M., 111, 112
Kweder, Sandra, 224
Ky, K. N., 162, 164

L

Lady Tasting Coffee, 82, 85–93
   case and questions for, 85–91
      coffee shop wager, 85–87
      tasting coffee, 89–91
      tasting tea, 87–88
   summary and overview of, 362
   teaching notes for, 91–93
      classroom management, 92–93
      introduction and background, 91
      objectives, 91
      student misconceptions, 91–92
   web version of, 93
Lake Ontario fish, reproductive effects of PCBs in, 235, 261–266
Lange, Bob, 295, 296
Lauer, T., 28
Learning, 26
   active, 27–28
   constructivist model of, 31–32
   in cooperative groups, 28, 29
   critical thinking and domains of, viii, ix
   evaluation of, 371–377
   problem-based, 28, 30–31, 367, 372–373
   situated, 31–32
   as a social process, 31–32
Lecture method, x, 24
Legalization of marijuana for medicinal use, 185, 213–218
Lemons, P. P., 375
Lindberg, David, 1
Litaker, Wayne, 15
Long, D., 124

Love Potion #10: Human Pheromones at Work, 130, 147–157
   case and questions for, 147–152, **149, 150**
   summary and overview of, 363
   teaching notes for, 152–157
     classroom management, 153–157
     introduction and background, 152–153
     objectives, 153
     student misconceptions, 153
   web version of, 157
Lucas, James, 336
Lyon, Bruce, 100–103
Lysenko, T. D., 129

M

Mandatory HPV vaccination of girls, 185, 225–230
Marijuana, medicinal use of, 185, 213–218
Marks, J. G., Jr., 124
Mars exploration, 184, 187–195
Marshall, Barry, xi, 37, 70–79
Mass hysteria, 59–60, **60**, 66
McCarthy, Jenny, 238, 242
Measles-mumps-rubella (MMR) vaccine and autism, 234, 237–245
Media. *See* Science and media; Science and media cases
Medicinal Use of Marijuana, 185, 213–218
   case and questions for, 213–215
     "Terminal Cancer," 213–214
     "The Story of the Lotus Eaters," 215
   summary and overview of, 364
   teaching notes for, 215–218
     classroom management, 216–218
     introduction and background, 215–216
     objectives, 216
     student misconceptions, 216
   web version of, 218
Memory Loss in Mice, 82, 95–98
   case and questions for, 95–96
   summary and overview of, 362
   teaching notes for, 96–98
     classroom management, 97–98
     introduction and background, 96
     objectives, 96–97
     student misconceptions, 97
   web version of, 98
Mendola, Pauline, 262–264
Menstrual cycle and PCBs in Lake Ontario, 235, 261–266
Mental Rotations Test (MRT), 289, 290
Metacognition, 29
Miasma model of disease, 47–48
Michaelsen, Larry, 373
Michel, Helen, 5
Mierson, Sheella, 30
Miller, Jon, x
Misconceptions of students, xi, 2–3. *See also specific cases*
MMR (measles-mumps-rubella) vaccine and autism, 234, 237–245
Mom Always Liked You Best, 23, 82, 99–107
   case and questions for, 99–102
     biologists' method to attack problem, 100
     conclusions of study, 102, **102**
     data analysis, **101,** 101–102
     statement of problem, **99,** 99–100
     what is measured, 100
   summary and overview of, 362
   teaching notes for, 103–106
     classroom management, 104–106
     introduction and background, 103
     objectives, 103
     student misconceptions, 103–104
   web version of, 107
Moon to Mars: To Boldly Go...or Not,

Index

    184, 187–195
    case and questions for, 187–193
        prelude to space, 187–188
        public hearing, 188–193
    summary and overview of, 364
    teaching notes for, 193–195
        classroom management, 194–195
        introduction and background, 193
        objectives, 193–194
        student misconceptions, 194
    web version of, 195
Moore, B., vii
Moore, Joseph Earle, 332
Mother's Milk Cures Cancer?, 268, 271–279
    case and questions for, 271–276
        discovery, 271–272
        from lab to pharmacy, 275–276
        research, 273–274
        responses of scientific community, 272–273
    summary and overview of, 365
    teaching notes for, 276–279
        classroom management, 277–279
        introduction and background, 276
        objectives, 276–277
        student misconceptions, 277
    web version of, 279
"Mozart Effect," 130, 159–165
    case and questions for, 159–162
        enhanced performance, 159–160, **161**
        outlines of experiment, 161
        replication, 162
        research report analyses, 162
    summary and overview of, 363
    teaching notes for, 162–165
        classroom management, 164–165
        introduction and background, 162–163
        objectives, 163
        student misconceptions, 163

    web version of, 165
MRT (Mental Rotations Test), 289, 290
Mycotoxins, 61, 64, 66–67
Mystery Disease, **12, 13,** 15
Mystery of the Blue Death: John Snow and Cholera, xi, 35–36, 45–55
    case and questions for, 45–52
        Broad Street pump, 51–52
        mystery of blue death, 48–49, **49**
        sanitation and Victorian London, 46–47
        Snow and origins of anesthesiology, 47
        Snow's early life, 45–46
        solving the mystery of blue death, 50–51
    summary and overview of, 361
    teaching notes for, 52–55
        classroom management, 53–55
        introduction and background, 52–53
        objectives, 53
        student misconceptions, 53
    web version of, 55

N

NAACP (National Association for the Advancement of Colored People), 321
NASA (National Aeronautics and Space Administration), 187–195
Nash, Jack and Lisa, 312
National Academy of Sciences, 205
National Aeronautics and Space Administration (NASA), 187–195
National Association for the Advancement of Colored People (NAACP), 321
National Cancer Institute (NCI), 273, 282, 284

Index

National Center for Case Study Teaching in Science website, xi, 375. *See also web versions of specific cases*
National Commission for the Protection of Human Subjects of Biomedical and Behavioral Research, 317
National Institutes of Health (NIH), 233, 271, 273, 284, 302, 318
National Research Council (NRC), 26
National Science Education Standards, xi, 26, 369–370
National Science Foundation (NSF), 27
Native Yew Conservation Society, 282
Nature of science (NOS), 11–19
    case studies for teaching of, 14–18
        Breast Implants, 17
        The Dragon in My Garage, 14, **14**
        Family Photos, 16–17
        The Fish Kill Mystery, 15–16
        Mystery Disease, 15
        Prayer Study, 18
        Times Beach, 18
    case study alignment with concepts of, **12**
    case study questions for NOS scientific knowledge, **13**
    definition of, 11
Nature of Technology, The, 183
NCI (National Cancer Institute), 273, 282, 284
Nelson, C. E., 54
Neve, Rachael, 95
Newton, Isaac, 233
NIH (National Institutes of Health), 233, 271, 273, 284, 302, 318
Nobel, Alfred, 268
Nonsteroidal anti-inflammatory drugs (NSAIDs), 220
Norton, Gail, 180
NOS. *See* Nature of science
NRC (National Research Council), 26
NSAIDs (nonsteroidal anti-inflammatory drugs), 220
NSF (National Science Foundation), 27
Null hypothesis, 90
Nuremberg Code, 334, 337

O

Occam's razor, 130
Of Pandas and People, 197, 199
Office of Protection from Research Risks (OPRR), 318
On the Mode of Communication of Cholera, 48
Oppenheimer, Robert, 268, 270
OPRR (Office of Protection from Research Risks), 318
Orion Crew Exploration Vehicle, 187
Orphan drugs, 279

P

Paclitaxel (Taxol), 268–269, 281–288
Parker, R., vii
Pascal, Blaise, 372
Pasteur, Louis, 37
Paul, Richard, 27
PBL (problem-based learning), 28, 30–31, 367, 372–373
PCBs in the Last Frontier, 82–83, 109–112
    case and questions for, 109–111
        global transport, 110
        PCBs, **109,** 109–110
        riddle solved, 110–111
        significant difference, 110
    summary and overview of, 362
    teaching notes for, 111–112
        classroom management, 112
        introduction and background, 111
        objectives, 111
        student misconceptions, 112
    web version of, 112

Index

PCD (pre-implantation genetic diagnosis), 309–314
Pedagogical content knowledge, 21
Peer evaluation, 372–375, **374**
Peer review, 2, 234, 251–252
Peptic ulcers, bacterial theory of, xi–xii, 36–37, 69–79
Perry, Rick, 185, 225–229
Perry, William, 24, 54
Peugeot, Renee, 296, 302
Pfiesteria piscicida toxin, 15–16
Pharmaceuticals
    Amanda's Absence: Should Vioxx Be Kept Off the Market?, 185, 219–224
    Cancer Cure or Conservation, 268–269, 281–288
    How a Cancer Trial Ended in Betrayal, 269, 295–302
    Medicinal Use of Marijuana, 185, 213–218
    Mother's Milk Cures Cancer?, 268, 271–279
Pheromones, 130, 147–157
Photographic evidence, 16–17
PHS (U.S. Public Health Service) Division of Venereal Diseases, 330–336
Placebo effect, 131
Plato, 26
Plotkin, Mark, 282, 287
Poison Ivy: A Rash Decision, 83, 121–128
    case and questions for, 121–127
        effectiveness of jewelweed treatment, **126**, 126–127
        encounters with poison ivy, 121–122
        finding a treatment, 123–124
        mystery of blisters unfolds, 122–123
        scientific experiment, 124–126, **125**
    summary and overview of, 362
    teaching notes for, 127–128
        classroom management, 127–128, **128**
        introduction and background, 127
        objectives, 127
        student misconceptions, 127
    web version of, 128
Pollen allergies and climate change, 234–235, 247–252
Polychlorinated biphenyls (PCBs)
    in Alaskan lakes, 82–83, 109–112
    in Lake Ontario, reproductive effects of, 235, 261–266
Prayer-Gauge Debate, The, 169
Prayer Study: Science or Not?, 130–131, 167–172
    case and questions for, 167–169
    nature of science concepts aligned with, **12, 13,** 18
    summary and overview of, 363
    teaching notes for, 169–172
        classroom management, 172
        introduction and background, 169–171
        objectives, 171
        student misconceptions, 171
    web version of, 172
Pre-implantation genetic diagnosis (PCD), 309–314
Precautionary principle, 18
Pregnancy and PCBs in Lake Ontario, 235, 261–266
Prejudice and racism
    Bad Blood: The Tuskegee Syphilis Project, 269–270, 329–339
    Studying Racial Bias: Too Hot to Handle?, 269, 315–327
Problem-based learning (PBL), 28, 30–31, 367, 372–373
Problem solving, 22, 24
Progressive disclosure method, 257–258
Pseudoscience, xii, 169–171, 184, 242, 363.

Index

See also Unusual claims
Public health
    Ah-choo! Climate Change and Allergies, 234–235, 247–252
    Bad Blood: The Tuskegee Syphilis Project, 269–270, 329–339
    Mystery of the Blue Death: John Snow and Cholera, xi, 35–36, 45–55
    Sex and Vaccination, 185, 225–230
    Tragic Choices: Autism, Measles, and the MMR Vaccine, 234, 237–245
Public hearing cases, 364, 368
Publication of scientific discoveries, 233

Q

Quinault Indian Nation, 281–284

R

Racial prejudice
    Bad Blood: The Tuskegee Syphilis Project, 269–270, 329–339
    Studying Racial Bias: Too Hot to Handle?, 269, 315–327
Radetsky, Peter, 276
Ransohoff, D. F., 76
Rauscher, F. H., 162, 164
Reagan, Ronald, 2
Records of Salem Witchcraft, 61
Religious beliefs, 2
    And Now What, Ms. Ranger? The Search for the Intelligent Designer, 184, 197–206
    Prayer Study: Science or Not?, **12, 13,** 18, 130–131, 167–172
Reproductive effects of PCBs in Lake Ontario, 235, 261–266
Research ethics, 269, 289–293
Respect for persons, 317
Revkin, Andrew, 253, 257
Rhine, J. B., 135, 136, 138
Ridley, Matt, 183, 233
Rising Temperatures: The Politics of Information, 235, 253–259
    case and questions for, 253–255
    summary and overview of, 365
    teaching notes for, 255–258
        classroom management, **256,** 256–258
        introduction and background, 255–256
        objectives, 256
        student misconceptions, 256
    web version of, 259
Roberts, Eugene, 95
Rofecoxib (Vioxx) withdrawal from market, 185, 219–224
Role-play cases, 364–366, 368
Rubric for assessing claims and evidence, 377

S

Sagan, Carl, xii, 14, 129, 169, 234
Salem Witchcraft Papers, 58
Salem's Secrets: On the Track of the Salem Witch Trials, xii, 36, 57–68
    case and questions for, 57–64
        classroom extensions, 63–64
        data interpretation, 61–63, **62**
        ergot poisoning, 61
        mass hysteria events, 59–60, **60**
        Salem's secrets, 57–59
        societal frame, 63
    summary and overview of, 361
    teaching notes for, 64–68
        classroom management, 66–68
        introduction and background, 64
        objectives, 65
        student misconceptions, 65
    web version of, 68
Salk, Jonas, 270

Index

Schatz, Irwin J., 335
Schaudinn, Fritz, 330
Schmitt, Harrison, 189
Schulman, Lee, 21
Science, 1–9
    Berkeley model of how it works, **4**, 4–7
        case study of asteroids and dinosaurs, 5–7
        vs. how it really works, 7–9, **8**
    limitations of, 2
    misconceptions about, 2–3
    nature of, 11–19
    as a way of knowing, viii, 81
Science, technology, engineering, and math (STEM) education, x, 27
Science and media cases, xii, 233–235
    Ah-choo! Climate Change and Allergies, 234–235, 247–252
    Eating PCBs from Lake Ontario, 235, 261–266
    Rising Temperatures: The Politics of Information, 235, 253–259
    summary and overview of, 365
    Tragic Choices: Autism, Measles, and the MMR Vaccine, 234, 237–245
Science and society cases, xii, 183–185
    Amanda's Absence: Should Vioxx Be Kept Off the Market?, 185, 219–224
    The Case of the Tainted Taco Shells, 184–185, 207–212
    Medicinal Use of Marijuana, 185, 213–218
    Moon to Mars: To Boldly Go...or Not, 184, 187–195
    And Now What, Ms. Ranger? The Search for the Intelligent Designer, 184, 197–206
    Sex and Vaccination, 185, 225–230
    summary and overview of, 364

Science for All Americans, 26
Science journals, 233–234
Science on Trial, 17
Scientific enterprise, 11
Scientific knowledge, 11
Scientific literacy, x, 27, 151
Scientific method, viii–ix, x–xi, 1–9. *See also* Experimental design cases
    concept map, **170**
    unusual claims and, xii, 129–131
Scientific process, 177, **178**
Scotchmoor, Judy, 1
Scott, David, 189
Search for Extraterrestrial Intelligence (SETI), 191
Seelye, Katharine, 253
Selecting the Perfect Baby: The Ethics of "Embryo Design," 269, 309–314
    case and questions for, 309–312
    summary and overview of, 366
    teaching notes for, 312–314
        classroom management, 313–314
        introduction and background, 312–313
        objectives, 313
        student misconceptions, 313
    web version of, 314
Semmelweis, Ignaz, xi, 35, 39–43
SETI (Search for Extraterrestrial Intelligence), 191
Sex and Vaccination, 185, 225–230
    case and questions for, 225–229
        arguments against mandatory HPV vaccination, 227–229
        governor's case, 226–227
        Texas tempest, 225–226
    summary and overview of, 364
    teaching notes for, 229–230
        classroom management, 230
        introduction and background, 229
        objectives, 229

student misconceptions, 229–230
web version of, 230
Sexually transmitted diseases (STDs), 225–230
Shaman's Apprentice, The, 287
Shaping the Future: New Expectations for Undergraduate Education in Science, Mathematics, Engineering, and Technology, 27
Shaughnessy, J. B., **161**, 164
Shaw, G. L., 162, 164
Silicone breast implants, **12, 13**, 17
Situated cognition, 31–32
Skepticism, ix, xii, 22–24, 129
Skin cancer, trial of BCX-34 in, 269, 295–302
Skirrow, Martin, 37, 75
Smithsonian Institution, 282
Snow, John, xi, 35–36, 45–55
Snyder, Harry W., Jr., 296, 302
Social psychology, 83, 113–120
Societal issues. *See* Science and society cases
Socrates, 26
Solomon, David, 273
Space exploration, 184, 187–195
Spanos, Nicholas, 61
Staphylococcus aureus, 72
Statistical experiment, 82, 85–93
STDs (sexually transmitted diseases), 225–230
Steele, K. M., 162, 165
Stem cell research, 311
STEM (science, technology, engineering, and math) education, x, 27
Stossel, John, 18
Studying Racial Bias: Too Hot to Handle?, 269, 315–327
case and questions for, 315–322
ethics and conduct of research with human subjects, 316–318, **319**
how IRB evaluated proposal, 318–321
research proposal, 315–316
response of supporters of proposed research, 321–322
summary and overview of, 366
teaching notes for, 323–327
classroom management, 324–327, **325**
introduction and background, 323
objectives, 323
student misconceptions, 323–324
web version of, 327
Sulzberger, Arthur O., 258
Svanborg, Catherine, 271–279
Syphilis project in Tuskegee, 269–270, 316, 329–339
Szent-Györgyi, Albert, 77

T
Taxol (paclitaxel), 268–269, 281–288
Teaching notes for cases, xi. *See also specific cases*
Terrestrial Planet Finder, 192
Terry, D. R., 375–376
Thanukos, Anna, 1
Think-pair-share method, 257
Thomas Aquinas, 199
Times Beach, **12, 13**, 18
Tobias, Sheila, x
Tragic Choices: Autism, Measles, and the MMR Vaccine, 234, 237–245
case and questions for, 237–242
the choice, 237–239
the conference, 240–241
the connection, **239**, 239–240, **240**
the consequences, **241**, 241–242
summary and overview of, 365
teaching notes for, 242–245
classroom management, 243–245

introduction and background, 242
objectives, 242–243
student misconceptions, 243
web version of, 245
Truzzi, Marcello, 2, 130
Turner, Frederick Jackson, 190
Tuskegee syphilis project, 269–270, 316, 329–339
Type I and Type II errors, 90, 293
Tyser, R. W., 376

U

Understanding Science: How Science Really Works website, 1
Unusual claims cases, xii, 129–131
    The Case of the Ivory-Billed Woodpecker, 131, 173–181
    Extrasensory Perception?, 130, 133–140
    Love Potion #10: Human Pheromones at Work, 130, 147–157
    The "Mozart Effect," 130, 159–165
    A Need for Needles: Does Acupuncture Really Work?, 130, 141–146
    Prayer Study: Science or Not?, 18, 130–131, 167–172
    summary and overview of, 363
Urushiol, 122–124, **125, 126**
U.S. Department of Health, Education, and Welfare, 317
U.S. Food and Drug Administration (FDA), 97, 152, 185, 218, 219–224, 279, 302
U.S. Forest Service, 281–284
U.S. Public Health Service (PHS) Division of Venereal Diseases, 330–336

V

Vaccines
    Sex and Vaccination, 185, 225–230
    Tragic Choices: Autism, Measles, and the MMR Vaccine, 234, 237–245
Validity of evidence, rubric for assessment of, 377
Vena, J. E., 264
Venter, Craig, 233
Vioxx withdrawal from market, 185, 219–224
Vision for Space Exploration, 187–188
Vonderlehr, Raymond, 332–333
Vygotsky, L. S., 31

W

Wakefield, Andrew, **240**, 240–242, **241**, 244–245
Warren, J. Robin, xi, 36–37, 69–79
Wassermann blood test, 330
Water contamination and cholera, 36, 46–55
Watson, James, 233
Watson-Glaser Critical Thinking Appraisal, 376
Web versions of cases, xi. *See also specific cases*
Whitehead, Henry, 36, 51
Whitman, Christine Todd, 253–254
WHO (World Health Organization), 247
Willingham, Daniel, viii
Witch trials in Salem, xii, 36, 57–68
World Health Organization (WHO), 247

Z

Zechmeister, E. B. and J. S., **161**, 164